"David Grinspoon has written the book I wanted to write, and he's done it so very well that I'll be forever thankful I never got to it! Covering subjects that span from the Epicureans of ancient Greece to the SETI Institute, from the Copernican revolution to the latest in astrophysics, the origins of DNA, crop circles and alien abductions, David is willing to address questions about how we got here that other science writers are unwilling to touch. He discusses aspects of the revolutionary field of astrobiology from a deeply scientific perspective and delves into some more bizarre publicly held ideas, sympathetically but without compromising his scientific grounding. David's style is so direct, so personal, and so punctuated with delightful humor that reading this book feels like a living room conversation."

—RUSSELL *"Rusty"* SCHWEICKART,
NASA Astronaut

"An exuberant, provocative look at the possibility of extraterrestrial life, what it might be like, and what it might mean. . . . Wisecracks, philosophical musings, and personal anecdotes make his text as lively as it is authoritative. The best look at this subject since Carl Sagan's *Cosmic Connection*." —*Kirkus Reviews* (starred review)

"Grinspoon takes us on a thrilling ride through the history of well, everything. . . . With metaphor, analogy, and clear, entertaining writing, Grinspoon helps us understand how we got here. . . . [A] terrific book." —*San Diego Union Tribune*

"Superb. . . . Everything is here—theories of planetary formation and evolution, the origin of life, the evolution of complex life, and even the evolution of intelligence and technology. . . . David Grinspoon has succeeded marvelously at producing a comprehensive, enjoyable overview of astrobiology, the epitome of multi-disciplinary research. . . . Anyone, even a professional scientist, who wishes to become familiar with contemporary astrobiology should read this book. It's a prime place to become more than casually acquainted with one of the hottest, most interesting fields of science." —FRANK DRAKE,
Astronomy magazine

"A book that balances solid science with offbeat, often hilarious detours. . . . Entertaining." —*Discover*

"*Lonely Planets* is an entertaining and thought-provoking book. . . . David Grinspoon provides a masterful synthesis of the history, science, philosophy, and even theological implications of extraterrestrial life. . . . His lively and engaging writing is well-suited to the broad range of subjects encountered here." —*Science*

"In his low-key, conversational language, Grinspoon asks provocative questions about modern science's unyielding rigidities. . . . Definitive proof that life on this planet is intelligent *and* funny."
—*Entertainment Weekly*

"Grinspoon seasons the discussion with witty anecdotes, personal experiences, and reminders of what has been demonstrated and what is still theoretical. Fun to read, Grinspoon comes across like a buddy in a bar, trying out ideas over a beer or few. He deserves a large readership." —*Booklist*

"In *Lonely Planets*, David Grinspoon brings together what has never before been synthesized: the history, science, culture, and politics of the search for life in the universe. Along the way, you will not escape his practical and often humorous observations of the quest; he is a planetary scientist as well as a dreamer, borne of the space age."
—NEIL DEGRASSE TYSON, astrophysicist and Director,
Hayden Planetarium, New York City

"David Grinspoon's *Lonely Planets* is a thorough and thoroughly readable review of our chances for finding life on other worlds, and what this may mean to us. Breezy yet deep, fun to read, and thought-provoking."
—BEN BOVA, President Emeritus of the National Space Society
and prolific author of science fact and fiction

"Less a popularization than a personable chat on life, the universe and everything. . . . Grinspoon handles the wide variety of material necessary for a coherent narrative with great aplomb." —*Publishers Weekly*

"Grinspoon tackles E.T. in a style that will satisfy science nerds and English majors alike."
—*Wired*

"Grinspoon tells engagingly the story of humankind's long fascination with the possibility of extraterrestrial life." —*Scientific American*

"Outstanding . . . Grinspoon does a great job of explaining what we know so far, how we know it, and why we have quite a ways to go before the story is complete. . . . Keen scholarship, witty observations, and thought-provoking banter from a working scientist who, like Sagan, can really write. . . . What sets Grinspoon apart from other scientists who do a good job writing for a lay audience, i.e., Lawrence Krauss and Brian Greene, is the attention he pays to the limitations of science. . . . There are a number of noteworthy scientists now vying for the crown of premier science popularizer worn with such distinction by Sagan. With the addition of *Lonely Planets* to his earlier *Venus Revealed*, Grinspoon stakes his own strong claim to that crown.
—*San Jose Mercury News*

"An easy to read, thought-provoking look at humanity's unending fascination with extraterrestrial life." —*Gannett News Service*

"Grinspoon steps into the Cosmos-sized shoes of the late Carl Sagan with *Lonely Planets*, the best, most entertaining examination of the possibility of other life in the universe since Sagan's best work. . . . It's fascinating as pure information and made positively delightful by Grinspoon's willingness to be playful. . . . His writing is rife with a kind of uppity humor. He's like a teenager who's just decided he wants to be an astrobiologist, thrilled and humbled but still full of attitude." —*Boulder Daily Camera*

"*Lonely Planets* is a Lonely Planet guide to the universe, packed with useful information served up with a wry, amusing twist . . . in a breezy, engaging style unlike anything you'll find in most popular science books. . . . Mr. Grinspoon steps deftly into Mr. Sagan's mantle . . . on this mind-altering trip through the cosmos."
—*Dallas Morning News*

L o n e l y   P l a n e t s

Lonely

Planets

The Natural Philosophy of Alien Life

DAVID GRINSPOON

An Imprint of HarperCollinsPublishers

HarperCollins books may be purchased for educational, business, or sales promotional use. For information please write: Special Markets Department, HarperCollins Publishers Inc., 10 East 53rd Street, New York, NY 10022.

First Ecco paperback edition published 2004

Designed by Jessica Shatan Heslin

The Library of Congress has catalogued the hardcover edition as follows:

Grinspoon, David Harry.
    Lonely planets : the natural philosophy of alien life / David Grinspoon—1st ed.
        p.   cm.
    Includes bibliographical references and index.
    ISBN 0-06-018540-6
        1. Life on other planets.   I. Title

  QB54.G74 2003
  576.8'39—dc21   2003048509

ISBN 0-06-095996-7 (pbk.)

04 05 06 07 08 ❖/RRD 10 9 8 7 6 5 4 3 2 1

For my parents

Evelyn Betsy Grinspoon

and

Lester Grinspoon

with love and gratitude

Penetrating so many secrets,
we cease to believe in the unknowable.
But there it sits, nevertheless, calmly licking its chops.

—H. L. MENCKEN

Contents

Foreword:
It Came Out of the Sky

We are all in the gutter, but some of us are looking at the stars.
— OSCAR WILDE

On a tranquil late afternoon in early January 2004, the sky split open and an alien ship dropped out. In a tired, rusty desert land where nothing more than a dust devil had stirred for a hundred million years, the monotony was shattered and a thundering, glowing ball of light rushed toward the ground. Suddenly, at about two hundred feet, the visitor inflated like an angry puffer fish, growing to many times its original size, and then, seconds later, landed with a mighty "whump!" bouncing as high as a four-story building. After twenty-eight more bounces—each one raising a fearsome cloud of dust that slowly drifted off—it came to rest on a desolate, sandy plain scattered with worn and broken rocks.

Nineteen days later, on the other side of the world, a twin vessel made a similarly strange, bouncing entrance, rolling to a stop in a small crater sunk into a vast flat wasteland of salt-crusted rocks sprinkled with metallic, berrylike spheres. Each visitor quickly began to transform itself, deflating its landing cocoon to reveal a small hibernating creature within. Extending wheeled legs, mechanical eyes, and other peculiar sensory limbs, each slowly crawled off its now defunct landing pod. The Martian arrival had begun.

Back on Earth, just two months later, in late March 2004, hundreds of scientists pursuing alien life congregated in a hastily constructed NASA facility in northern California—a colossal white tent with semi-translucent siding, illuminated by rows of massive searchlights. Armed government guards checked ID of all who wished to enter. At first glance it resembled some top-secret *X Files*–type government installation, but a peek inside dispelled that impression. Instead of emotionless space-suited functionaries intent on inscrutable experiments, the giant hall was filled with a motley assortment of nerds (myself among them) and student nerds-in-training. Fashions ranged from suits and ties to sandals and shorts. Information-packed posters hung in long rows, displaying the latest scientific results on "astrobiology"—the study of extraterrestrial life. A platoon of headsetted journalists, chasing us around with microphones and cameras, completed the scene. This was the Third Astrobiology Science Conference, held at NASA's sprawling Ames Research Center, spread along the southwestern shore of San Francisco Bay—a tentful of carbon-based, water-loving, marginally intelligent organisms gathered on the thin skin of planet Earth, to prognosticate about the possibilities of life beyond.

We had a lot to talk about. NASA's two Mars Exploration Rovers had made their spectacular bouncing landings only two months prior and had already made fantastic discoveries that had recharged and refreshed the perennial debates about life on Mars.

Of course Mars was all the rage, so I was somewhat surprised, but delighted, to have been invited to the conference to speak about one of my pet ideas: "Sympathy for the Devil: The Case for Life on Venus." It was fun to play Lucifer's advocate for the astrobiology community and attempt to sell Venus's overlooked charms to this skeptical but far-reaching audience.

In a speculative field like astrobiology, complacency, overconfidence, and unsupported consensus are all serious dangers. After all, our field is still lacking in any actual bona fide extraterrestrial research volunteers. So outside ideas, however ultimately wrong-headed they may prove, are welcomed, as long as they can be supported with plausible arguments that don't break too many of our agreed-upon rules. (What are these rules? Why do we agree upon them? Should we? Read this book.) My conjectures about possible microbial life in the clouds of Venus were deemed to pass this test, and so they were invited into the tent, joining the more "conventional" notions of life underground—in possible Martian hot springs and the buried seas of Jupiter's icy moons. (See chapters 11 and 12 for more on possible Venusian life.)

This gathering under a tent, though perhaps not nearly as exotic as a secret government alien research lab, was definitely not your mom's science conference. The two other programs for which I was enlisted that week were a panel called "Ethics of Exploration" and a public debate in which a group of scientists and science fiction writers argued over the possibility and desirability of "Terraforming Mars" (the future engineering of Mars to be more like Earth).

Now hold on. Science fiction? Ethics? At a science conference? Yes, if the science is astrobiology. In addition to the "strictly physical" questions we wrangle with—such as making life to begin with, transporting it between worlds, and keeping it alive in a wide range of planetary environments—questions about life in the universe inevitably spill over into other realms. Intellectually, astrobiology itself is a rather big tent, somewhat hastily constructed, in which the Earth, space, and life sciences commingle with wild speculation, a dash of philosophy, and even a splash of spirituality.

The most eagerly awaited moment of the weeklong conference came Tuesday afternoon, when Steve Squyres, principal scientist for the Mars rovers, gave us an update on the activities of his two little Martian robot geologist puppies, *Spirit* and *Opportunity*. There was an air of celebration at this session, as a community that has known recent and repeated failure enjoyed a great success. This cockamamie bouncing–on–air bags landing scheme had worked once before with the *Sojourner* rover in 1997, but we all had a lot more riding on this attempt than just the equipment. During the weeks prior to the landings, there had been a palpable nervousness, fueled by the fact that we still weren't sure what went wrong with our last attempt to land on Mars in 2000 and tempered somewhat by the thought that this time we were sending two identical, carefully tested rovers and at least one of them ought to work.

This time they both did. Each survived the bouncy landing without a hitch, and at the time of the Astrobiology Science Conference, each was inching across one of the thousand unexplored deserts of Mars, scratching and poking among the ruddy dirt and ancient rocks, shaking loose buried secrets, snapping pictures all the while.

Squyres—rail-thin, angular, and as always sporting jeans over cowboy boots—was looking very bright-eyed for someone who'd been living on Martian time for the last three months. As far as I could tell, he hadn't changed a bit since we first crossed paths in the summer of 1978 as students at Cornell. I ran into him in the hallway before his talk, and though at that moment he was the coolest person in the solar system, he didn't act with one ounce of self-importance. He recounted the latest rover findings and the fun he was having, as if he were just an old colleague at a meeting telling me about his latest pet project. Which he was, but . . . his pets were on Mars, and they were on the move. Steve took the stage, to thunderous applause, meant for the triumphant little robots as well as for their driver. These are the moments we live for. It is thrilling to be back on Mars.

He began with a spirited recap of the rovers' initial forays on Mars and a preview of their possible futures. Then he got right to the good stuff—the possible stuff of life on Mars. What had we found? When I think about it I still get so excited I can hardly talk—or type. All my life, and my professional career, I've been enthralled with the possibility that through planetary exploration we can learn something definite about whether we have living company in the universe beyond Earth. Well, the universe had just dropped us a big hint. We found rocks on Mars that were formed of sulfate salts. The only way we know of to make that kind of rock is through the evaporation of salty seas from a place that must have been soaking wet for significant periods of time.

Why are we so hyped-up about finding sea-formed rocks on Mars? Well, as far as we know at present, life needs water. On Earth, where there is water there is life. Over the last few decades, circumstantial evidence had been building for large quantities of surface water in the Martian past. Orbital photographs revealed shapes strongly suggestive of watery rivers and lakes. Yet, there has always been the nagging possibility that we were searching so hard for signs of the familiar that we were misinterpreting the photos and maps, mistaking the action of lava, wind, ice, or some other unknown carver for the work of our beloved water. But the rocks don't lie. Now, at last, we've sampled the ground itself, and the evidence is no longer circumstantial. We've found the smoking gun (which in this case is a dripping Super Soaker) of past habitability. We now know there was other wet ground, beyond the Earth, in our solar system. Right next door. Buckets of rain once ran like salty tears over the face of our little red brother Mars. This discovery proves that Mars is indeed an important place for astrobiology exploration—a place where many kinds of Earth life could once have survived—so why not Martians? The idea that we might really find fossils of bygone creatures on the Red Planet can no longer be regarded as far-fetched.

Among other things, this will be a major shot in the arm for our desires and plans for future missions, providing the encouragement (and most likely the funding) we need to keep going, to send new machines there that can look for fossils or chemical traces of past life.

Soon we will want to return Martian samples to Earth. With the right Mars rocks in our own laboratories, we will be able to more definitively test the idea that life once graced our red planetary neighbor.

The raised prospect of new missions to—and from—Mars heightened the exigency of our ethical discussions, topics that just a year ago seemed more academic. How much should we care about—and spend to guard against—the possibility that we might contaminate Mars with microbes from Earth, or even the slight but disquieting chance that we could bring something back from Mars that might enjoy snacking upon our own biosphere? John Rummel, NASA's planetary protection officer, was at the conference, addressing these issues.

Does the news that Mars once had the conditions for life increase the threat of contamination? Perhaps not. Many of us believe that whatever biology once graced this rusty world disappeared long ago, along with the sputtering geology and the evaporating seas. The Mars rovers are wandering places where almost nothing has happened in uncountable eons. Seen up close, these landscapes verify our belief that the surface of Mars is incredibly ancient. There is nothing in the new photos to suggest any recent action, beyond the frequent bursts of dusty winds. Most of the geological activity is long gone, leaving a surface freeze-dried, ossified, and sculpted in places into bizarre forms not seen on Earth, because you couldn't find a place on our planet that has been left to the wind alone for a billion years. The rovers haven't found much that changes our views of present-day circumstances on Mars. Their biggest discoveries are about conditions in the deep past, including the enticing possibility of ancient life.

Even the finding of once-soggy ground, as spectacular as it is, is not revolutionary. It doesn't overturn our current notions about Mars—in fact, it confirms them. Yet, while the rovers were grabbing all the attention, about a week before the conference, another report had come in from Mars that could have truly revolutionary implications. The European Space Agency's *Mars Express*, which arrived in Mars orbit on Christmas Day 2003, had caught a whiff of something in the air. Something that didn't belong there. Something that might indeed be a sign of life there today. The announcement received much less press attention than the rovers, which, after all, were taking cool pictures. On board the orbiter, an infrared spectrometer—which precisely dissects the radiation leaking from the planet into a million distinct colors—had detected a most unexpected trace gas in the Martian air. Feeble signs of methane had been found on Mars. Methane is CH_4, a carbon bonded to four hydrogens.

Chemically, it is out of place in an atmosphere like that of Mars, which is composed almost completely of carbon dioxide (CO_2). Finding methane on Mars is like finding a gazelle strolling unnoticed through a pack of hungry lions.

To me this announcement was shocking, and it seemed even more unreal than any of the strange postcards sent home by the rovers. I used to say that methane on Mars would be one finding that could change my mind about that planet being a perfectly dead world. Why? Because as life evolves on a planet, the atmosphere evolves along with it. Chemically, the two become intimately coupled. On Earth, the oxygen we breathe, the protective ozone layer, and, yes, the trace of methane in the air are all chemical by-products of 4 billion years of biology. Life in turn has molded its chemistry to cleverly utilize the atmosphere pervading our world. It may be this way on all planets with life and air. If so, then a close study of a planet's air will always reveal life or the lack of it.

More specifically, life is a process that consumes energy and produces gases that are "out of equilibrium" with the rest of the atmosphere. These gases don't last long because they react quickly with their surroundings. Methane does not belong in the atmosphere of Mars, just as oxygen doesn't belong in the atmosphere of Earth. Oxygen here is a product of green plants. What (or who) is producing the methane on Mars?

When I first heard about the methane observation, I was highly skeptical, but at the same time I felt my pulse quicken. Could this really be the faint breath of underground colonies of Martians?

We've been wrong about Mars many times. Mars is so like Earth in some respects, and so close at hand. As if it is our only friend in a large, empty universe, we sometimes project too much onto poor old Mars. We are hungry for signs of life, and this hunger is dangerous. In science the desire to find a certain answer can lead us astray. I describe a few of these wrong turns in chapter 3 of this book, such as the late 1950s "discovery" of chlorophyll—the green stuff in green plants—on the Red Planet. This sensational announcement (in *Science* magazine) was greeted without skepticism at the time because of the prevailing view that the seasonal color changes observed through telescopes were caused by vegetation. (We now know that they are caused by wind-blown dust.) The chlorophyll was later discredited, its "fingerprint" shown to be caused by compounds of deuterium (heavy hydrogen) in our own atmosphere. Half a century earlier, the scientific community had been briefly enthralled,* and then greatly embarrassed, by Percival Lowell's claims of finding intelligently designed canals crisscrossing the map of Mars. The canals either conveniently disappeared just as our telescopes and cameras improved enough to truly see them, or (more likely) they were never there at all.

Given this history of false starts and retreats, we've learned not to lightly declare that the Martians have at last been found. Or have we? In the spring of 2004 the teams reporting on the methane detections all openly speculated on underground Martian life as a likely source. Such speculation is in part facilitated by an attitudinal pendulum within science that has again swung toward acceptance of the possibility of life on our neighboring planet.

The methane claim was quickly bolstered by several independent observations. In addition to the detection by *Mars Express* in orbit, it has now been seen by two different groups of ground-based observers using some of the best telescopes on Earth, in Hawaii and Chile. So the methane, it seems, is there. But does it really mean life on Mars?

When you actually look at the numbers, the evidence is not immediately convincing because the quantity of methane is so tiny. Seen in the infrared, methane has a distinct and strong signature. In an atmosphere of carbon dioxide it stands out like a blood stain in a fresh snow bank. So, having scrutinized Mars in the infrared for decades, why haven't we found it before? The signal is, in fact, very weak, implying that the methane is extremely scarce. The data suggest something like ten parts per billion (ppb) methane in the Martian air, so for every billion molecules of carbon dioxide (CO_2) there are, apparently, ten molecules of methane (CH_4). That's hardly a methane mother lode. Yet, *there must be a source*. And that is the part that sets our minds spinning. On Earth the main sources of methane are biological ones. Methane doesn't last long in our air either, but bacteria living in rice paddies and in the guts of cows (for example) supply a constant trace. Could underground bacteria on Mars be the culprits here?

Given "Sagan's law" that "extraordinary claims require extraordi-

---

* For a couple brief decades, that is.

nary evidence," we are obliged to search for other, more mundane explanations before we trumpet the news (once again) of finding life. How do we know whether we should accept, or rule out, an alternative explanation for something as potentially important as methane on Mars? Many times a "back of the envelope" calculation is sufficient—an exercise in which we plug in rough but reasonable estimates for the important quantities and see if the answer we get is in, or at least near, the ballpark.

For example, a friend e-mailed me, suggesting that maybe the tiny residue of methane is simply leaking from the small collection of derelict spacecraft we've left on the planet. Yet this can quickly be ruled out with the nearest envelope: given that the entire Martian atmosphere weighs about 2.5×10^{16} kilograms, or about 25,000 trillion kilos, this means that ten parts per billion methane, as small as that sounds, still adds up to about 90 million kilos of methane. So, if we had 90,000 spacecraft on Mars (as opposed to about a dozen), each weighing a thousand kilos and each composed entirely of methane gas (not a rec-ommended construction material), then this could work as an explana-tion. Envelopes are good for *reductio ad absurdum* arguments, which tell us where not to waste our time.

A more promising possibility is volcanic venting. On Earth volcanoes burp great quantities of methane into the atmosphere. However, Earth is a volcanically active planet at present and Mars, overall, is not. My favorite candidate for a nonbiological source is the steady rain of mete-ors. Organic material falls from space all the time on Mars, as on all other planets. We don't know the precise rate at which this space gunk is entering Mars's atmosphere, but we can make some reasonable infer-ences based on the observed rate on Earth and applying what we know about orbits and the gravitational reach of Mars. When I put this all together on the back of a nearby envelope, I find that the amount of organic carbon landing on Mars each year is likely close to the needed supply rate for the observed methane. Decaying organic matter is a classic source of methane—think swamp gas. My oversimplified calcu-lation tells me that the infall of meteors, and the subsequent release of organic gas as they break down in the atmosphere, could provide the right trickle of methane. No bugs required.

Is the methane falling from the sky? Maybe. Maybe not. With future space missions we'll eventually be able to measure the rate at which meteors supply Mars with organics. Much sooner than that, we'll have better observations of the methane. Where, exactly, in the atmosphere does it reside? Is it concentrated near the surface (implying an underground source)? Is it coming from specific locations? The latter is the most important observation we can make in the near future. If we find it venting from certain surface features, this destroys my "falling from the sky" idea and gives us important clues to the conditions belowground that may be fostering the chemical activity—be it geological or biological—supplying the doomed molecules.

As is often the case at scientific conferences, the most interesting discussions took place in the hallways between sessions. There was much chatter about the hot new methane observation but little consensus on the right interpretation. Should we take it as a sign of life? Nobody was arguing that the idea was ridiculous. So here we had a possible new sign of life on Mars (discovered, I was aware, in the few short months since *Lonely Planets* was first published) being taken seriously by this skeptical crowd. This stunning development heightened my awareness of the fluid state of our ideas about life in the universe and the breathtaking pace of discovery in the solar system.

Whether or not the methane turns out to be the breath of Mars bugs, our attempts to explain it will certainly further our ideas about how to detect life elsewhere. It often happens in science that our difficulty in understanding a new observation exposes the weaknesses in our previously agreed-upon ideas, ushering in a period of confusion that ultimately leads to new insights. The difficulty we are having in interpreting the methane discovery exposes the inadequacy of the well-accepted "disequilibrium equals life" protocol for identifying planets with life. *How much* disequilibrium does it take to signal the presence of life? This leads to some fascinating questions about the relationships between planets and life. Can a planet be a little bit alive? Or is there some threshold amount of biological activity required for a robust biosphere?

In my view, the methane is probably not a sign of life. (I'd sure love to be wrong about this.) For reasons I describe in this book, I don't expect that the signs of life on a planet with an atmosphere will be feeble or subtle. I believe that if life has survived on a planet for many billions of years (as it has on Earth and must have on Mars for there to be life today), then it will have become deeply intertwined with the atmosphere of that world in a way that will make the atmospheres of living worlds distinct—flagrantly distinct—from those of nonliving worlds.

As of this writing, both rovers are still crawling over new Martian vistas and calling home daily with their latest dispatches. They have both remained remarkably healthy in the killing Martian cold, but will they survive the coming winter? *Spirit* has developed a bum wheel, and *Opportunity* is tempting fate with daring exploits in a scary-looking crater. They may roam for another year, or die in the coming weeks, but either way they have, by any measure, far surpassed our expectations. I wish them well.

Up above, in Martian orbit, *Mars Express*, *Mars Global Surveyor,* and *Mars Odyssey* are all still operating, and plans are being drawn up for a more ambitious Mars rover mission, to launch in 2009 and search more directly for signs of past organic life. It's a busy time on Mars, and—life or no life—we need to keep exploring. Mars seems to be telling us that it once had conditions in which living creatures could have thrived. So how could we not go and look for fossils? Can you imagine—the chance to study the traces left by evolution on another world and compare them with the shells and bones of Earth?

Meanwhile, the *Cassini* spacecraft has just entered Saturn orbit, returning its breathtaking first close images of the mighty rings, with their picture-perfect waves and ripples. These gorgeous patterns look so much like mathematically simplified computer models that I am now completely convinced that the Pythagoreans were right: God is math. In October 2004, *Cassini* will make its first close encounter with the enigmatic, organic-rich moon Titan. Later, it will release the *Huygens* probe, which, in January 2005, will descend through the clouds and crash (or splash) onto Titan, photographing and sampling the air all the way down. Titan, as I describe herein, is a long shot for some unknown kind of extreme cold–adapted life, but a safer bet as a place that will teach us more about organic evolution in an environment that in many ways resembles the young Earth on the eve of life.

Back on the home planet, recent outbreaks of UFO sightings have been reported in Iran and Mexico, but nobody outside the world of dedicated "UFOlogists" seems convinced. Our radio searches for signs of intelligent life continue to grow in power and reach, but as yet the aliens have not called in.

Most discussion of life elsewhere focuses on the possibility of finding simple, microbial life. Yet SETI (Search for Extraterrestrial Intelligence) was also represented there under the astrobiology big top. Seth Shostak, from the SETI Institute, reported on the completion of Project Phoenix just two weeks before our conference, which listened to the radio emissions from 750 stars between 1995 and 2004, finding, alas, only noise and no signals. He pointed out that since Earth has been "leaking" radio and television signals for sixty years now, there are a thousand stars that are close enough to have noticed us and responded by now. None have. Perhaps this doesn't tell us much. Our galaxy alone has some hundred billion stars, so this sample represents less than a millionth of a percent. At least we know that not every single star in our galaxy is occupied by creatures who instantly answer any faint radio signal with a powerful reply, though in our case this could be due to the content of our programming.

Shostak was upbeat. SETI researchers are famously vague when it comes to making predictions of success. The standard line is that we could hear something any day, or it could take centuries or millennia, but we should keep on searching, because we can. So it came as a surprise to hear Shostak make a much more specific prediction—that SETI will succeed within twenty years if it is to succeed at all. Given the exponential increases in our listening power, he suggested, within two decades we will have the capability to search for signals from a large fraction of the stars in our galaxy. So, he said, we are certain to find them soon if they are there at all. It is always refreshing to hear something new at a SETI meeting, but I found myself wondering if this prediction doesn't somehow give us too much credit.

Even farther out there (in a good way) were the speculations of Steven Dick, NASA's chief historian, who raised the intriguing, perhaps disturbing, possibility that we may live in a "postbiological" universe, where extraterrestrial intelligence need not imply extraterrestrial life. This could come about if, on most planets, biology gives way to machines that outlive, outthink, and outevolve their slimy organic precursors. Dick suggested that it may be the machines who inherit the universe, and that in most places this may already have happened. Does intelligent life always cede its future to intelligent machines? Does it always seek out other life? Or does it usually destroy itself in an orgy of shortsighted technological cleverness? Is it always compelled to move beyond its home planet to colonize other worlds?

Questions about the behavior of intelligent life elsewhere inevitably lead back to questions about our own nature and future. Our evening debate on the wisdom and feasibility of the future terraforming of Mars led back to discussions of environmental ethics and the human role on Earth. What responsibilities, to any indigenous life-forms and to the life of Earth, including our descendants, do we carry with us as we explore Mars and contemplate going there to live? If we purposefully alter Mars to increase its ability to support life, will it be a desecration or a restoration, a salvation or a contagion? Do we display obnoxious hubris even to ask the question?

Joining us on the terraforming panel, virtually at least, by satellite hook-up from his home in Sri Lanka, was one of the science fiction heroes of my youth, Arthur C. Clarke. Unfortunately, the satellite connection wasn't working very well, and the comments of Sir Arthur (who, by the way, invented the communications satellite) were almost completely unintelligible. We all would have liked to hear what Arthur had to say on the matter, but his main role for the evening became one of providing amusement to the audience by interrupting the rest of us at awkward moments with little electronic bleeps, blurps, and word fragments, like some intermittent alien signals tantalizingly close to pure noise. Perhaps somebody up there was having some fun, reminding us that for all our talk about future high-tech wonders, our early twenty-first-century technology is still full of bugs. Why worry about terraforming Mars when we can't get a reliable phone connection between Sri Lanka and California? They can put a man on the moon but . . . In fact, during the entire evening, the only thing I was sure I clearly heard Clarke say about terraforming was, "Well, I think we should ask the Martians first." I agree. We do need to ask the Martians, or at least seek them out, so we can be bloody certain that there is nobody already living there before we start redecorating the place to be more like home.

On Earth this is a time of war and terror. But it's been a good year in the rest of the solar system. The real story of our time, in an evolutionary sense, may not be who started or lost this or that sorry war. This is when we take our first steps off Earth and gain the ability to seek out cosmic companionship. Yet, any aliens watching our behavior at the moment could not be blamed if they were to recommend against our being invited to join the galactic league of planets. Just remember, as Doris Lessing's dad used to say to her (as described in the dedication in *Shikasta*, "If we blow ourselves up, there's plenty more where we came from."

David Grinspoon
Denver, Colorado
July 2004

Preface

Here's how this book is supposed to start:

"My fellow humans, we stand here today at the edge of a new age of cosmic discovery that will transform all of our lives. Recent breakthroughs have sparked a scientific revolution in the search for life in outer space. Any day now, we may meet with success and find proof that we are not alone in the universe."

And, indeed, it's true. Numerous recent findings have helped to ignite a resurgence in scientific interest in the study of extraterrestrial life. These include possible fossils found in a rock from Mars, the first discoveries of worlds orbiting distant suns, evidence for the largest liquid-water ocean in the solar system underneath the icy surface of Jupiter's moon Europa, and an astonishingly wide range of newly discovered organisms living in extreme terrestrial environments previously believed to be uninhabitable. Together, these announcements have encouraged renewed hopes for finding alien life and helped to fuel a movement that some have called "the astrobiology revolution."

But, in researching this book, I have repeatedly been struck by the great similarity between our current ideas about alien life and those that were expressed decades and even centuries ago. Today, our researches and ruminations are informed by much new information. Still, a book summing up everything we *know* about alien life would contain only one word: *nothing*. I've managed to add an additional 150,000 by following our quests for aliens through history, speculative science, philosophy, and fantasy. After all, if Jerry Seinfeld can do a sitcom about nothing, why can't I write a book about something we know nothing about?

For me, extraterrestrial life has been a recurring theme from my days as a teenage space-head to my more recent employment as a professional planetary scientist funded to study astrobiology. Along the way I picked up academic degrees in the two least practical things I can think of: philosophy and planetary science. The topic of alien life allows me to finally combine all of this "useless and pointless knowledge."

The first popular-science book devoted to the question of extraterrestrial life was written in 1686. In the preface to his *Conversations on the Plurality of Worlds*, the French poet and philosopher Bernard le Bovier de Fontenelle wrote, "I've tried to treat Philosophy in a very unphilosophical manner; I've attempted to bring it to the point where it's neither too dry for men and women of the world nor too playful for scholars."

In his day scientists were still philosophers, science was still "natural philosophy," and belief in a cosmos full of planets inhabited by intelligent creatures was becoming widespread among European scholars. Like Fontenelle, I've been unphilosophical in places. Many scientific ideas and truths, to be expressed accurately, must be couched in endless caveats and qualifications. When I write, a little imaginary scientific colleague is always pouncing on my shoulder, telling me to clog the science at every turn, whispering in my ear, "Provide more detail," "Show *how* we know that," and "Don't you dare step out on that limb." I've largely tried to ignore that little monster, lest the book become too freighted with detail and fall from your hands.

I've organized the book into three parts: "History," "Science," and "Belief." In the first section I give a brief history of beliefs about ETs. Our changing images of extraterrestrials over the eons have reflected our evolving sense of ourselves and how we humans fit into the universe. The history I tell is selective and largely intended as a setup for what follows, to allow us to examine some modern ideas about extraterrestrial life in a historical context. This part is told in chronological order. Several topics and events that I mention here, I discuss in more detail later in the book. After laying down this rhythm track with the major beats of history, we are free to wander in time without getting lost. The rest of the book is nonchronological, and I sometimes jump back and forth in time chasing the thread of an idea.

When I changed offices a year ago, sacrificing some floor space for a view of the Rocky Mountain Front Range, the burly moving guy wrestling my filing cabinet full of heavy scientific reprints onto a dolly inquired, "Can I ask you a question? When I read in the paper that scientists agree on this or that, I wonder who decides what leading scientists all agree on, or what world scientific opinion is. How does that really work? Is there a commission?" It was a great question. In the

"Science" section I present a snapshot of present scientific thought about ET life. Part of what I've tried to do here is to hold up a mirror to the scientific process, in an attempt to illuminate how we decide what is true. My wife once compared me and my fellow planetary scientists to kids on a playground. Everyone is excited about a certain game for a while and all crowd around. Then somebody at the other end of the playground says, "Hey, check this out!" and all run over and start playing the new game. In these pages I try to portray the collective thought process of the scientific community, as ideas ripple through, are tossed around, put to the test, and are embraced or rejected. Sometimes they are embraced without being put to the test.

This is not a comprehensive treatment, as I have no wish to write a fifteen-volume *Encyclopedia Galactica*. Nor do I wish to be superficial, so I have been highly selective. I skim lightly over topics that have been well covered in other recent books and dive more deeply into areas that I feel have been neglected or mistreated elsewhere, or where I think I have something new to say. More comprehensive treatments of many topics can be found in the notes on sources and suggestions for further reading, or my sporadically updated on-line chapter notes at funkyscience.net. Copious additional illustrations for each chapter in this book can also be found at this site.

This book is highly opinionated and biased in numerous ways. In a raw field like planetary science or astrobiology, any researcher worth her grant money has opinions about contentious issues that are not held by all of her colleagues. I do not shy away from expressing my own nonconsensus views, but I will try to point out when I am doing so and even endeavor to describe the opposing views and explain why some researchers hold these erroneous opinions. ☺ I don't claim to be objective, unbiased, or correct about everything. This is a combination of what you'd hear if you sat in on one of my undergraduate lectures and what I'd tell you if we got talking over a beer afterward. Hopefully, I'll at least keep you entertained.

One of the themes of this work is the long, often uneasy relationship between astronomy and biology, the two scientific fields that must get in bed together if we want to make real progress in understanding the potential of this universe to create life in other places. After a century of flirtation, they started going steady in 1960 with a tentative, insecure union called exobiology. Then, after a thirty-five-year courtship, they finally took the plunge in the late 1990s in a marriage called astrobiology.

Some scientists have been studying the question of extraterrestrial life for decades, but until recently it was not considered entirely respectable, and it could even be a risky career move. The astrobiology movement represents a shift in attitudes among a scientific community that previously regarded this study with suspicion or derision. This was fueled by some exciting new discoveries, but also by a heightened awareness within NASA that the public and the media respond more to stories about alien life than to anything else we do. Here, I try to present a first-hand description, from within my own field of planetary science, of this large and rather abrupt change in attitude about the scientific search for ET life.

Last summer, at a friend's wedding reception in California, I ran into David Morrison, the head of space sciences at NASA's Ames Research Center and a leader in the new astrobiology movement. He asked me about my new book, and after I briefly described the work in progress he said, "It sounds like you're writing a book about astrobiology." I tried to explain to him why that wasn't exactly the case, but we were both heavily into the wedding spirit, so I probably didn't do an articulate job of it. There are several books about astrobiology—some of them quite good—and they all begin by saying: "Poised on the edge of the most momentous breakthrough in human knowledge, scientists have sparked a revolution that is sweeping the nation like a dance craze." Yet, the way I see it, astrobiology is the newest name for an old quest. Here I will try to put the present moment—and the belief that we are hot on the trail of aliens, and witnessing the start of a new scientific revolution—in a wider historical frame.

To me, the study of ET life is as interesting for what it reveals about our own biases and hidden assumptions as it is for what it reveals about life in the universe. We strain the boundaries of good science when we extrapolate to the rest of the cosmos based on our one example of a planet with life. We come across many questions that are great fun to contemplate. Could there be a world ruled by intelligent plants? Life on a gas world like Jupiter? Planets that are much *better* suited for life than Earth? Sure. Why not? Such questions force us to refine our views about intelligence and evolution and push us to define life in a universal sense, even though all we know is life on Earth.

Because I pay special attention to the limits of science, in a sense this is not strictly a science book but a work of *natural philosophy*. By using this term, I want to encourage a certain perspective on the science, an attitude where we keep ourselves honest by frequently questioning the framework of assumptions we use. I discuss some new ideas, currently on the shifting boundary between science and natural philosophy, that may be helping us to derive a less Earth-bound view of what it means for a planet to become alive.

Science is attempting a noble new assault on the question of our cosmic aloneness. But the question encompasses far more than just science. Astrobiology, I believe, is leading the way in helping the scientific community to, once again, think like natural philosophers, harkening back to a time when science was not distinct from philosophy, when the universe was not carved up into the turf of separate disciplines and subdisciplines each speaking its own specialized language, and when even the lines between our study of the physical universe and our spiritual quests were not so finely drawn.

After I've lulled you into submission and taught you to respect my authority as a scientist, then hopefully you won't notice when, in the "Belief" section, I start crawling farther out on various unsupported limbs, where the juiciest fruit is often found. In this section I allow myself more freedom to discuss my own beliefs. There is such a thing as scientifically informed intuition, and I rely more on this inexact tool in the last section of the book. My explanations and justifications are inevitably looser than those found in the "Science" section. I'm saying this now to give myself license, so look out.

Here I discuss our efforts to theorize about, and even communicate with, intelligent aliens living on planets circling distant stars. I also grapple with the widespread beliefs that aliens are already here studying us or perhaps even infiltrating our societies.

True confession: The whole time I have been writing this book I have had as a companion looking over my shoulder a three-and-a-half-foot-tall, large-headed, green alien with big black eyes. He is not flesh-and-blood or even silicon-and-plasma but a squeaky-squeezy plastic inflatable hanging by a string from the ventilation pipe, yet he serves to remind me that, at least as a cultural phenomenon, aliens are indeed among us. I don't attempt a comprehensive survey of the history and current phenomena of ufology, but I offer a montage of my impressions and experiences with some true believers.

After I discuss some of the more fringe ideas about aliens that permeate modern culture, I speculate on some future possibilities. What might intelligent life become, eventually, on Earth or elsewhere, and what are the implications, both scientific and spiritual, of these far-future evolutionary possibilities for the ultimate role of life and mind in our universe?

. . .

In writing this book I feel as though I've been abducted by aliens and abruptly returned to Earth after two years, hoping my friends, family, cats, and scientific and musical colleagues will remember who I am. A great many people have helped me on this interrupted journey.

For helpful comments on earlier drafts I thank Mark Yanowitz, Jason Salzman, Rebecca Rowe, Mark Bullock, John Spencer, Lester and Betsy Grinspoon and the Rev. Dr. Jeff Moore. Jake Bakalar provided detailed and insightful editing and Kevin Zahnle (a.k.a. Thrak) contributed perceptive and amusingly exasperating commentary on the entire manuscript. Fortunately for me, Kevin found all six mistakes in the first draft.

I thank my colleagues at the Southwest Research Institute for putting up with my strange schedule and being generous with their insights and expertise. In particular Hal Levison, Alan Stern, Robin Canup, Bill Ward, Henry Throop, Clark Chapman, and Luke Dones have enlightened me on various topics touched on in this book. Needless to say, none of them are to be blamed for any mistakes or opinions.

For conversations, correspondence, suggestions and sources I thank Steven Dick, Anthony Aveni, Harry Cooper, Larry Klaes, Andy Chaikin, Glen Webster, David Deamer, Jim Head, John Lewis, George Musser, Ken Croswell, Ben Bova, Amir Aczel, Peter Grinspoon, Josh Grinspoon, Tim Ferris, Carl Pilcher, Nick Schneider, Dorion Sagan, Penny Boston, John Bally, Larry Esposito, Fran Bagenal, Bruce Jakosky, Don Hunten, Sean Solomon, Dirk Schulze-Makuch, John Scalo, Alexander Zaitev, Martha Hausman, Chris O'Brien, Ginny Sutherland, Tom Donahue, Nicolai Kardashev, Philip and Phyllis Morrison, Frank Drake, Peter Ward, Don Brownlee, Chris McKay, Simon Conway Morris, Ronald Weinberger, John Mack, Tim Pickard, Dennis Overbye, Guillermo Lemarchand, Doug Vakoch, Jacques Vallee, Jack Mustard, Athena Andreadis, Andy Spencer, Shaun Brooks, Rick Griffith, Bob Pappalardo, Guy Consolmagno, John Rummel, Mike Meyer, David Morrison, Chris Chyba, Jeff Kargel, Damon Santostefano, Helen Thorpe, Peter Heller, and Dan Sjogren.

Borin Van Loon provided inspired artwork that graces some of these pages.

Thanks to Jim and Harriet Campbell for the peace and hospitality of their Willow Spring Bed and Breakfast in Colorado's magical San Luis Valley. NASA and the National Science Foundation have generously supported my research into comparative planetology. For keeping the Funky Science office lurching along, I thank Holly Holloway, Tom Arriola and, in particular, Antony Cooper for his unflappable calm, competence, generosity, and creativity.

My brilliant and thoughtful agent Tina Bennett patiently guided me through the entire book process, from proposal to publication. At Ecco/HarperCollins I thank Dan Halpern, Julia Serebrinsky, and Gheña Glijansky for invaluable advice, perseverance, and faith.

Most of all, I thank my wife Tory Read for her wit, charm, intelligence, artistry, and soul, for reading through numerous drafts, providing skillful editing, love, patience, and laughter, and for making whatever planet we happen to be on the opposite of lonely.

David Grinspoon
Denver, Colorado
March 2003

PART I

History

1 | Spirits from the Vasty Deep

We have all felt this impulse in our childhood as our ancestors did before us, when they conjured goblins and spirits from the vasty void, and if our energy continue we never cease to feel its force through life. We but exchange, as our years increase, the romance of fiction for the more thrilling romance of fact.

—PERCIVAL LOWELL

I can call spirits from the vasty deep.
Why, so can I, or so can any man, but will they come when you do call for them?

—SHAKESPEARE, *Henry IV*

PROLOGUE: BRUISED BY AN ALIEN

It was a dark and stormy night—and already a weird one. My friend Damon and I trudged around through a snowstorm in the Meatpacking District, hunting for a spoken-word/hip-hop/acid-jazz event someone had said we had to see, while the wind whipped the streets into soft, majestic canyons. We were ants lost in a liquid-filled snowy Manhattan, and somebody up there was giving it a good shake. After hours of increasingly blind and frozen searching, we ducked into a corner bar where a jazz quartet was only slightly mangling Coltrane's "Out of This World" and sat down to regroup and have a drink. At one point we looked up at the TV and there was Rudy Giuliani dancing with the Rockettes wearing fishnet stockings and high heels. It was a little unsettling.

Soon it was several drinks later, the band was finishing Miles Davis's "So What." and our waitress was ready to end her shift. She leaned over our table and asked just what we were blathering on about and what we were doing in New York. We had been baiting each other—as we have been doing since the eighth grade—into some twisted science fiction scenario that seemed good at the time. Of course she was much more impressed with Damon, who is both cuter and a film director, than me, a scientist and a "writer" (yeah, right!). She started talking about her acting experience and aspirations.* She was not obnoxious or pushy, just friendly, and we welcomed the diversion. Eventually, perhaps just to be polite, she asked me what kind of scientist I am. What I *said* was "I study planets and I'm writing a book about aliens," but what I was *thinking* was "I wonder what her story will be." And then she told us.

(Cue spooky, New Age music.)

One night about a year ago she had stayed out late and had a few drinks herself—she doesn't remember how the night ended. There was a strange interval of missing time, and she woke up the next morning with an elaborate marking on her right outer thigh. It was a large, stick-figure discoloration about six inches tall. It looked just like a bruise, but it didn't hurt like one. And just like a bruise, it faded. The design appeared to show some sort of helmeted and antennaed space-creature. I asked her if she could draw it for us, and she did, right there on the back of our bar tab. She even signed it "Jillian." Here it is:

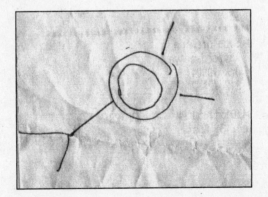

---

*This is how you know you're not in Kansas anymore. In Denver the waiters are semi-employed musicians. Actors would be waiting tables in New York or L.A.

Because you never know, I asked her permission to use the drawing and the story in this book. She agreed without hesitation. She smiled but didn't seem to be putting us on. This clearly intelligent, articulate, and apparently undisturbed woman was certain that she had had some kind of alien encounter. Damon asked her if quaaludes were involved but she swore that it was nothing like that.

Given the Giuliani vision, not to mention Jillian's story, I would have been inclined to think I had just hallucinated the entire evening, except the next morning when I woke up, there it was. No markings on my thighs, but a bar tab in my wallet with an alien on the back.*

One thing I've learned is that when it comes to aliens, everywhere you go, somebody's got a story. This actually solved a problem for me. I already knew that I wanted to write about science and our beliefs about aliens. To me, this is familiar territory. But how would I include stories about UFOs, abductions, cattle mutilations, crop circles, and so forth, the phenomena that are widely associated with the topic of extraterrestrials everywhere except in the pages of scientific journals? There are so many stories out there. It would be futile to try to be proportionately representative, yet you'd have to be blind not to see that aliens are all around us. Without really trying I've picked up my share of alien paraphernalia: beach towels, glow pops, rolling papers, magnets, a little green dancing statuette, and even a pipe-smoking-alien lawn gnome. Much of this alien lore is tongue-in-cheek, but some not entirely, and some leaves the tongue just hanging out there flapping loose in the breeze. Fortunately, many on all sides of the UFO debates approach the question with a proper dose of humor.

But what balance to strike? After all, dammit Jim, I'm a scientist, not a comparative sociologist. Anything I have to say that is of any interest to you, fair reader, is more likely from the perspective of a working scientist, not a UFO dilettante.

Since it seems that virtually everyone has something to say on the subject, a strong belief, an opinion, or a must-read source of esoteric evidence, I decided that rather than go to the aliens, I'd let them come to me. After all, the cultural airwaves are saturated with alien signals, on all frequencies, in all directions. All you need to do is unfurl your antennae, turn

---

*And they say that there are never identifiable artifacts left over from alien encounters. Well, I still have it if anyone wants to run isotopic tests.

on your signal analyzers, and let her rip. So, I thought, I'll just pay attention to all of the transmissions passing through my little region of space.

It's pretty hard to find someone who doesn't believe in aliens, although that can mean very different things to different people. Some folks are convinced that NASA has photographed—and covered up—elaborate cities on the surface of Mars. Others believe that little ET creatures make nocturnal visits to the bedrooms of innocents, kidnapping them, doing strange experiments, and then returning them to their beds with important lessons for humanity. Legends of crash sites with alien bodies and smashed saucers abound. The government has hidden and dissected them, the story goes, in order to "reverse engineer" fantastic energy and propulsion technologies that could set us all free, if only the truth were released. On the other side of the rainbow, and closer to my neck of the woods, are scientists, "astrobiologists," who, through some reverse engineering of our own evolution and biochemistry, have convinced themselves that any life on other worlds must be just like ours.

These diverse beliefs are all modern responses to an ancient question.

THE QUESTION

Hello?

Is anyone out there?

The question persists, ringing through the void like an electromagnetic prayer. It may be innate—an instinct for self-discovery built into the cosmos, a reflex reaction to conscious awareness, springing autonomically to mind like air drawn into a lung.

The question goes way back. We've been wondering, speculating, fretting, hallucinating, and prognosticating about aliens about as long as anyone can remember. Strange creatures, variations on the human theme, have always inhabited our fantasies and nightmares. Ever since the lost, distant time when we became self-aware, waking slowly from our ape dreams, pausing on some East African savanna to stare down in amazement at our flexible fingers or up at the silent stars, we've had the capacity and the inclination to wonder whether there were others like us elsewhere.

Our ideas about where this "elsewhere" might be have evolved along with our sense of where "here" is. There is an old, somewhat magical sense of the term *other worlds,* where physical existence and location in space are irrelevant. These presumed realms of existence that somehow parallel or mirror ours have created infinite space for all manner of imagined creatures not of this Earth. As we peer further back in time, we see the fantastic off-world creatures permitted by modern science morphing back into gods, angels, dragons, underworld demons, and animal spirits.

To dream up modern, scientifically sanctioned extraterrestrials, we first had to think of ourselves as terrestrials. We had to conceive of our home planet as a limited part of a larger universe containing other similar places. The Epicureans of ancient Greece started us down this road. Their universe contained an infinite number of worlds, many of them inhabited. Epicurus himself (341–270 B.C.) said, "We must believe that in all worlds there are living creatures and plants and other things we see in this world. . . ."

Did the Greeks mean what we mean by *other worlds?* Yes and no. Their other worlds were not mere abstractions, they were actual physical places like Earth. But they had nothing to do with stars, planets, or anything we observe in the sky. Indeed, the Greeks thought of the stars as being not so far away, and the idea that the sun might be just another star, if it occurred to anyone, must have seemed absurd. Their other worlds were somewhere beyond the stars, too far away for us to ever see. The history of ideas about extraterrestrial life is one of shifting ground between physical and metaphysical realms. In this geography, the other worlds of the Epicureans occupied an interesting middle ground—places that were very real but forever concealed from us.*

The Epicureans reasoned that the sheer vastness of the cosmos made all things, including other inhabited worlds, not only possible, but inevitable. The Epicurean Metrodorus wrote, "To consider the Earth the only populated world in infinite space is as absurd as to assert that in an entire field sown with millet only one grain will grow."

This idea, that a universe containing an infinite amount of stuff not only allows but requires the existence of anything we can think of, is often called the *principle of plenitude.* As I will show you, this principle has often reappeared in more modern guise.

Not all the Greeks agreed on this. Plato and Aristotle opposed the Epicureans and argued against the existence of multiple worlds. Aristotle had a scheme that explained the entire physical structure of the universe,

---

*Modern cosmology, with its "multiverse" containing infinite other universes that we can never observe, has returned us to a similar picture.

but in the process ruled out other worlds. According to his doctrine of "natural place," everything is composed of the four elements earth, air, fire, and water. Heavy stuff falls toward the center of the universe and light stuff rises. All motions—rocks tumbling down a cliff face, leaves blowing in a breeze, a billowing column of smoke, and water cascading over a cataract—are explained by the settling of earth and water, and the soaring of air and fire. This theory held sway in European thought for over a millennium both because it worked pretty damn well and because of Aristotle's aura of untouchable authority. In fact, we have found much of what Aristotle was groping for in our periodic table of the elements and laws of gravity.

So what was Aristotle's problem with other worlds? He didn't think of gravity as a force that draws every object toward all others. Everything was pulled only toward the center of the Earth, which was also the center of the universe. If any other worlds were out there, they would come crashing down on our heads. Besides, how would these worlds hold themselves together if their earth and water were drawn not toward their own centers, but toward *the* center, down beneath our feet? Aristotle also believed that the physical laws governing heaven and Earth were fundamentally different. In this divided cosmos, everything beyond the atmosphere was made of completely different matter from all things terrestrial, so there could be no life like ours out there.

Both theories—Epicurean and Aristotelian—are elegant and logical, but they lead to opposite conclusions about other worlds and extraterrestrial life. This, in a nutshell, is the problem with philosophizing in the absence of evidence. Without observations to ground our theories in the real world, we can produce flawless arguments to reach any conclusion we want. Physical theories constructed without observation, however ingenious they may be, are ladders standing in quicksand. Whether we're really doing much better than this with our modern scientific studies of life elsewhere, or perhaps just fooling ourselves into thinking that we are, is one of the questions I'll wrestle with in this book.

ON THE REVOLUTIONS

We couldn't really worry about life on other planets before we realized that we live on a planet, and for that we needed a revolution—the Copernican revolution. That's the name we give to the radical metamorphosis from a geocentric (Earth-centered) to a heliocentric (Sun-centered) worldview. This shift in perspective was the first and hardest jolt in a series of sudden awakenings to just how pint-size and peripheral we—and our entire planet—are in the mind-numbing vastness of creation.

Except for a few historical oddities that never gained any following, all models of the cosmos up into the sixteenth century placed the Earth in the bull's-eye at the center. That's no surprise. If you simply believed your eyes and your sense of balance, and you did not have the benefit of modern astronomical observations, you would hold these truths to be self-evident: that our world is immobile and the "dome of night" is as it appears: a perfectly spherical realm in which all celestial bodies travel in paths that circle us. Any theory that questioned this picture challenged lifetimes of intuition and common sense.

Along came Nicolaus Copernicus, a visionary Polish astronomer who advocated a Sun-centered cosmology mostly for aesthetic reasons—it was a more pleasing design for the solar system. In his book *De Revolutionibus Orbium Coelestium (On the Revolutions of Heavenly Orbs)*, he proposed that the Earth orbits the Sun along with the five other visible planets.

It must have sounded crazy at first. Today we are brought up immersed in a Copernican world. We learn as soon as we can understand the words that the Sun appears to travel through the sky because we stand upon a spinning Earth. We are continually exposed, through catalog covers, movie posters, Web sites, and the evening news, to the sight of our blue-and-white home planet spinning in space. Given this subliminal indoctrination, it is hard for us to imagine the frightening disorientation that Copernicus's theory produced in the minds of his contemporaries. I'm not sure what the equivalent theory would be today, but its author might be ridiculed as a New Age flake, dismissed as a dreamer or pitied as a babbling, delusional acid casualty. It would have to be a theory that suggested that much of our shared reality is illusory. Copernicus took contemporary common-sense conceptions (about "the world," the Sun, up and down, the human place in the grand scheme, and what it means simply to stand still and watch the stars), turned them on their head, and sent them spinning.

Because it sparked a complete shift in worldview that has long since been accepted by all but a few Flat Earthers and Republican senators and was the first step in a drastic descoping of human importance and influence in the universe, *De Revolutionibus* is a good candidate for

Most Radical Book Ever Written. Copernicus himself never had to deal with the upheaval caused by his revolution. He received the first printed copy of his new book in May 1543, on the day he died.*

Astronomers didn't immediately embrace the new system of Copernicus. It didn't actually do a good job of predicting the motions of the planets—supposedly its major selling point. That was because Copernicus, like Aristotle nineteen hundred years before him and every philosopher in between, assumed that all shapes and paths in the heavens were drawn in circles. In the Copernican system, the Sun occupied the exact center of perfectly round concentric orbits on which Earth and the other planets traveled. And so it remained, until a mad scientist named Kepler came along and squished those circular orbits into oval-shaped ellipses.

LOVE 22 AND KEPLER'S LAWS

Johannes Kepler was a late-sixteenth-century philosopher/freak who walked the fine line between genius and delusion. He had a lifelong conviction that a secret, simple mathematical order lay hidden just beneath the confusing, chaotic surface of the universe. He found it hard to find steady work and, like many astronomers of his day, kept a day job as a court astrologer, casting fortunes for the rich and famous.†
With a seamless blend of mysticism and science he pursued his search for the numerological and geometrical designs of creation.

The more I learn about Kepler's actual life and work, as opposed to the filtered version we are taught (and then teach) in Astronomy 101, the more he reminds me of guys like Love 22. When I was in college in Providence, this jovial crazy person named Love 22 hung around Thayer Street preaching the gospel of the number 22 to all who would listen. He lived in a red-white-and-blue converted school bus, and though his material possessions were few, he knew the secret of cosmic harmony and wisdom. It all had to do with the number 22. Wearing his trademark Uncle Sam uniform, he handed out $22 dollar bills and showed how the number 22 is hidden in the names of presidents, prophets, and all phrases of spiritual wisdom. He ran a perpet-

---

*In 1686, Bernard le Bovier de Fontenelle wrote of Copernicus's death, "He didn't want to rebut all the contradictions he foresaw, and he skillfully withdrew from the affair."
†Today we just write grant proposals.

ual campaign for president and governor, on the Love 22 ticket (Love for Gov!). Love was quite the comedian but he seemed sincere. My friends and I thought that he was rather sweet and enjoyed talking to him.*

There is only one Love, but there are many like him. There is a personality type—and you've got to have it to go in for this kind of existence—that is remarkably impervious to the fact that virtually everyone thinks you're out of your mind. These self-appointed misunderstood geniuses are convinced they've discovered some system of knowledge that humanity needs. I've met them handing out pamphlets in cafés in San Francisco, Tucson, Providence, Cambridge, Boulder, Ann Arbor, and Madison, trading wisdom for cash to buy food or wine or to Xerox more pamphlets. Often, more than money, they want a sympathetic ear. An earnest fellow in Boston once showed me mathematically detailed plans for faster-than-light starships and time machines.

Because I've published articles in popular-astronomy magazines, I get letters from people all around the world with elaborate theories of everything. I don't throw them out. I keep them in a file labeled *Kook*. Maybe somewhere in the kook files of the world's astronomy writers is an obscure tract containing the seeds of the next Copernican revolution. Kepler, the father of planetary physics, if he were alive today, might well be living in a converted school bus on the outskirts of some college town peddling mystical pamphlets, living off donations, and spouting cosmic wisdom to anyone who would listen.

Kepler is a missing link between the two modern sources of belief in aliens. The man who worked out the mathematical laws of planetary motion was motivated largely by a desire to cast more accurate horoscopes. Today, two separate strains of believers about alien life coexist in our culture: rationalist scientific followers of SETI (the Search for ExtraTerrestrial Intelligence) and mystical, New Age UFO believers. The roots of science and pseudoscience are completely intertwined in Kepler's work. Like a modern scientist, he was seeking the simple patterns underlying apparently complex phenomena. Like a modern New Ager he was obsessed with numerological coincidences and convinced they had cosmic significance.

---

*I hadn't heard of him in, well, nearly twenty-two years, but today, thinking about Kepler, I searched for Love on the Web and learned that he's in Key West, still doing his 22 bit with what seems to be a larger comedy factor than I remember.

Kepler believed in the Copernican system, for reasons that were essentially mystical. The Sun should be at the center of everything, he felt, because it is the symbol of God and the source of heat and light. In his restless, obsessive quest to explain the proportions and motions of the planetary orbits, he crafted innumerable schemes, most of which seem today to be elaborate, colorful nonsense. He wondered why there were six planets (Earth plus the five visible to the unaided eye). He wanted to find the significance of this number and an explanation for the five distance intervals between the planets, a simple geometry that would make it all fit together and reveal the plan of the creator. At age twenty-four, in a fit of inspiration, he thought he found the answer.

He seized upon the fact that there are five "perfect solids" (pyramid, cube, octahedron, dodecahedron, and icosahedron) and also (get this) five unexplained distances between the planets. Coincidence? He didn't think so. He constructed a model of the solar system with the five perfect solids stacked tightly inside one another, like a cubist set of Russian dolls. When he discovered that the relative sizes of the shapes in this model are exactly the same as the size ratios of the planetary orbits, it blew him away. This was the secret structure to the universe he had been searching for. "The delight that I took in my discovery," he wrote, "I shall never be able to describe in words."

Today Kepler's solar system model, like most of his other discoveries, is seen as a wacky and amusing dead end. Yet, Kepler considered this model, not "Kepler's laws" that we teach in every astronomy course today, to be his greatest achievement. This design for the solar system and the rush he got from its discovery inspired a lifelong, and ultimately successful, quest for the laws of planetary motion. Though his genius was profligate, undisciplined, and borderline crazy, his keen intellect was less bound by convention than that of his contemporaries.

Kepler was bothered by the failure of the Sun-centered solar system model to predict planetary motions accurately. The planet Mars, in particular, strayed from the sky path prescribed for it by the Copernican model. In his determination to save the Copernican system, Kepler tried innumerable mathematical schemes to make it work, often obsessing maniacally for months on a new idea, only to toss it out and start on another. Finally, in a classic example of out-of-the-box thinking (in this case the box is round), he calculated the motions that Mars would exhibit if its orbit were not circular but egg-shaped, elliptical. Eureka!

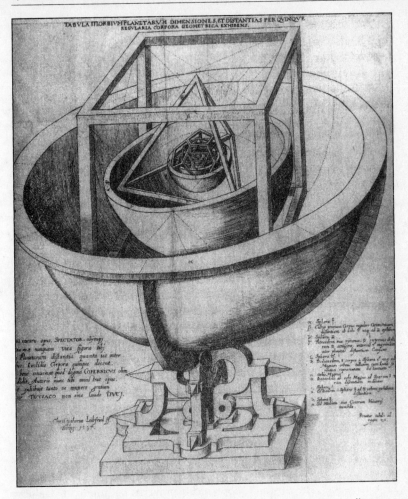

Kepler's "new age" model of the solar system, which he saw as his greatest discovery.

Suddenly it all worked. Mars and the other planets moved exactly as predicted once Kepler liberated them from Aristotelian circles and allowed them to follow elliptical paths in a Sun-centered system.

It worked. But was it real? Were the planets—Earth among them—really moving around the Sun in this manner? Could our world, our rock-solid, all-encompassing Earth, truly belong to the same class of objects as those ethereal little lights roving the night sky? The answer was not long in coming.

THROUGH THE LOOKING GLASS

In January 1610, Galileo Galilei swung his crude telescope skyward, smashing the perfect, crystalline celestial spheres of Aristotle,* and knocking the Earth off its immobile, biblically enshrined pedestal. Galileo's early observations of Venus, Jupiter, and the Moon were nails sealing the coffin of the pre-Copernican worldview.

Studying Venus, Galileo saw what anyone with a small backyard telescope and the patience to watch for a few months can see today: the evening star is approaching and receding from Earth. He realized that Venus is shining by reflected sunlight and, from Earth's perspective, passing alternately in front of and behind the Sun. This only makes sense if Venus and Earth are both traveling around the Sun.

Turning his glass toward Jupiter, Galileo discovered that the giant planet was attended by four tiny companions that tag along on its orbit, rearranging themselves night after night. He had found the moons of Jupiter, the first new worlds. The existence of moons orbiting Jupiter showed that not everything travels around the Earth. This spelled doom for the old Earth-centered cosmos of Aristotle.

The surface of our Moon, viewed through Galileo's telescope, displayed a complex topography of shadows, pits, and mountains. This was not the flawless, smooth sphere required by Aristotle's dichotomy between a perfect, spiritual celestial realm and an imperfect Earth. The Moon's "flaws" suggested to Galileo that it was a world like Earth. Suddenly, it didn't seem at all preposterous that the other planets might be Earth-like. The abstract Copernican universe became real. Galileo concluded that the other planets are worlds, and that "the world"—our Earth—is merely one of many planets circling the Sun.

Galileo caught hell from the Church. In what has become a modern myth of science's collision with biblical authority, he was brought before the Inquisition, forced to recant his Copernican beliefs, and lived out his days under house arrest.

---

*Aristotle's spheres were made of crystal because they had to be solid to hold up the Sun, Moon, planets, and stars, yet transparent since we can see through them.

PLANET-HOPPING JESUS

Largely because Aristotle was invulnerable, early Christian scholars almost unanimously denied the existence of other worlds that might be occupied by rational beings. In the tale of Genesis, God creates the Earth for human habitation, and other worlds are not mentioned at all. Like hand in glove, this human-centered narrative fits snugly into Aristotle's cosmos in which perfect, untouchable heavens envelop an Earth that is unique, central, separate, stationary, and inferior.

Furthermore, the possibility of intelligent creatures on other worlds presented paradoxes for anthropocentric Christianity. If Jesus died for our sins alone, would intelligent aliens on other planets be damned by his neglect? Or are they free from sin? If so, why did we get such a raw deal? If not, was Christ a planet-hopper who managed to be incarnated on all such worlds?

St. Augustine, widely recognized as one of the greatest thinkers of Christian antiquity, argued that if other worlds were inhabited by humanlike creatures, each would need a Savior, which was impossible because Christ was singular. Several scholars, however, found clever loopholes through which to admit intelligent extraterrestrials into a Christian universe. The most common argument was that other worlds would not need a redeemer because mankind's sin was so original. More specifically, aliens could not be sons of Adam and did not inherit his sin, so they were off the hook.

Aristotle's hold on the Christian imagination began to loosen when some scholars pointed out that a universe with only one world implied limits on the creative powers of God. In 1277, Etienne Tempier, the bishop of Paris, issued a proclamation declaring Aristotle's terrestrial/celestial dichotomy a heresy. This precipitated a sea change in attitudes toward other worlds and alien life. Many Christian scholars began breaking from Aristotle, and numerous treatises were published arguing that God could make as many worlds as he damn well pleased. He is, after all, God.

Was the existence of alien life forbidden by the uniqueness of Christ's incarnation or required by God's omnipotence? In 1440 Nicholas of Cusa, a German ecclesiastic, wrote *Of Learned Ignorance,* a widely celebrated book that exuberantly rejected Aristotle's hierarchical, Earth-centered cosmology, advocating in its place a universe bustling with life on every star. But Cusa was not scorned by the Church hierarchy for his belief in life elsewhere. On the contrary, after writing *Of Learned Ignorance* Cusa was made a cardinal. So why did the Church celebrate Cusa and, 150 years later, condemn Galileo?

There are several reasons. First, Galileo was somewhat of a tactless boor—a quality often left out of the Galileo myth—and his obnoxiousness helped seal his fate. Perhaps if he had put the right spin on his new discoveries, Rome would have showered him with praise and rejoiced in the addition of new worlds to God's creation. Instead, he seemed to go out of his way to piss off the Church authorities with his know-it-all comments on Scripture. He might have fared better if he had kept a lid on it and not told the clerics how to interpret the Bible.

In his *Dialogue concerning the two Chief World Systems* (1632), Galileo popularized his findings and proselytized for Copernicanism. In this book, the character who played the role of doubting the Copernican system was a pompous ass with the unflattering name Simplicio. In an impolitic move that well illustrates his arrogance, Galileo had Simplicio give voice to the anti-Copernican views of Pope Urban VIII, mirroring the pope's words so closely that His Holiness became convinced that Simplicio was created to mock him. The infuriated pope was all too eager to preside over Galileo's sentencing.

Galileo was also a victim of bad timing. He challenged authority at a time when the Church was threatened by the Reformation. Even worse, Galileo's world-shaking telescopic discoveries were made before heretic monk Giordano Bruno's ashes had cooled. Bruno, a Dominican friar who was condemned and burned at the stake in Rome on February 17, 1600, believed in an infinite cosmos filled with life virtually everywhere—on planets, stars, meteors, you name it. He is often mentioned in the same breath with Galileo as another martyr for Copernicanism and science in general. In reality, his colorful advocacy of other worlds and alien life was seen by his persecutors as a minor offense compared to his sorcery, pantheism, and denial of Christ's divinity. Bruno was murdered by the Church, first and foremost, for espousing superstitious mumbo jumbo and devil worship, and less so for promoting the new astronomy. If they wanted to, New Age mystics or satanists could claim him as a martyr with at least as much veracity as scientists do today.

Bruno was one of the first to advocate that each star is a sun with its own retinue of orbiting planets inhabited by intelligent creatures. Yet, he based this conviction on metaphysical principles and mystical visions rather than observation or physical theory. Bruno couldn't have cared less about evidence, measurement, or the intricacies of planetary motions. His adaptation of Copernicanism was a convenient co-opting of a recently published theory to support his belief in an infinite number of inhabited worlds—a belief derived from a spiritualistic faith in the unity of the cosmos. If anything, Bruno did harm to the progress of science (and certainly to poor Galileo) by encouraging the Church authorities to associate Copernicanism with flagrant anti-Christian agitation. Surely some of the wrath that the Church vented on Galileo was really meant for Bruno, who refused to recant, reaffirming his beliefs and taunting his persecutors with his dying breaths as flames engulfed his body and freed his soul to travel among his infinite worlds.

Personality and timing aside, Galileo's biggest problem was simply that he had found the goods. The stark reality of his evidence suddenly made Copernican beliefs much more threatening. Before Galileo's telescope opened a window to a new reality, cosmological questions were all hypothetical. Discussions of other worlds seemed as abstract and immune from verification as arguments about how many angels could dance on the head of a pin. Now, there were actual planets that you could see in the sky, and their existence implied that the Earth itself is in motion, contrary to the received truth found in Scripture. Telescopes and planets are mentioned in the Bible no more than particle accelerators and quarks, but with a little digging and creative interpretation, those who shrank with horror from the new, less human-centered universe could find scriptural objections to back up their fears.

The difficulty of the transition from biblically received knowledge to observational cosmology is well represented in a scene in Bertolt Brecht's play *Galileo*. A group of learned astronomers have called upon Galileo to express their concern over his claims of finding new worlds. He invites his skeptical visitors to simply have a look through the telescope and see the new worlds for themselves. He is confident that, once they have seen with their own eyes, they will drop all objections. Fearing trickery or sorcery, they refuse to look.

Forced to take a stand by Galileo's observational successes and rhetorical excesses, the Church decided to put the kibosh on Copernicanism, but it was too late. Word was out. Telescopes are easy to manufacture. Soon observers all over Europe were marveling at the moons of Jupiter and mapping the mountains of the Moon. Cusa and Copernicus had laid the dry timber and Galileo had provided the spark. A wildfire of rampant Copernicanism ripped through seventeenth-century Europe, and though the Church leaders spread fear to douse the flame, they could not stamp it out.

| Plurality of Worlds

Every great scientific truth goes through three states:
first, people say it conflicts with the Bible;
next they say it has been discovered before;
lastly they say they always believed it.
—LOUIS AGASSIZ

Where the telescope ends, the microscope begins.
Which of the two has the grander view?
—VICTOR HUGO, *Les Misérables*

AFTER THE REVOLUTION

The Copernican revolution opened the floodgates, and the modern debate over life on other planets began in earnest. The original Copernican revolutionaries had approached ET life cautiously, but their followers went wild. Advocacy of the Copernican solar system became identified with a universe filled with planets and intelligent life, a "plurality of inhabited worlds." Throughout the seventeenth, eighteenth, and nineteenth centuries, the phrase *plurality of worlds* was used to describe the idea of a densely inhabited cosmos, and "pluralists" believed in it.

Many of Galileo's followers and defenders confused the issues of Copernicanism and plurality. In the minds of the most ardent proponents of each, life on other worlds validated the Copernican system and vice versa. They would stand or fall together. The logic must have been irresistible: if Earth is just another planet, and other planets are like Earth (as far as could be told with seventeenth-century telescopes), then why shouldn't the planets be Earth-like in every way, including inhabitants? This is called the argument by analogy, and it has reappeared in various forms right up to the present as a rationale for belief in extraterrestrials.

When we knew next to nothing about the other planets, the scant observations that did exist were interpreted as implying the existence of extraterrestrial life. The facility with which seventeenth-century natural philosophers put such a hopeful spin on their interpretations should serve as a cautionary tale to modern scientists itching to find evidence of alien life.

Kepler's belief in an advanced civilization on the Moon was based, at least in part, on careful observations. Recognizing the importance of air and water for life, he found evidence for both on our Moon. He thought that the dark areas were water and the bright areas dry land. In support of this he noted that the bright areas have rugged surfaces.

Kepler decided that the Moon was not only habitable but densely inhabited. Because the lunar "spots" (he did not call them craters) are perfectly circular, he judged they must be cities: "When things are in order, if the cause of the orderliness cannot be deduced from the motion of the elements or from the composition of matter, it is quite probably a cause possessing a mind."

Kepler couldn't imagine a natural process that created such perfectly circular forms, so he concluded that rational creatures were responsible. Nearly four hundred years later, Arthur C. Clarke wrote *2001: A Space Odyssey*, in which astronauts on the Moon dig up a monolith underneath the crater Tycho (named after Kepler's despotic boss) and come to the same conclusion: this doesn't look natural, so someone must have built it and put it there.

Now we know that the lunar craters are not cities. We've been to the Moon. We understand how the energetic explosions caused by high-velocity cosmic collisions produce beautifully circular impact craters on all planets with solid surfaces.

If we find ordered structures without a known "natural" cause, is this an indication of extraterrestrial intelligence? Kepler thought so, and modern theorists of SETI (the Search for ExtraTerrestrial Intelligence) agree.* Yet, as Kepler's inference of lunar cities illustrates, failing to deduce the "cause of the orderliness" may be due to the ignorance of those attempting the deduction. Looking for surprising order in nature

---

*Proponents of "intelligent design" creationism make a similar argument, replacing ET with God.

is not a bad approach for seeking extraterrestrial intelligence. But Kepler's mistake should remind us to beware lest the limitations of our own intelligence cause us to find erroneous evidence of intelligence elsewhere.

At the time of the Copernican revolution, the discovery of the "new world" by European explorers was recent history. This strengthened pluralist arguments by analogy. Vast realms of our own planet, previously unknown to Europeans, were found to be densely inhabited by diverse creatures both familiar and exotic and, most importantly, by other humans.* In the wake of these findings, widespread life on the other "new worlds" of the post-Copernican solar system seemed just as likely.

In addition to showing the planets to be worlds, the telescope also revealed a seeming infinity of unknown stars, which many observers believed to be suns. Teleological reasoning—the logic that things must be created for a purpose—was rampant in the seventeenth century. It was widely believed that the Earth, Sun, and Moon were created for our habitation, comfort, and pleasure. All of those innumerable suns must have been created for *someone*.

CURIOSITY AND POOR EYESIGHT: FONTENELLE SPREADS THE WORD

In 1686, Bernard le Bovier de Fontenelle, a poet, novelist, and natural philosopher who would later become secretary of the French Academy of Sciences, wrote *Entretiens sur la Pluralité des Mondes* or *Conversations on the Plurality of Worlds*. This book was an instant best-seller and international sensation. Excitement about pluralism was building, and Fontenelle both rode and helped to propagate that wave. Writing in a playful, whimsical style, he produced what has been described as the first popular-science book. To this day it is still a good read, and not just for a peek into the mind of a seventeenth-century pluralist and popularizer. It is a work of sweeping imagination written in provocative, witty prose with a bit of an edge. Fontenelle predicts spaceflight, discusses the habitability of other planets in our solar system and beyond, and offers vivid descrip-

---

*Whether the disastrous consequences of this "discovery" for those human inhabitants augurs potential dangers of contact with extraterrestrials is a question I'll return to in later chapters.

tions, centuries before the Apollo project, of the Earth as seen from space. He even urges his readers to consider what the inhabitants of Jupiter might think about Earth: "Even if they saw our Earth on Jupiter and knew about it there, still they wouldn't have the faintest suspicion that it could be inhabited. If anyone were to think of it, heaven knows how all Jupiter would laugh at him. It's possible we're the cause of philosophers being prosecuted there who have tried to insist that we exist."

Conversations consists of a sequence of five dialogs, on five consecutive moonlit evenings, between a learned philosopher and an uneducated but sharp-minded marquise. The philosopher is convinced that the heavens are full of inhabited worlds. The marquise initially doubts this pluralist vision, and as she is gradually won over, Fontenelle skillfully defuses the doubts in the minds of his readers. By using a female character, Fontenelle implicitly advocates the notion that women can handle physics and philosophy—a radical notion in seventeenth-century France—while infusing *Conversations* with a delightful flirtatious quality.

Placing his wild conjectures in the voices of his characters, Fontenelle leaves the reader uncertain if the author himself actually believes what his philosopher argues. This frees him to offer dangerous speculations, as in his hints that the planets might be where they are just by chance instead of a creator's design. Aware of the Church's continuing objections to pluralist ideas, Fontenelle used a standard disclaimer, asserting that the aliens of his speculations were assuredly not men, did not descend from Adam, and were therefore irrelevant to biblical concerns. Nevertheless, *Conversations* was placed on the Catholic index of banned books one year after publication. But it was a hit with the public.

Passionate about his beliefs but never dogmatic, Fontenelle pokes fun at everything, including himself and his profession: "All philosophy is based on two things only: curiosity and poor eyesight; if you had better eyesight, you could see perfectly well whether or not these stars are solar systems, and if you were less curious, you wouldn't care about knowing, which amounts to the same thing. The trouble is, we want to know more than we can see."

It is remarkable how many modern scientific arguments about alien life were anticipated by Fontenelle. Sure, we have a lot more data now from astronomy, biology, paleontology, and geology. Yet, at its core, scientific belief in aliens rests on most of the same leaps of faith now as in the seventeenth century.

After convincing the marquise of the correctness of the Copernican system, the philosopher proposes that there is intelligent life on the Moon. He likens our view of the lunar surface to a Parisian in the towers of Notre Dame gazing at Saint-Denis in the distance: "Everything one can see of Saint-Denis strongly resembles Paris; Saint-Denis has steeples, houses, walls, and it might resemble Paris in that it's inhabited as well. All this will make no impression on my townsman; he will obstinately maintain forever that Saint-Denis is uninhabited because he has seen nobody there. Our Saint-Denis is the Moon, and each of us is a Parisian who has never gone outside his city."

"We'll never know the people on the Moon," the marquise declares, "and that's heartbreaking." The philosopher, unafraid to speculate about spaceflight, responds, "The art of flying has only just been born; it will be perfected, and someday we'll go to the Moon. Do we presume to have discovered all things, or to have taken them to the point where we can add nothing? For goodness' sake, let's admit that there'll still be something left for future centuries to do."

The witty marquise counters this by asking, if such spaceflight is possible, why haven't the people on the Moon already come to Earth? The philosopher responds, "The Europeans weren't in America until after six thousand years. It took that much time for them to perfect navigation to the point where they could cross the ocean. Perhaps the people on the Moon already know how to make little trips through the air; right now they're practicing. When they're more experienced and skillful we'll see them, with God knows what surprise."

This last exchange presages the modern scientific debate about extraterrestrial visitation. The marquise's question has been rediscovered as Fermi's Paradox: if spaceflight between inhabited worlds is possible, surely we should have been visited by now. One popular modern answer to this is identical to the philosopher's response: it takes a long time to develop spaceflight and travel between worlds. They're just not here yet. Note that in Fontenelle's day, six thousand years was considered a good estimate for the age of the universe. The numbers have changed, but the arguments have not.[*]

By the third evening, having convinced the marquise that the Moon is inhabited, the philosopher retreats to a careful agnosticism on the question: "You should never give more than half your mind to beliefs

---

[*]I'll come back to Fermi's Paradox in a later chapter.

of this sort, and keep the other half free so that the contrary can be admitted if it's necessary." He then gives physical arguments, based on telescopic observations, suggesting the Moon may be uninhabitable. Noting that observers on the Moon could easily deduce the existence of water on the Earth by watching the motion of clouds, he points out that, by contrast, the features of the moon are fixed and motionless: "By this reasoning, the Sun doesn't raise any vapors or mists above the Moon. So then she's a body infinitely more solid and hard than our Earth, whose most volatile elements separate easily from the rest and rise up as soon as they're stirred into motion by heat. The Moon must be some mass of rock and marble where there's no evaporation, and furthermore, evaporation is so natural and so necessary where there are waters, that there can't be any waters if none is taking place. Who then are the inhabitants of these rocks which can produce nothing, and of this land which hasn't any water?"

"What?" cries the marquise. "Don't you remember that you assured me there were seas on the Moon that one could distinguish from here?"

"I'm sorry to say that's only a conjecture," answers the philosopher. "Those dark places that are taken for seas are perhaps only great cavities. At the distance we are, it's understandable not to guess quite accurately."

Frustrated by the philosopher's flip-flopping, the marquise quips, "That's a great deal of ignorance based on a very little science." I wonder what she would think of a modern astrobiology conference.

Later that evening the philosopher tries to convince the marquise that all the other planets are inhabited, once again using reasoning that still has a modern ring. Fontenelle was well aware that the telescope was not the only seventeenth-century Dutch optical invention that had opened up new realms of the cosmos to humanity. In the 1670s Antonie van Leeuwenhoek, the unsung Galileo of the microscope, reported the discovery of a multitude of invisible, microscopic organisms squirming in every drop of ordinary water. If there is abundant and ubiquitous life hidden from our senses, Fontenelle reasoned, why shouldn't this be true on cosmic as well as microscopic scales?

"Take all of these species of animals newly discovered, and perhaps those that we easily imagine which are yet to be discovered, along with those that we've always seen, and you'll surely find that the Earth is well populated. Nature has distributed the animals so liberally. . . . Can you believe that after she had pushed her fecundity here to excess, she'd been so sterile toward all other planets as not to produce anything living?"

Today we've identified a great many more of those species "yet to be discovered." We've found strange critters cavorting around undersea volcanoes and microbes slumbering miles beneath the ground. In an echo of Fontenelle, these modern discoveries are trumpeted in newspaper headlines as hopeful signs for alien life.

Fontenelle even discusses the significance of exotic life-forms, today dubbed extremophiles. He proposes a hidden biosphere of rock-eating organisms living deep underground on other planets: "Even in very hard kinds of rock we've found innumerable small worms, living in imperceptible gaps and feeding themselves by gnawing on the substance of the stone. Imagine how many of these little worms there may be, and how many years they've subsisted on the mass of a grain of sand. Following this example, even if the Moon were only a mass of rocks, I'd sooner have her gnawed by her inhabitants than not put any there at all."

Fontenelle argued for infinite variety among intelligent life-forms on other planets, mirroring the bountiful diversity of life on Earth. He imagined that life elsewhere would take turns that we cannot predict, as it adapted to local environmental conditions. He endowed his aliens with new senses and other capabilities unknown to us.

In response, his marquise declares, "My imagination's overwhelmed by the infinite multitude of inhabitants on all these planets, and perplexed by the diversity one must establish among them; for I can see that Nature, since she's an enemy of repetition, will have made them all different."

On the fifth evening, the philosopher makes the final leap, placing inhabited worlds around the distant stars. He wonders about the diversity of solar systems, something that we are today just starting to figure out. The marquise asks, "May these systems not, despite this similarity, differ in a thousand ways? After all, a basic resemblance doesn't exclude infinite differences."

The philosopher responds, "Definitely. But the difficulty is to figure it out. What do I know? One vortex has more planets revolving around its sun, another has fewer. In one there are subordinate planets which revolve around the larger planets; in another there are none. Here they're gathered around their sun like a little platoon, beyond which a great void extends . . . elsewhere they travel around the edges of the vortex and leave the middle empty. I don't doubt that there may be some vortices deserted and without planets. . . . What do you want from me? That's enough for a man who's never left his own vortex."

The final conversation ends with a discussion of the birth and death of stars. When the marquise expresses sorrow that some stars expire, the philosopher offers the comforting thought that new stars might also be born. Foretelling one of the most important insights of twentieth-century astronomy, he asserts, "I also believe that the universe could have been made in such a way that it will form new suns from time to time. Why couldn't the proper matter to make a sun, after having been dispersed in many different places, reassemble at length in one certain place, and there lay the foundation of a new system?"

Here Fontenelle rejects the ancient notion of the immutability of the heavens, instead seeing continual change among the stars and hinting at the modern notion of galactic ecology, the great cycle in which new generations of stars are continually reincarnated from the ashes of their ancestors.

Throughout his long life (February 1657 to January 1757—one month short of a century) Fontenelle published numerous revisions of *Conversations,* each time bringing it up-to-date with the latest astronomical observations—adjusting the diameter of Venus or the distance to Saturn in accordance with the most recent measurements.

ASTROTHEOLOGY

The seeds of pluralism, spread in the seventeenth century by Fontenelle and his contemporaries, took root and grew in the next. General receptivity to the concept of alien life was fostered by the growth of natural theology—the quest to learn about God through the study of nature. Natural theologians were not afraid, like Galileo's inquisitors, to look through the telescope. What could reflect more fully the glory of the creator than a universe filled with inhabited planets?

Between 1700 and 1750 at least a dozen books were published advocating pluralism, mostly on the basis of natural theology. The most successful popularizer of this view was the Reverend William Derham, who published *Astro-Theology, or A Demonstration of the Being and Attributes of God from a Survey of the Heavens* in 1714. Derham asked, "What is the use of so many Planets as we see about the Sun, and so many as are imagined to be about the Fixt Stars? . . . The answer is, that they are Worlds, or places of habitation."

Derham's natural theology removed the potential conflict between science and religion for those with pluralist inclinations. Encouraged by the freedom to speculate that this reconciliation allowed, scientists and mathematicians began to deduce the characteristics of life on other worlds, sometimes with questionable results.

In 1735, German mathematician Christian Wolff published a formula for calculating the size of creatures on other planets. Assuming that (1) height is proportional to the size of the eye, and (2) that the square of the diameter of the pupil is proportional to available light intensity, he calculated that the average height of those living on Jupiter was precisely 13.57 feet! This calculation was cited for more than a century by both sides in the pluralist debate, and Wolff's method was used to compute sizes of organisms on several other planets.

Today, of course, this seems arbitrary if not downright loopy. We can regard it as a good example of how scientists sometimes use math inappropriately, sticking equations where they don't belong in an effort to make our answers seem more "scientific," more exact, and thus more legitimate. If the assumptions behind our calculations are absurd (as Wolff's now seem with the hindsight of three centuries), then our results, beneath their shiny mantle of quantitative credibility, are nothing but garbage. Which of our current efforts will someday be placed in this same pile?

WORLDS WITHOUT END

The daring and liberating pluralist vision captivated Enlightenment poets and philosophers. Immanuel Kant, one of the last Enlightenment philosophers, is widely known as a giant of philosophy who forcefully questioned the sources of all knowledge. Kant defined the Enlightenment as the "emergence of man from his self-imposed infancy." As a young man, Kant had turned his attention to astronomy with equally impressive results. In his *Universal Natural History and Theory of the Heavens,* written in 1755 when he was thirty-one years old, he made several important and lasting contributions to cosmology. Like the first tadpole bravely gazing above the surface of the pond and daring to contemplate frogdom, Kant had a vision of a universe many orders more vast than anyone had previously conceived.

Kant, standing five foot two and weighing less than a hundred pounds, was a shrimp of a man with a Galaxie 500 mind. Though he didn't see much of planet Earth during his stay here, never in his life traveling more than fifty miles from his hometown of Königsberg, Kant gave humanity an unbelievably enlarged universe. Pondering the diffuse glowing band we see at night and call the Milky Way, he made a tremendous conceptual leap, deducing that this view is an optical illusion born of our position *within* the Milky Way. Kant realized that if our Sun is embedded within a flattened disk of innumerable suns, then the far shores of this disk, too distant for individual stars to be seen, would appear to us as the milky band we see stretched across the night. Kant had found us the Milky Way galaxy. Indeed, he found the other galaxies as well.

Contemporary thought held that the numerous "spiral nebulae," diffuse smudges of light spread randomly through the sky, were enlarged puffs of gas or dust as distant from us as the stars. Kant, noting their elliptical shapes, proposed that they were other entire "Milky Ways," much farther away than the stars we can see, each a far-flung disk composed of numberless stars. He even speculated that these other Milky Ways were themselves bound together in larger groups held together by gravity—what today we call clusters of galaxies.

Kant and his fellow Enlightenment philosophers saw themselves as leading the human race in a courageous jailbreak from a prison with walls built of superstition and prejudice; they were freedom fighters for reason. In his galaxies, Kant found the perfect physical manifestation of the Enlightenment quest to cast aside comforting illusions and unquestioned assumptions in search of deeper, wider truths. "The cosmic space," he wrote, "will be enlivened with worlds without number and without end."

It would be nearly two hundred years before astronomers could confirm Kant's intuition, and in some ways we have still not caught up with him. The distant galaxies are so far away that, unlike the planets of our solar system or even the "fixed stars" in our own galaxy, it is literally difficult to imagine traveling to them or communicating across the interminable, empty intergalactic stretches. Even science fiction writers, who are encouraged to bend the rules to help the rest of us probe the edges of the possible, rarely imagine intergalactic travel or cross-talk. When *Star Trek* writers need an incredibly far-off land as a plot device, they choose a remote quadrant of our own Milky Way galaxy. Whether they're accidentally marooning *Voyager* in the distant "gamma quadrant" or twisting space into a wormhole reach to *Deep Space Nine*'s "delta quadrant," these visionaries actually restrict themselves to a minuscule patch of universe. They never venture even as far as nearby Andromeda, which at only 12,000,000,000,000,000,000 miles is a stone's throw away compared to any other galaxy.*

It is as if, even in a fictional universe where faster-than-light "warp drive" is accepted as a quotidian transportation option, it would strain credulity to postulate extragalactic visitation or communication. So, in a sense, the discovery of external galaxies throws us back conceptually to the other "worlds" of the ancient Greek Epicureans—a universe composed almost entirely of places that are real, but forever out of reach. The difference, of course, is that we can *see* the galaxies, even if we still don't quite dare dream of traveling to them. And now we know we live in one.

Kant also came up with one of the first physical theories that explained the origin of planetary systems. In Kant's scheme, stars begin life as rotating clouds of gas that spin faster as they collapse under the weight of their own gravity. These flattened, spinning clouds create rings of material around a young sun that eventually coalesce into orbiting planets. Kant had conceived a crude picture of what would later be called the nebular hypothesis, understood today as the way solar systems form.

Kant ended his book with detailed speculations on the nature of extraterrestrial creatures. Among these were the seeds of an idea that persisted, in various forms, for centuries: that there is a progression in stages of evolution with distance from the Sun, so that planets closer to the Sun are younger and more primitive than Earth and those farther out are more highly evolved. This framework led Kant to believe that the creatures on Mercury and Venus are dimwits, and that the Jupiterians and Saturnians are geniuses, with earthlings occupying "exactly the middle rung . . . on the ladder of beings." Later in life, embarrassed by these youthful conjectures, he insisted that the more speculative final chapters be censored from reprints of *Universal Natural History and Theory of the Heavens*. But he never lost his pluralist convictions. In his 1781 masterpiece, *Critique of Pure Reason*, written when he was fifty-seven years old, he wrote, "I should not hesitate to stake my all on the truth of the proposition—if there were any

---

*With your unaided eyes from a dark site, you can easily see Andromeda, a faint, soft oval smudge floating off the knees of Cassiopeia.

possibility of bringing it to the test of experience—that at least some of
the planets which we see are inhabited. Hence I say that I have not
merely the opinion, but the strong belief, on the correctness of which I
would stake even many of the advantages of life, that there are inhabitants in other worlds."

Encouraged by Enlightenment philosophers, astronomers in the eighteenth century increasingly adopted pluralism. Among them was
Edmund Halley, the British Astronomer Royal who is most famed for
Halley's Comet and who argued that all planets "are with Reason sup-
pos'd habitable."

A LUXURIANT GARDEN: PLURALISM GOES MAINSTREAM

At the age of nineteen, astronomer William Herschel was forced to flee
his native Germany when the French invaded his hometown in 1757.
He settled in England and became one of the most famous scientists of
the Age of Reason. Systematically mapping the heavens with a telescope, he was in many ways the first modern astronomer. It is difficult
to determine how much of his reputation was really due to the work of
his brilliant, diligent, and unsung sister Caroline, who helped build his
telescopes, performed many of his observations, and did most of his
calculations.

William and Caroline were both professional musicians who pursued
astronomy on the side, but together, in their spare time, they observed
and classified numerous stars, and they were the first to see that the
diverse colors and types of stars belonged to an evolutionary sequence.
They realized that stars have life cycles with visibly different stages.
William likened our brief glimpse of stellar evolution to a glance at a
garden with plants at all stages of growth. In *The Construction of the
Heavens,* he wrote, "The heavens . . . are seen to resemble a luxuriant
garden which contains the greatest variety of productions in different
flourishing beds . . . is it not almost the same thing whether we live successively to witness the germination, blooming, foliage, fecundity, fading, withering and corruption of a plant, or whether a vast number of
specimens, selected from every stage through which the plant passes in
the course of its existence, be brought at once to our view?" This insight
set the stage for our modern evolutionary view of cosmic phenomena.

William was obsessed with building the largest telescopes in order to see farther than anyone ever had.* Many of his instruments were more impressive in size than optical quality, but at the age of forty-three he stumbled upon the planet Uranus, the first "new" planet ever discovered by humans. Like the first discoveries, in our own time, of new planets beyond our solar system, Herschel's new world encouraged pluralist speculation. Up until that moment, for all we actually knew, there might be only six planets in the whole universe. Finding Uranus confirmed the widespread conjectures that other planets were out there, beyond the ones we can see with the unaided eye. Herschel capitalized skillfully on the naming rights to his new planet, calling it the Georgian planet. King George III rewarded him with an annual pension of two hundred pounds, which allowed him to quit his day job as a musician and devote himself full-time to his astronomical passion.†

Herschel's zeal for bigger scopes and better views was driven by a lifelong belief in the existence of life elsewhere. He once wrote that if he had a choice, he'd prefer to live on the Moon. Herschel was determined to find definite evidence of life on Earth's nearby companion. He made drawings detailing his observations of lunar forests and cities. Some of Herschel's reports of lunar vegetation, turnpikes, canals, and even pyramids read like today's tabloid headlines you might find yourself reading furtively in the checkout line at the grocery. An amateur astronomer hoping to be accepted as a professional, William refrained from publishing these spectacular findings out of concern for his reputation, but to Herschel the discovery of Uranus was important chiefly because it provided the financial means to support his more important studies of lunar civilization.

Today, Herschel's interpretations of planetary observations seem clearly biased by wishful thinking. He played up similarities between Earth and Mars and ignored obvious differences to support his belief that "its inhabitants probably enjoy a situation in many respects similar to ours." But his insights into stellar evolution and his discoveries—especially of the planet Uranus—serve as lasting reminders that investi-

---

*Freud would not be born for another century.
†I'm surprised that modern planet discoverers haven't resorted to hitting up modern royalty for naming rights. "Planet Bill Gates" would be easier to remember than "the planet orbiting HR3522."

gations driven by fervid beliefs can sometimes produce spectacular and lasting scientific results.

SYSTEM OF THE WORLD

The most important mathematician of the late eighteenth century was Pierre-Simon Laplace. Math tutor and confidant of Napoléon—who dubbed him "the Newton of France"—Laplace made lasting contributions to probability theory, calculus, electricity and magnetism, and especially *celestial mechanics*—the theory of the motions of heavenly bodies. In his *System of the World,* published in 1796, he developed his "nebular hypothesis," which explained the origin of the solar system from a contracting, flattened, rotating disk of gas. Kant had painted a similar picture forty-one years earlier, but Laplace put mathematical flesh on the bones of Kant's idea, presenting a masterful derivation showing that such a spinning disk would inevitably result from a cloud collapsing in space.

Today, the physics we use to describe solar system formation is directly descended from Laplace. One important implication of the nebular hypothesis is that all stars, if they form in the way that ours did, should be born with planets. For nearly two hundred years pluralists had been saying that other stars must have planets. This, however, was based only on a metaphysical principle: the argument by analogy. Laplace came along and showed how this belief is supported by a sophisticated scientific theory. His nebular hypothesis was a major shot in the arm for pluralism.

Like most scientists of his day, Laplace believed in a plurality of inhabited worlds. Of the planet Jupiter he wrote, "It is not natural to suppose that matter . . . should be sterile upon a planet so large." Laplace described the startling insignificance of humanity and the Earth in the vastness of space, but he urged his contemporaries to rejoice in the power of the human intellect, rather than despair at our pathetically diminished stature:

"Man appears upon a small planet, almost imperceptible in the vast extent of the solar system, itself only an insensible point in the immensity of space. The sublime results to which this discovery has led may console him for the limited place assigned him in the universe."

The success of mathematics and physics in explaining so much of the material world emboldened men like Laplace to believe that there was a physical, mechanical explanation for everything, and to question the role of God in creating and maintaining the cosmic dance. When Napoléon, after having read *System of the World,* asked Laplace why there was no mention of God in his work, he replied, "Citizen First Consul, I have no need of that hypothesis."

By 1800, the international scientific community was nearly unanimous in holding pluralist beliefs. Religion was also finally adapting to the new consensus reality. It became more common to use theological arguments to argue for pluralism than against it!

Among Enlightenment intellectuals, belief in extraterrestrial life became a major source of doubt in Christianity. The plurality of worlds was used to argue against the reality of the Incarnation of Christ. American revolutionary hero Thomas Paine made this point without pulling his punches in *Age of Reason* (1794):

"From whence then could arise the solitary and strange conceit, that the Almighty, who had millions of worlds equally dependent on his protection, should quit the care of all the rest, and come to die in our world, because they say one man and one woman had eaten an apple! and, on the other hand, are we to suppose that every world in the boundless creation had an Eve, an apple, a serpent, and redeemer? In this case, the person who is irreverently called the Son of God, and sometimes God himself, would have nothing else to do than to travel from world to world, in an endless succession of death, with scarcely a momentary interval of life."

In the obvious conflict between pluralism and certain aspects of Christian doctrine, the tables had been turned.

3 | A Wobbly Ladder to the Stars

The discussions in which we are engaged belong to the very boundary regions of science, to the frontier where knowledge ends and ignorance begins.
—WILLIAM WHEWELL, 1853

No shortage of explanations for life's mysteries. Explanations are two a penny these days. The truth, however, is altogether harder to find.
—SALMAN RUSHDIE, *The Ground Beneath Her Feet*

NOTHING FROM WHICH TO REASON

Over the course of the seventeenth and eighteenth centuries, belief in extraterrestrial life traveled from the fringes to the mainstream. Pluralism went, in the words of science historian Michael Crowe, "from being a belief of a few to a dogma taught in scientific textbooks and preached from pulpits." A backlash was inevitable. In 1853, William Whewell, the master of Trinity College, Cambridge, wrote one of the first polemics against life elsewhere that was based on scientific evidence. He was the first to use a line of argument that has since been offered many times and that has most recently been resurrected as the Rare Earth Hypothesis. Whewell's arguments against the existence of intelligent life elsewhere were based on the following:

1. Empirical evidence indicates that the other planets in our solar system have conditions so different from Earth's that they must be uninhabitable by any life we can imagine.

2. There was no evidence of Earth-like planets orbiting other stars. Therefore Earth may be completely unique in the universe.

3. The history of life on Earth, as revealed by geology, shows that intelligent life is only a recent phenomenon here. During most of cosmic history, complex life did not exist even on this planet. Therefore, arguments by analogy are weakened.

Whewell concluded, "The belief that other planets, as well as our own, are the seats of habitation of living things, has been entertained, in general, not in consequence of physical reasons, but in spite of physical reasons; and because there were conceived to be other reasons, of another kind, theological or philosophical, for such a belief."

Whewell's book provoked outrage, as belief in extraterrestrial intelligence was deeply entrenched in scholarly and popular culture throughout the nineteenth century. He became an antipluralist lightning rod. Mocking the perceived arrogance of Whewell and his hypothesis, the renowned American astronomer Maria Mitchell wrote, "They say in Cambridge that Dr. Whewell's book, *Plurality of Worlds,* reasons to this end: The planets were created for this world; this world for man; man for England; England for Cambridge; and Cambridge for Dr. Whewell!"

Mitchell's mid-nineteenth-century scientific judgment was that "there is nothing from which to reason. The planets may or may not be inhabited."

THE EVOLUTION REVOLUTION*

In 1859, six years after Whewell's book reignited the pluralism debate, another book touched off an even greater firestorm—one that smolders still. Charles Darwin's *On the Origin of Species* permanently changed our view of the relationship between Earth and its life and, ultimately, between life and the cosmos. Darwin was a pluralist, and his theory of evolution did for our species what the Copernican revolution had done for our planet: it knocked us off our pedestal. Darwin and his followers showed that we are part of a continuum of life in which all species are descended from common ancestors. Evolutionary theory gave us a new answer to the question of why the Earth seems so ideally suitable for

---

*With a tip of the hat to Lancelot Link, Secret Chimp.

life. Our planet only *seems* to have been created as the perfect environment for man, Darwin suggested, because we have evolved by natural selection to best take advantage of conditions found here.

The implications for the "plurality of worlds" debate were both obvious and profound. Pluralists saw in the theory of evolution by natural selection the seeds of a universal process of life-making and evolving. Teleological arguments, long a mainstay of pluralist belief, were abandoned and replaced by evolutionary arguments. No longer were the other worlds clearly made for someone else, as this world was made for us. The new rationale was that other planets must be inhabited because natural selection would fashion living beings to take advantage of local conditions throughout the cosmos.

Darwin's radical theory generated ripples that were felt in all areas of natural science. What started as a new idea about Earth and its inhabitants was soon being applied to life elsewhere, and even the life of the cosmos as a whole. Astronomers began to think in terms of cosmic evolution: Did the universe itself evolve through distinct phases, perhaps culminating in the evolution of life?

Biology and astronomy were not the only fields in rapid transition. During this same period, advances in chemistry showed the stuff of life to be less mysterious, less distinct from the rest of nature. "Organic" matter was found to be composed of complex compounds of the element carbon.

In the 1860s, it finally became possible to investigate experimentally another of the metaphysical pillars on which belief in plurality had rested: the assumption of the uniformity of the cosmos. This became subject to test by the new methods of *spectroscopy:* precise measurements of the intensity of light at each separate color, or wavelength. Viewed in this carefully dissected light, each chemical compound has its own specific signature, and we can determine what something is made of by examining the light it shines or reflects. Stick a spectroscope on the end of a telescope and voilà: you've got a probe for studying the composition of distant planets and stars. When Sir William Huggins reported his pioneering spectroscopic work showing that distant stars are made of the same stuff as the Sun, he wrote that this provided "an experimental basis on which a conclusion, hitherto but a pure speculation, may rest—viz. that at least the brighter stars are, like our sun, upholding and energizing centres of systems of worlds adapted to be the abode of living beings."

As the nineteenth century drew to a close, these advances in biology, chemistry, and physics had removed several barriers to belief in life on other worlds. If life here is a natural consequence of universal physical processes, and our solar system is nothing special, then why shouldn't the universe be crawling with all kinds of creatures?

LOWELL'S MARTIANS

Into this atmosphere of great scientific receptivity to pluralism strode Percival Lowell, a brash, wealthy Bostonian on a mission. His scientifically informed fantasy of a Martian civilization would capture the world's imagination, temporarily advancing the pluralist cause but ultimately setting it back for most of the century.

Upon graduating with honors from Harvard in 1876, Lowell gave a commencement address on Laplace's nebular hypothesis. Before turning to astronomy full-time in his forties, he traveled repeatedly to Asia in pursuit of arcane knowledge of Eastern religion. From these journeys he gained a mystical belief in the unity of the cosmos that influenced his devout pluralism. "Each body," he concluded, "under the same laws, conditioned only by size and position, inevitably evolves upon itself organic forms."

Shortly after returning to Boston in 1893, Lowell learned of the telescopic observations of the Italian astronomer Giovanni Schiaparelli, who, in 1877, had drawn maps of Mars with unprecedented detail, showing a network of straight lines that he called canals. Schiaparelli's canals sparked the imagination of Percival Lowell, who began to conceive a way of confirming his cosmic vision of abundant life and intelligence. It did not take him long to throw together a well-financed (with Lowell family money) expedition of Harvard astronomers to Arizona in search of an ideal high-altitude site to build a new observatory. He chose a forested hill at seven thousand feet on the outskirts of Flagstaff, where today the Lowell Observatory remains one of the premier American institutes of planetary research. Charismatic, and fabulously wealthy, Lowell had the means to build an observatory, but he lacked the patience and objectivity of a great observer. He arrived in Flagstaff to begin his observations in May 1894. Less than a year later he was already popularizing incredible discoveries and a radical new view of Mars.

Lowell confirmed the presence of a geometric pattern of Martian canals covering the entire globe. The dark areas, however, were not watery seas as Schiaparelli and others had concluded. Rather, he determined, they were deserts. He developed an elaborate interpretation of the current Martian condition, in which the canals had been built by a race of intelligent Martians vastly superior to humans in intellect and technical capabilities. Mars, once much like the Earth, had lost its oceans. The Martians, trying to preserve life on their dying world, had constructed the canals to carry spring melt from the polar caps to cultivated fields on the rest of the planet.

Lowell publicized this theory in a series of papers in scientific and popular journals, in a well-attended lecture tour, and in his book *Mars,* published in December 1895. Some critics noted the suspicious similar-

Percival Lowell's view of Mars, showing widespread deserts and a planetwide irrigation system.

ity of Lowell's Martian observations to his pre-observational views of cosmic evolution. Indeed, on the eve of his departure for Flagstaff he had proclaimed to the Boston Scientific Society, "Investigation into the condition of life on other worlds, including last but not least, their habitability by beings like or unlike man . . . is not the chimerical search some may suppose. On the contrary, there is strong reason to believe we are on the eve of a pretty definite discovery in the matter." Other skeptics showed that a planetwide irrigation system fed by the tiny polar caps would not work.

But the public ate it up. Such was the power of Lowell's will and imagination, and the skill of his oratory, that he took the whole world with him on an elaborate, decades-long Martian fantasy ride. His books were best-sellers, and his sensational lectures were standing-room-only affairs, with frenzied throngs spilling onto the street. At the beginning of the twentieth century, Lowell's advanced, canal-building Martians were all the rage on Earth.

The scientific community was sharply divided over the issue. The furious disagreements, lasting for decades, mostly centered on the question of whether the canals were artificial or "natural." Most accepted that they existed. Numerous careful and renowned observers the world over also saw the canals.

At the time, that seemed like a powerful independent verification, but now it stands as a warning of the traps we can set for ourselves when we push science too far. When we overinterpret sketchy data at the limits of current abilities, the gaps in our data may be filled by our desires, by the power of suggestion, and by the undeniable force of consensus in forming opinions.

Finally, in the 1920s, the debate ebbed. Improved telescopes and photographic techniques showed that the question of interpretation was moot. The canals were not the invention of advanced Martians trying valiantly to save their planet from global change. They were created by a turbulent atmosphere, an active imagination, a charismatic individual, and a pervasive will to believe. They are simply not there.

LOWELL'S LEGACY

Lowell went to his grave in 1916 firmly convinced of the reality of his Martian civilization. Today, he is remembered as the man who, by force of personality, led the world on a wild-goose chase that ultimately proved to be one of the most embarrassing episodes in the history of science. It's unfortunate, because his Martian obsession is only a highly visible dark patch in a legacy with many bright spots. Lowell was a successful popularizer of astronomy who established an observatory of lasting value. His efforts to derive comprehensive theories of planetary evolution paved the way for modern planetary science. Indeed, though I find myself loath to say it (speaking as a comparative planetologist), he may have been the first comparative planetologist.

Lowell's turn-of-the-century views of planetary evolution now seem an interesting mix of nineteenth- and twentieth-century ideas. His description of Mars was influenced by Kant's nebular hypothesis in which the planets formed sequentially with distance from the Sun, so that the farther ones were older and those closer in were younger. Mars was ancient, dried-out, and over-the-hill, a vision of Earth's sad future, and Venus was like a youthful, prehistoric Earth. The red planet had evolved much further than Earth both geologically and biologically, losing its water as Earth would at some time in the distant future. Martian evolution had progressed eons further, producing much more advanced sentient life.

But advanced age was only one reason Lowell offered for the differences between Mars and Earth. He proposed a second explanation that has proved more durable. His statement that Mars, "being smaller, aged more fast than the Earth," anticipated one of the results of modern comparative planetology. We have now established that the planets formed all at the same time, not one at a time in sequence with distance from the Sun. But we have also become increasingly aware of how a planet's size influences physical evolution. A small (compared to Earth) planet like Mars has less gravitational hold on its atmosphere and oceans, and less internal heat to sustain prolonged, replenishing geological activity. Mars did lose most of its water early on and has since suffered billions of years of geological dormancy. Thus its biological potential seems largely buried in the distant past. Lowell's portrait of Mars as a world dry and old before its time, too small to hold its air and water, is mirrored in our modern views.

Lowell boosted interest in extraterrestrial life to an all-time high among both the scientific community and the public at large. This surge in awareness would outlive Lowell and his theories by many decades, impacting the pluralist debate throughout the twentieth century. The picture of Mars as a reasonably Earth-like place with a moderate climate, water running on the surface, canals, and living creatures implanted itself permanently in the popular psyche.

Science fiction writers conjured freakish images of Martian life, which mingled with the science in the public imagination. *The War of the Worlds,* H. G. Wells's 1897 tale of interplanetary invasion, was written in response to the 1894 "discovery" of a civilization on Mars. It was Lowell's dried-out, dying Mars from which the superior, malevolent Martians launched their attack, seeking Earth's greener pastures. Since then a never-ending stream of fictional aliens have invaded, trying to steal our water, our food, our women, our men, our cattle, and our precious bodily fluids.

Lowell's Mars persisted in fiction long after his theories were banished from science. Water-filled canals are found in Ray Bradbury's *The Martian Chronicles* (1950), where the alien invaders are humans from Earth, in Arthur C. Clarke's *The Sands of Mars* (1952), and in Robert Heinlein's *Stranger in a Strange Land* (1961). I must admit that, having absorbed all of these stories in my science-fiction drenched youth, I find it a bit hard to give up entirely on Lowell's canals—even though high-resolution spacecraft imaging has thoroughly destroyed any rational hope of finding them. For the public, the canals never entirely went away. They just gradually morphed into the dried-up river channels revealed by the *Mariner 9* spacecraft in 1971.

Among scientists, however, when solid evidence inevitably caught up with Lowell's Mars—revealing a desiccated, frozen, desert world with a thin, unbreathable atmosphere, and no canals or other signs of civilization—the backlash was extreme. The pluralist cause was discredited so severely that it has only recently recovered.

Scientific belief in alien life has always rested on some combination of observational evidence and other, nonscientific reasons to believe, which, inasmuch as they are conscious, may be classed as metaphysical and, inasmuch as they are not, psychological. The Lowell affair left pluralism with its pants down, exposing with embarrassing clarity the roles that wishful thinking and herd mentality can play in scientific claims about life elsewhere.

Belief in intelligent extraterrestrial life suffered a precipitous fall from grace in the post-Lowell decades. A more general belief in life on Mars, however, persisted. The new view, held by virtually all leading astronomers in the 1920s and 1930s, was that although Mars had no

H. G. Wells brought the world a new archetypal fear: the menace from space.

intelligent life, its surface was covered by vegetation that caused the margins of the broad dark regions to shift with the seasons.

In 1928, a symposium published in the *New York Times Magazine* headlined "Eminent Astronomers Give Their Reasons for Belief That Life Exists on the Great Red Planet" quoted many prominent astronomers, including Harlow Shapley at Harvard and Henry Norris Russell at Princeton, endorsing the likelihood of vegetable life and the possibility of primitive animal life on Mars. Astronomers held this view until close-up spacecraft pictures supplied a reality check in the 1960s.

Advances in spectroscopy gradually drove home the reality of Mars—an incredibly cold and dry world with a thin atmosphere devoid of oxygen. Yet, other measurements episodically resurrected hopes for life. As recently as 1957, Harvard astronomer William Sinton reported in *Science* magazine that he had found the spectroscopic signature of chlorophyll, and "this evidence, together with the strong evidence given by the seasonal changes, makes it seem extremely likely that plant life exists on Mars." Sinton made these observations from a telescope at Lowell Observatory.

CONVERGENT OR CONTINGENT?: BIOLOGICAL PESSIMISM

Twentieth-century advances in biology, chemistry, and astronomy so drastically redrew the map on which the pluralism debate was held that it was given a new name. The phrases *plurality of worlds* and *pluralism,* which smacked of seventeenth-century natural philosophy, disappeared, and the more scientific-sounding *extraterrestrial life debate* came into vogue.

Setting the tone for modern biology's involvement in this debate was none other than Alfred R. Wallace, Darwin's codiscoverer of the theory of evolution by natural selection. Wallace, provoked by the popularity of Lowell's Martian theories, studied the probability of intelligent life evolving on other planets. In his 1903 book, *Man's Place in the Universe,* he published his conclusion: intelligent life is unique to Earth.

This declaration set up an intellectual conflict between biology and astronomy over the question of extraterrestrial life. Astronomers, wowed by the sheer number of stars and planets where life could possibly evolve, were generally the optimists. Biologists, impressed by the arduous, and perhaps unique evolutionary journey of life on Earth, were the pessimists.* Twentieth-century astronomers were getting used to the idea of cosmic evolution and developing a picture of stellar and galactic life cycles in which one phase followed inevitably from the next. They often assumed that life would spring forth naturally on other planets at a certain phase of their development, as it apparently had here.

Biologists, struggling unsuccessfully to find a theory of life's origins on Earth, concluded that such a development was difficult and unlikely. They were more inclined to believe that life was unique to Earth. Arguments in favor of extraterrestrial intelligence have primarily been based on physics and astronomy (with a dash of metaphysics and wishful thinking). Arguments against have largely been based on biology (with a pinch of Earth-centered parochialism).

Wallace drew attention to the role of *contingency* in biological evolution. Since evolution of intelligence here required an absurd number of lucky breaks, he argued, the probability against its ever occurring again trumps even the unfathomably large number of places in the universe where it might happen. Plenitude is defeated by the unlikely, contingent path of evolution. Ever since, many biologists have echoed this, asking, "What are the chances?"

This biological stance is best refuted not by astronomy but by another biological argument. Wallace's pessimistic reckoning of intelligence's prospects takes no account of *convergent evolution*. Evolution is remarkably good at finding solutions to the problems of survival in diverse environments, and the same evolutionary invention often occurs separately in very different species.

The eye of the octopus is a common example. It is remarkably similar in structure and function to the human eye, but we did not evolve from octopi, nor they from us. Nor do we and octopi share an ancestor that had eyes. Evolution independently found the same design for a visual organ that helps you get around whether you've got eight legs and suckers or two legs and toes. Similar examples abound in nature. Another frequently cited one is the similar streamlined shape of marine mammals, such as dolphins, and large fishes. These creatures are unrelated but evolution found them the same solution for swimming swiftly through Earth's seas.

---

*Yes, the terminology I use reveals a bias that life elsewhere is a *good* thing. Isn't it?

Good designs that drastically increase the chances of survival often evolve separately more than once, and perhaps with some inevitability, among the diverse species of a biosphere. For this reason, it does not seem unreasonable that animal species evolving on another planet would have organs we would recognize as eyes. Can we say the same about brains? That's the big question.

The possibility that convergent evolution applies to the development of intelligence completely changes the probability arguments. If intelligence confers a huge survival advantage, then perhaps it will inevitably arise on other planets. Whether intelligence is such a trait is still a matter of significant debate.

PANSPERMIA: HERE, THERE, AND EVERYWHERE

Where does life come from? We now regard this question as an essential component of the scientific quest for extraterrestrial life. Prior to the twentieth century, however, it was only marginally a part of the debate, regarded as more of a philosophical question we could not hope to address scientifically. Forgetting the rest of the universe for a moment, the ultimate source of life on Earth remains a deep mystery. It is a mystery we need to solve if we hope to know something about life's distribution throughout the cosmos. Observing nature, you will quickly find the age-old, obvious answer: "Life comes from other life." And like a three-year-old responding to every answer with another question, we must ask, "But where did it come from originally?"

The explanation that sufficed for millennia—divine creation—ceased to satisfy. Neither Galileo nor Darwin had questioned this solution to life's ultimate riddle. But their followers, impressed with the ability of natural philosophy to resolve nature's greatest puzzles, began a systematic search for answers. Discounting literal interpretations of Genesis, which nineteenth- and twentieth-century scientists were increasingly willing to do, there are really only two possibilities. Life either somehow was brought to this planet from elsewhere—*panspermia*—or it arose here by natural processes from nonliving substances—*spontaneous generation*.

The French chemist Louis Pasteur (immortalized on every carton of pasteurized milk) sought throughout his life to find experimental evidence for the prevailing view that life somehow arose from inanimate

materials. He failed and, in doing so, believed he had proven the opposite. His experiments demonstrated that the seemingly "spontaneous"
growths observed in spoiled food were actually introduced from the
outside. (This also led directly to the food sterilization technique that
made Pasteur a household name.) Pasteur's results, published in 1860,
convinced many scientists that spontaneous generation does not occur.
So then, whence came life?

Perhaps it arrived here in meteorites. During the nineteenth century,
when scientists figured out that these strange, charred lumps of rock
and metal fall from beyond the Earth, meteorites became the subject of
intense study. Since the 1830s we have known that they contain organic
matter. Some scientists, assuming that organic materials can only be
produced by living organisms, took this to be both evidence of extraterrestrial life and the solution to the riddle of Earth life's origins. It came
out of the sky. Riding on meteorites.

No one argued this more eloquently than William Thomson—
a.k.a. Lord Kelvin—the Scottish physicist who invented the absolute-
temperature scale that scientists use to measure temperatures above
absolute zero in "degrees Kelvin." Thomson compared the origin of life
on Earth to the rapid blooming of a newly formed and initially barren
volcanic island. The seeds must drift in from elsewhere. He argued (cor-
rectly) that every year many tons of meteorites fall to Earth from space.
Some of these, he suggested, must be the fragments of planets once rich
with life:

"Hence and because we all confidently believe that there are at present, and have been from time immemorial, many worlds of life besides
our own, we must regard it as probable in the highest degree that there
are countless seed-bearing meteoric stones moving about through
space. If, at the present instant, no life existed upon this earth, one such
stone falling upon it might, by what we blindly call natural causes, lead
to its becoming covered with vegetation. . . . The hypothesis that life
originated on this earth through moss-grown fragments from the ruins
of another world may seem wild and visionary; all I maintain is that it
is not unscientific."

The idea that Earth life was seeded from elsewhere was refined and
advanced in the first years of the twentieth century by Svante
Arrhenius, a polymathic Swedish chemist who was often far ahead of
his time. He barely graduated from university in 1884, earning scorn

from his professors for unorthodox ideas about the electrical conductivity of solutions. These same ideas earned him the Nobel Prize in chemistry in 1903. That year, emboldened by success to venture farther out on the limbs where both the fruits and the dangers of speculation can be found, he published his theory of life's origin from outer space. His name remains the one most closely associated with the concept of panspermia.

Arrhenius agreed with Kelvin that life was seeded from space. He did not think, however, that meteorites were the most likely carriers. Life would be unlikely to survive either the violent collisions that produce meteorites or the heating and shock that occur when these rocks fall to Earth. Instead, he proposed that seeds were carried throughout the universe in tiny particles of dust. Noting the way a comet's dusty tail is blown away from the Sun by the gentle pressure of sunlight, he proposed that this "radiation pressure" was the force that distributed "living seeds" to the planets.

Arrhenius calculated that the Sun's radiation pressure could blow seeds from Earth to Mars in twenty days, to Jupiter in eighty days, and to the nearest star (Alpha Centauri) in nine thousand years. He argued that these interplanetary transit times were short enough for the seeds to remain viable: "In this way, life would be transferred from one point of a planetary system, on which it had taken root, to other locations in the same planetary system, which favor the development of life."

And what of the much longer travel times between the stars? The exceedingly frigid temperatures of interstellar space, he suggested, would freeze-dry the traveling seeds, preserving them to survive even these epic journeys.

Arrhenius developed his ideas on panspermia much further in his book *Worlds in the Making: The Evolution of the Universe* (1908), which used detailed theories to argue that life develops inevitably on numerous worlds in our solar system and others. For his sweeping synthesis of astrophysics, chemistry, and biology, Arrhenius could be considered the first astrobiologist.

The idea of panspermia, like one of Arrhenius's intrepid interstellar seeds, is hard to kill. But panspermia does not solve the problem of life's origin, it just removes it from Earth. Even if life came here from elsewhere, it still originated somewhere. Pushing it off into outer space merely relocates the mystery.

LIFE IS CHEMICAL

Another take on the origin of life arrived with the new discipline of bio-chemistry. As the extraterrestrial-life debate was waged between the optimism of physicists and the pessimism of biologists, it made sense that new approaches should arise from the field that bridged the conceptual gap between atoms and organisms.

Proteins (organic chemical components of all living cells) were first isolated from cellular material in the first years of the twentieth century. Around that time many simple biochemical reactions were duplicated in laboratory flasks, adding to a growing sense that life is, fundamentally, chemistry. This more sophisticated version of spontaneous generation renewed hopes of finding the key to life's origins in special brews of chemicals native to the primitive Earth.

A chemical origin of life became widely accepted after 1936 when the Russian biochemist Aleksandr Ivanovich Oparin published his landmark book *Origins of Life*. Oparin, no doubt influenced by the dialectical materialist philosophy permeating the Moscow air, postulated an inevitable historical process in which conditions on the young Earth caused the molecules of life to rise up and organize out of nonliving matter.

The prevailing view of Earth's earliest environment at the time included an atmosphere composed mostly of carbon dioxide (CO_2), similar to that which we already knew to exist on neighboring Venus. Oparin argued for a very different kind of ancient atmosphere, rich in methane (CH_4) and ammonia (NH_3), gases which had recently been detected in the atmospheres of Jupiter and Saturn.

This proposed change in the early atmosphere, from CO_2 to CH_4 and NH_3, has a crucial effect on the social behavior of carbon atoms. In an environment rich in hydrogen compounds such as methane and ammonia, called a *reducing* environment, carbon atoms will tend to grab on to each other, forming the giant carbon conga lines and group carbon hugs we call complex organic molecules. Carbon behaves very differently in an *oxidizing* atmosphere richer in carbon dioxide or oxygen (O_2). The carbon is seduced by oxygen's pull and ignores its own kind. Organic molecules don't stand a ghost of a chance.*

---

*Yes, it's true. Our precious oxygen is lethally toxic to the basic molecules of life. We'll return to this chemical irony in a later chapter.

Oparin described how, on an early Earth with a reducing environment, simple organic compounds formed and began reacting with one another. This led to "chemical evolution" in which the more stable (or "fit") molecules hang around, accumulating and evolving further. The result was a rich soup of chemicals that gradually increased in size and complexity until the organic molecules essential to forming the first living cells were abundant in the ponds and oceans of the juvenile Earth.

Origins of Life was a watershed in modern thought about life's beginnings, strongly influencing both astronomical and biological beliefs about the primitive Earth for the rest of the century. Although Oparin's book was strictly about Earth, the theory described the inexorable chemical development of life from conditions believed to exist generally on young planets. The cosmic consequences were inescapable.

In the 1930s, we knew precious little of the actual conditions on other planets in our solar system, and even less about the primitive environments on these planets way back when life on Earth began. It seemed probable that early conditions were similar on all planets, so Oparin's chemical evolution seemed like a universal life-generating theory.

In 1953, at the University of Chicago, Nobel laureate Harold Urey and his grad student Stanley Miller realized they could test Oparin's thesis experimentally. Urey, one of the fathers of modern planetary science,* created the subfield of cosmochemistry, in which we follow the chemical forms of matter through the stages of cosmic evolution. As the first to improve upon the nebular hypothesis using sophisticated chemical modeling, he cleverly deduced what the planets were made of when they condensed out of the solar nebula. He concluded that the early Earth was rich in methane, ammonia, hydrogen, and water—a picture similar to Oparin's.

Miller, then a beginning student, wanted to try simulating the natural creation of organic chemicals on the early Earth. Urey was skeptical, but he agreed to let Miller attempt a preliminary experiment to test the concept. The setup was simple: brew up some primitive air, zap it with simulated lightning, and see if anything happens. They mixed ammonia, methane, and water in a flask and sparked it up. After a few days they were both astounded to find their experimental flask full of an ugly, sticky, brown goo. The gunk turned out to be made of amino

---

*And my "intellectual grandfather": Urey was thesis adviser to my thesis adviser, John Lewis.

acids—the building blocks of protein, the stuff of life! This finding far exceeded the ambitions of their initial mock-up investigation, which is now inscribed in textbooks as the Miller-Urey experiment.

The astonishing result suggested that unremarkable conditions and processes on the primitive Earth would inexorably have produced the molecules of life. Of course, that was Oparin's original thesis twenty years earlier, but a bird in the lab is worth two on the page: experimental proof is more convincing than the most sophisticated theoretical conjecture. Miller and Urey had not actually created life in the lab, but by producing life's crucial building blocks from garden-variety chemicals, they removed what had seemed a fundamental barrier to the spontaneous generation of life from nonlife on Earth or elsewhere. Like Oparin, Miller and Urey did not at first discuss the extraterrestrial applications of their work. However, ammonia, methane, and water were known to be among the most abundant compounds in the universe. Embedded in the results of Miller-Urey was the clear implication that the steps to life here were the result of common cosmic processes.

The Miller-Urey experiment, by establishing that the origin-of-life question is subject to experimental inquiry, generated not only a flask of dark, promising sludge, but a cottage industry. Investigators trying to discover the essential early steps of life have endlessly varied the formula of the gaseous brew, following evolving ideas about the primitive atmosphere, and they've zapped these mixtures with all kinds of energy that might have been present on the young Earth, including ultraviolet radiation and simulated asteroid-impact explosions. To this day, the resulting brown goos are eagerly analyzed like the precious elixir of life. Just as all later theoretical discussions of the chemical origins of life on Earth or anywhere in the universe can be seen as refinements of Oparin's ideas, all experimental efforts in the field are variations on the theme begun by Miller and Urey.

These rigorous scientific results returned to the study of extraterrestrial life much of the legitimacy it had lost in the aftermath of the Lowell affair. Just as the nebular hypothesis predicted that planets were a natural byproduct of the formation of stars, Oparin's theory and the Miller-Urey experiment implied that life itself is a natural byproduct of the formation of planets.

4 | The Planets at Last

The desire to know something of our neighbors in the immense depths of space does not spring from idle curiosity nor from thirst for knowledge, but from a deeper cause, and it is a feeling firmly rooted in the heart of every human being capable of thinking at all.

—NIKOLA TESLA, 1901

Fly me to the moon
Let me play among those stars
Let me see what spring is like
On Jupiter and Mars©
—OSCAR HAMMERSTEIN

BEING THERE

The 1950s were our last age of interplanetary innocence. These were the final moments of a 350-year stretch of blissful ignorance between telescopes and spacecraft—between figuring out what the planets were and learning what they were like. Like a first date that leaves you smitten and full of hopeful fantasy, the planets were easiest to idealize when we knew little about them.

We knew enough chemistry to believe that life came naturally to planets with the right conditions, but we had only hints of the actual environments on other planets. Scientists wondered in print if we would find "astroplankton" on the Moon, or plants and animals on the surfaces of Mars or Venus. As we were poised to enter the universe, our ideas about what we would find there were still greatly influenced by wishful thinking and simple extrapolation.

The few clues we had were often interpreted to encourage hope for nearby life, even while growing spectroscopic evidence implied that the atmospheres of Venus and Mars were not at all Earth-like and suggested that both were severely lacking in water. Many scientists were willing to believe that this evidence was not conclusive and to suspend judgment about life on other planets until we could go and see for ourselves.

Since the 1960s, we've finally been able to travel to the other planets, sending robotic extensions of our eyes and noses to radio back their findings. Early results threw buckets of cold water on our dreams of comfortable, Earth-like environments with abundant life. Viewed up close, the warm oceans of Venus dissolved into a choking sulfuric incinerator. The vegetable patches of Mars were mirages of windblown dust on a frozen, sterile desert.

When I was a child, my imagination was fired up by pictures of new worlds being explored for the first time. The images and stories of real spacecraft exploration blended smoothly in my adolescent brain with the worlds of science fiction. I had seen Neil and Buzz jump down a ladder onto the bright, dusty Moon when I was in the fourth grade, so the voyage to Jupiter in *2001: A Space Odyssey* did not seem unreal. I dreamed of spaceships and extraterrestrials and thought about how I would be forty in the year 2000. The Future. I imagined that one day I would travel to other worlds, following the trail of alien life. It helped that my parents were socially enmeshed in the Boston scientific community, and friends like Isaac Asimov, Carl Sagan, and Fred Whipple regularly dropped by our house with news of the latest discoveries or setbacks. I followed the ups and downs of planetary exploration as closely as other world-changing developments, such as Vietnam and the breakup of the Beatles.

At age eleven, in 1971, I was gripped by the drama of *Mariner 9,* a turning point in our understanding of Mars. Between 1965 and 1969, three other American spacecraft had reached the Red Planet, photographing small areas of the surface as they sped past on brief "flyby" missions. These craft, *Mariner 4, 6,* and *7,* had shocked and disappointed with their pictures of a Mars that looked very much like the surface of the Moon. Hopes for life on Mars were dashed on the rocks of an ancient, cratered landscape that looked dead as doornails. There was no sign of recent geologic or atmospheric activity, let alone running water or vegetation. These early missions also confirmed that Mars has only a wisp of an atmosphere and is far too cold for liquid water. The polar caps were made not of water ice but of dry ice (frozen CO_2).

Thus even while humans stepped onto another world and Clarke and Kubrick carried our imaginations out to "Jupiter and beyond the infinite," hope for an inhabited Mars was at an all-time low. But, the three early *Mariners* had photographed only 10 percent of the surface up close. Perhaps there was still room for surprise. However, the general vibe in the early seventies was that Mars was old and dead like the Moon, an eternally lifeless place.

All hopes for life on Mars, and Martian exploration in general, rested with *Mariner 9,* which was to be the first human-built spacecraft to orbit another planet. *Mariner 9* promised, if it worked, to expose the nature of the Red Planet definitively by photographing the entire surface.

Mariner 9 reached Mars, entered orbit as intended, turned on its cameras, and saw . . . absolutely nothing! Mars was not ready to divulge his secrets quite yet and had chosen to shroud himself in a global cloud of obscuring dust. Mars has a habit of working itself into a tizzy of violent winds and thick dust clouds that encircle the entire planet every few years, but the global dust storm that greeted *Mariner 9,* the "great dust storm of 1971," was one of the most intense we've ever seen, causing some to wonder if Mars was hiding something.*

Slowly, after many weeks, the dust began to settle and Mars revealed itself from the top down, with the highest mountains peeking first through the settling pall. The first features to appear—four huge dark spots near the equator—gradually emerged as gigantic volcanoes. The largest of these, which turns out to be the largest volcano anywhere in the solar system, was named Olympus Mons—Mount Olympus, the home of the Greek gods. As the dust cleared further, a new Mars was revealed: not the uniform dead world seen by the early Mariners, but a complex, varied planet with vast, jagged canyons dwarfing any similar features on Earth; wide volcanic plains covering much of the northern hemisphere; polar caps ringed by intricate layered terrain; and what appeared to be large networks of dried-up river valleys covering much of the ancient southern highlands.

Parts of those antediluvian southlands are devoid of features other than craters, at least if you don't look too closely. By sheer dumb luck, all the Mariners of the sixties had completely missed the most interest-

---

*The face!

ing features on Mars. This experience taught us a lesson about the dangers of drawing global conclusions from incomplete coverage.

I remember Carl Sagan showing up at our house with glossy prints of brand-new *Mariner 9* images and kvelling over them proudly as if they were baby pictures. I caught his enthusiasm like an incurable disease. My parents let me tack one of these pictures up in my room, and though it has yellowed a little, I still have it. It shows the great volcanoes just emerged from the dissipating global cloak of dust—an image full of the promise of continued revelation.

The Mars of *Mariner 9* is, in many ways, the Mars we know today. It is not a dead world like the Moon, or a living world like the Earth, but caught somewhere in between. Though many parts of its surface are heavily pockmarked with craters, revealing billions of years of geologic inactivity, ancient floods have also left their mark. The atmosphere, thin as it is, supports vigorous weather and continued erosion by windblown sand. Breezes blow and seasons change.

Mariner 9 also gave us strong hints of past climate change on Mars. The ancient valleys appeared to have been carved by rainfall, but no rain can fall in today's thin, frozen air. When the rivers ran, the atmosphere must have been thicker and warmer. Why did it change, and what happened to all the water? Mars, it seemed, started out more like the Earth, but had somehow gone cold and dry (shades of Percival Lowell). Might Mars and Earth have been similar long ago, when life was getting started here? If so, perhaps life sprang up on both worlds. Given the impressive ability of evolution to adapt to changing environments, might Mars still support some kind of life? This new hope spawned by *Mariner 9* gave us the lift needed to launch the Viking program.

Viking was the most ambitious and expensive planetary exploration program to date. It consisted of two orbiters and two landers—all successful. All four spacecraft were crammed with scientific instruments, but the centerpiece of the program, the raison d'être of the missions and ultimate source of their lavish funding, was the search for microbial life in the Martian soil. Each lander carried a package of three biology experiments.

The Viking landers set down in the summer of 1976. Along with other space-heads the world over, I was transfixed by the first pictures materializing on TV monitors. My teenage friends and I were at least briefly distracted from sex, drugs, and rock 'n' roll as the panoramic photos of dusty, rock-strewn, dune-filled landscapes gave us our first

Panoramic view of the *Viking 1* lander site.

good sense of what it might feel like to stand on the surface of another planet, gazing at the horizon.

The Viking cameras and almost all the other instruments worked flawlessly. The mission was amazingly successful and greatly enriched our knowledge of the atmosphere and surface of Mars. But the biology experiments were a bust. Though some early, puzzling readings provided brief, exciting moments of hope, the sum of all the results was convincingly negative: there is no life on Mars. At least no life that we knew how to search for. At least not in the surface soils at the two locations where the Vikings landed.

Hopes of finding life on Mars were demolished for two decades. But the Viking biology experiments, while failing to nourish any Martian microbes, gave us plenty of food for thought. How do we design an experiment to look for life on another planet when we've only observed it on this one? It's not a simple proposition. The question forces us to think deeply about what life really is—about the essential features that would transcend the specific natural history of one world.

UNCOVERING VENUS: WORLD GONE WRONG?

During those years of the first reconnaissance missions, the rest of the solar system didn't prove any friendlier to life as we know it. *Mariner 2*, the first machine (from Earth) to successfully visit another world, flew by Venus in December 1962 and radioed back news that was disheartening, at least for carbon-based creatures on Earth looking for close company or a nearby vacation paradise. The surface of Venus is hot as a kiln and dry as bones. There, organic molecules would fare about as well as a snowball in hell. The warm, wet, richly inhabited world that had dwelled in our scientific fantasies for hundreds of years was not to be found on Venus.

The drastic differences between Earth and Venus pose a huge challenge for our young science of comparative planetology. We are close neighbors, almost the same size and apparently built of the same materials overall, yet we have followed very different evolutionary paths. Most strikingly, Venus lacks just those features of Earth that seem most crucial to the survival and comfort of creatures like us: lots of water and a climate in the right range to keep it liquid. The temptation is strong to regard Venus as a world gone wrong, since Earth is so right— at least for us. We generally assume, although we can't yet prove, that Venus and Earth were very much alike at the start. But Venus, closer to the warming of the young Sun, suffered a "runaway greenhouse effect" early on. The young Venusian oceans boiled off into space in a global-warming disaster of mythical proportions, leaving "Earth's twin" a dried-up, burnt-out shadow of her former self. Admittedly, this analysis is rife with Earth-bound bias. A sentient Venusian sulfur slug might have a different perspective, but solar system history is written by the survivors.

The atmosphere of Venus is tricky to explore, with its acid clouds, ferocious winds, and turbulence that would make a United flight into Denver feel like a pony ride. But if you think that's bad, try exploring the surface. Obscured by clouds whether viewed from Earth or from orbit, it is difficult to probe in person, or even in robot, since we don't yet know how to design machines that can survive there for long without frying. The Soviet Union's persistent and methodical planetary exploration program was much more successful on Venus than on Mars. Two craft, *Venera 9* and *10,* built like big, round diving bells packed with refrigerants, made it to the surface in 1975 and snapped several pictures before surrendering to heat death. These photos, taken a year before Viking, were the first ever returned to Earth from the surface of another planet. They depicted gently rolling scenes strewn with volcanic-looking rocks, a little loose dirt hinting at some form of erosion, and a dull, cloudy sky off in the distance.

I found these barren, warped, rocky vistas to be slightly repulsive yet also enticing. Their unsettling otherworldliness was enhanced by the strange bits of alien Russian space technology rimming the foreground, and by the unusual geometry of the pictures, in which the camera captured a curving swath that dipped close to the ground in the center of the frame but out to the distant horizon on the edges. It was not a land-scape that made you want to pack a lunch and bound across it, yet those shady, distorted rock fields were somehow compelling. Like a piece of a fading dream you want to remember, this vague, tantalizing glimpse of an unexplored world made me want to see more. Little did I know at that time (I was a sophomore in high school) that I'd spend years of my adult life trying to unravel the story of Earth's twisted sister, Venus.

We didn't get our next direct glimpse of the surface of Venus until 1982, when I was about ready to graduate from college. This time *(Venera 13* and *14)* the pictures were in color, and the rocks were cast in red by the murky sunlight filtering through the thick clouds and crushing atmosphere. The strange, ruddy quality of the light served as a further reminder that these volcanic vistas were not of this world.

To uncover the story of Venus we needed global maps. Normal orbital cameras using visible light are useless for that purpose, so, like bats in flight or whales navigating the dark ocean depths, we must use echolocation to map the contours of the cloud-covered Venusian terrain. In 1979, the American spacecraft *Pioneer Venus* entered orbit, bouncing radio waves off the surface and recording the echoes to assemble our first crude global maps. These maps were a tease. You could see a lot of interesting structure, but you couldn't really tell what you were looking at. When I was a student at Brown University in the early eighties, one of my first research jobs was to help analyze these indistinct but enticing maps. We could make out numerous circular features dotting the Venusian plains. Were these impact craters or volcanoes? A lot rested on the answer to this question, as we did not know if the surface of Venus was ancient and full of craters like the Moon and the southern half of Mars, or young, restless, and volcanically active like the Earth.

After a decade of this torturous game of blindman's bluff (during which I got a Ph.D. in Tucson and then a postdoctoral fellowship at Ames, a NASA research center south of San Francisco), we got another spacecraft into orbit. *Magellan,* launched in 1989, mapped Venus for four years using cloud-penetrating radar. With these greatly improved radar eyes we saw towering volcanoes, vast plains flooded with lava, and a surface intermediate in age between ancient Mars and youthful Earth. *Magellan* did for Venus what *Mariner 9* had done for Mars, giving us a first clear global view. In many ways, our state of understanding of Venus today is where our knowledge of Mars was in the seventies, after *Mariner 9* and before Viking. Venus is a bit easier to get to but a lot harder to explore. It will take some new, advanced technology for long-lived landers like those of Viking to survive the sweltering conditions on Venus. I expect to see it happen.

GAS GIANTS AND ICE MOONS

During the 1970s and 1980s our eyes were opened to the rest of the solar system by the epic travels of the Voyagers. Launched in 1977, the year after the Viking landings, the two Voyagers flew by Jupiter in 1979 and Saturn in 1980. One craft, the indomitable *Voyager 2*, made it to Uranus in 1986 and Neptune in 1989. These missions took decades of hard work and intense planning, but the excitement was distilled into brief, manic "encounters" lasting only several days each, as one of the Voyagers would race past one of these giant planets, feverishly snapping pictures of its cloudy surface and its entourage of moons. Then, having safely radioed the bounty home, the spacecraft would quickly recede into the lonely depths of interplanetary space, heading for the next new world. During each encounter, multiple worlds were transformed instantly from obscure telescopic subjects into concrete, detailed places. The stunning pictures from these bursts of revelation will be treasured by humankind forever.

For planetary scientists the Voyager encounters were peak, formative experiences, and the trajectories of those two spacecraft through the outer reaches of our solar system became entwined with the trajectories of our lives. When *Voyager 2* reached Jupiter, I was a nineteen-year-old undergraduate assisting the team of scientists who retrieved and analyzed the photos beaming back from deep space. At the Uranus encounter I was participating as a twenty-six-year-old graduate student, and at Neptune I was a postdoc pushing thirty.

These encounters became bonding experiences for our community, part scientific conference, part family reunion, part soap opera. Each time our beloved robot craft plunged through yet another new system of worlds there was a gathering of the tribes as scientists and reporters descended on the Jet Propulsion Laboratory (JPL) in Pasadena, California, where the pictures and other information came down. Friendships formed and solidified. Romances began and ended. Some of those who were instrumental early in the mission were no longer with us at the later encounters.

Politicians and entertainers would show up to join in the fun, satisfy their curiosity, or make political hay from the stunning success of the Voyager project. One surreal morning at the Neptune encounter, after staying up all night watching the first close pictures come down from Triton, Neptune's schizoid frozen moon, my colleagues and I staggered out into the too bright California sunlight, and I could swear we stumbled upon Vice President Dan Quayle (who is not a rocket scientist) trying to milk the occasion by delivering an astonishingly insincere speech to a politely inattentive crowd.

Voyager revealed an outer solar system much more varied than we had expected and expanded our ideas about where we might find life. The most delightful surprises involved the myriad diverse moons orbiting these giant planets. The life stories of these small worlds turned out to be more complex and interesting than we had surmised. Surprisingly, several of them showed signs of recent geological activity.

The *Galileo* spacecraft orbited Jupiter from 1996 to 2003, dropping further hints suggesting (though not yet proving) the existence of a liquid-water ocean beneath the surface of Jupiter's moon Europa. This icy moon became a major focus of our remaining hopes for alien life within our own solar system.

MAD SCIENCE

Say what you will about the seventies. Maybe you still think disco sucks, but at least we were setting sail for the planets on a regular basis. After the stunning successes of planetary probes launched between 1961 and 1978, planetary exploration lost momentum in the eighties. The Soviets' winning streak at Venus continued into the early eighties, as they landed more cameras and sent balloon-borne instruments below the clouds, but no American planetary craft was launched toward any planet for an entire decade. This was due in part to the aftermath of Viking. Though the search for life on Mars was considered a long shot by most of the scientists involved, they had overhyped this angle to sell the mission to Congress and the public. As a result, the failure to find life was perceived as a failure of the mission, and that added to the difficulty in getting new missions funded.

But there was a larger problem. Whether we like it or not (and many of us do not) Cold War competition gave planetary exploration its first big push. Our fantastic voyages had largely been funded and supported as part of an elaborate form of saber rattling. The technology needed to launch a lunar or planetary probe is not dissimilar to that needed to deliver an ICBM. What better way to inspire missile envy than by nailing another planet? Our science had been enabled by the desire of governments locked in the MAD embrace of the Cold War to show off their destructive potential. By the late 1970s these demonstrations had served their purpose, and Apollo was only a memory. The governments that had supported our science to back up their threats were losing interest.

The most dangerous competition on this planet provided the impetus for our first liberating leaps beyond it. That these initial interplanetary voyages were financed and driven, financially and technically, by a potentially suicidal contest carries with it all of the moral contradictions of modern science.

Yet, our first pictures of Earth from space fostered a powerful new sense of the unity and rare beauty of our planet. Then the uniqueness of Earth was brought home to us with our first photos of the other planets. It is almost impossible not to have a global perspective while doing planetary science. Even during the height of the Cold War, Russian and American planetary scientists cooperated on their exploration plans and shared data across the geopolitical fault lines. It doesn't make sense for one planet to have two isolated space programs.

Along with the planetary perspective comes a realization that many of our problems and opportunities are global. The Earth has many lands but only one atmosphere, and we are all in it together. Shifts in global consciousness must be impossible to perceive accurately while they are in progress, but it is not inconceivable that a new planetary identity, which might be decisive for long-term human survival, is slowly dawning (painfully slowly). Certainly, efforts to communicate with intelligent extraterrestrials do not make much sense unless they are made on behalf of all humans. Merely contemplating the possibility of finding other life makes obvious our deep identification with all Earth's inhabitants.

It's ironic that the technology that enables you to beat gravity and see your planet whole can also threaten global self-destruction. One wonders if elements of this same drama might not be playing out elsewhere in the galaxy. If we are going to anthropocentrically worry about others on distant planets hastening their own demise with newfound techno-toys, we might also hope that the leap into space will have a similar liberating and unifying effect on their alien souls.

WORLDS BEYOND

Anyway, after a decade-long malaise, having kicked the monkey of Cold War aggression off our backs, we've now demonstrated that, hot or cold, we don't need war to explore. Planetary exploration was resurrected in the 1990s. In 1989, *Magellan* and *Galileo* were both launched toward Venus, where *Magellan* went into orbit and started its radar mapping. *Galileo* made a close flyby, getting a gravitational kick from Venus for its long journey out toward Jupiter.

The heartbreaking loss of *Mars Observer,* launched in 1992 (it went silent as it reached Mars in August 1993, probably due to a fuel-tank explosion), was an inauspicious start to a new series of Mars missions. But then, on July 4, 1997, *Mars Pathfinder,* swathed in air bags, made its daring bouncedown to land safely in an ancient flood zone. *Pathfinder* delivered the cute *Sojourner* rover, which was a scientific and public relations success, as it crawled around in the red dirt, snapping photos and sniffing rocks, showing the world that we were back. This triumph was followed by *Mars Global Surveyor (MGS),* which began its orbital mapping in 1998. With the *MGS* cameras we can see surface details a couple of meters across,* and its laser altimeter has provided us with fantastically detailed global topo maps that reveal the possible sites of ancient Martian oceans.

In October 2001, *Mars Odyssey,* a new orbiter, reached the Red Planet and detected vast fields of water ice mixed into the upper few feet of rock and soil surrounding the polar caps. New infrared images from orbit have revealed a complex layered structure in some areas, suggesting new complexities to the planet's past geologic history that have not yet been puzzled out. The search for signs of life-supporting environments, in the past or even present, is once again the primary goal of our entire revamped Mars exploration program.

In 1997, *Cassini,* a sophisticated, schoolbus-size spacecraft, departed for Saturn. *Cassini* began orbiting the ringed planet in July 2004 and in January 2005 will drop off the *Huygens* probe at Saturn's moon

---

*As opposed to about a kilometer with *Mariner 9* and fifty meters with Viking.

Titan. Titan is of special interest for possible extraterrestrial life. Voyager revealed abundant signs of interesting organic chemistry beneath Titan's thick, hazy, nitrogen-rich atmosphere. Studying this environment may help us understand the origin of life on Earth or, just possibly, reveal a new kind of cold-loving biology. Launches planned in the next few years include several new Mars missions, new Mercury and Venus orbiters, the first flyby of Pluto, and several new missions to study asteroids and comets.

The real planets provided a rude awakening from our dreams of Earth-like neighbors, but we're now learning to love them as they truly are. Just the fact that the planets were where we thought they'd be and our traveling machines actually reached them and worked has got to be the most solid confirmation of the scientific and technological revolutions of the past four centuries.* The resurrection of planetary exploration, in the decade after the Cold War ended, shows that human curiosity and wonder can occasionally transcend the deadly conflicts that have so often fueled our greatest bursts of innovation.

But what about worlds beyond our Sun and planets? Ever since Bruno and Fontenelle, most thinkers have concluded that our universe contains copious extrasolar planets. Throughout huge transformations in knowledge and worldview, scientists have consistently believed, without any direct evidence, that the Sun's family of planets is not unique among the overwhelming profusion of stars filling the night sky. But people believe all kinds of things, and neither analogy nor consensus makes it so.

In 1995 another cherished belief crossed the threshold from reasonable conjecture to observed fact: we started to find actual planets beyond the solar system. The first definitive detection of an extrasolar planet was made around a star called 51 Peg. After the first detection, planet hunters stalked their quarry with renewed vigor. Before you knew it, distant planets were popping up all over the place. As of this writing, in the eight years since the first detection, more than one hundred extrasolar planets have been found.† This is truly something new—a definitive, affirmative answer to an ancient question.

---

*Stuff that in your socially constructed pipe and smoke it!
†Such is the pace of current discovery that whatever number of planets I plugged in here in the final edit of galley proofs will be wonderfully obsolete by publication date.

Though in some ways the twentieth century saw the end of our astronomical innocence, the recent confirmation that our galaxy is rich in
planets has granted us a reprieve. Cosmology and extragalactic astronomy have banished us to the periphery of the universe, and spacecraft
exploration has replaced naive fantasy with cruel fact for the nearby
planets of our own system. But these brand-new worlds provide an
enlarged landscape for our speculative dreams. Like the planets of our
own solar system after the Copernican revolution and before planetary
exploration, we know these worlds are there, but we know little about
them. We can let our imaginations run wild.

ASTROBIOLOGY: NEW HYPE AND NEW HOPE

We tend to think of extraterrestrial life as a modernistic or futuristic
idea that rests on the findings of twentieth-century science and may
find verification in twenty-first-century discoveries. We think of it as
something for which humanity has had to be slowly prepared, while
gradually getting used to the disorienting and humbling knowledge that
we are very small and may not be so special in the universe.

Yet, as I've described, it is an old idea. Our seventeenth- and
eighteenth-century scientific heroes, the people we name spacecraft
after today, were almost all confirmed believers in a fertile, densely
inhabited universe. They based this belief on empirical observations of
other planets combined with metaphysical extrapolations. Today, our
rationale for believing in life beyond the Earth still involves a mix of
observation and extrapolation by analogy and plenitude. What is different now is the data. We've pretty much ruled out life on the Moon.
At one point we had ruled out life on Mars, though it has recently—at
least temporarily—been ruled back in. The boundary between that
which we can directly observe and that which we must deduce has
receded. But we still have to take that leap beyond the top rung of the
ladder of the known to reach for our conclusions.

Belief in aliens, among scholars at least, actually suffered some of its
lowest moments during the twentieth century. The search for extraterrestrial life was a semi-taboo subject among scientists. The public never
lost interest, but for decades professional astronomers risked disapproval
or career repercussions if they pursued the question as a major research
topic. Twentieth-century biologists mostly ignored the question.

Then in the late 1990s the field was revitalized, enjoying newfound respect and funding. It even got a new name: what had previously been referred to as exobiology, bioastronomy, cosmobiology, xenobiology, or exobotany was now rechristened *astrobiology*. NASA administrator Dan Goldin declared that the alien quest should now be the cornerstone of our approach for studying the entire physical universe. A NASA Astrobiology Institute was founded in Silicon Valley, and the money began flowing. Researchers in diverse fields are suddenly discovering that it pays to consider yourself an astrobiologist.

Why the official change of heart? Several recent discoveries have rekindled hope for a living universe. In August 1996, a year after the first extrasolar planet was discovered, President Clinton made a startling announcement. American scientists had found fossils in an ancient meteorite from Mars.* Even though many scientists soon challenged this interpretation of the microscopic, wormlike structures, it got us all thinking "Why not?" Worldwide front-page headlines gave the field a boost that lasted long after serious doubts were raised about the biological origins of the possible Martian microfossils.

That same year *Galileo* entered Jupiter orbit and returned sharp pictures of Europa, providing new circumstantial evidence for a possible life-giving underground ocean. Around the same time several new kinds of bizarre life-forms were discovered on Earth, living under conditions previously thought to be deadly. Our conceptual limits on life's domain widened. This harmonic convergence of discoveries was seized upon by advisers and officials at NASA, who were also discovering that casting our space research in terms of the search for cosmic company could supply a much needed PR lift.

Thus was born the "astrobiology revolution," which I will discuss in more detail in the next section of this book.

ACROSS THE GREAT DIVIDE

Dreams of alien life will not die. Almost as if we need to believe that we are not alone in the universe, or even somehow know we are not alone, this idea is resurrected in every era in new forms reflecting the attitudes and convictions of the day. Today, beliefs about extraterrestrials are polarized in a way that mirrors our culture's divergent attitudes about

---

*Shades of the panspermia of Kelvin and Arrhenius.

science. We scientists view the question as part of our turf, as the proper subject of astrobiology. In recent decades, however, another train of belief has been gathering steam on a parallel track. Although science has been on a roll of confidence and power since the Enlightenment, antiscientific voices are increasingly audible on the cultural airwaves. Creationism, faith healing, astrology, postmodern relativism, and New Age spirituality are among those beliefs that mainstream science regards as dangerous superstitions threatening the rational basis of our society. Many adherents of these beliefs view science as equally dangerous, threatening our very survival with an amoral, materialistic, antispiritual attitude and an out-of-control pursuit of new technology. Nowhere is the gulf between expert scientific opinion and popular folk beliefs greater than on the subject of aliens.

Millions of rational adults currently believe that UFOs are alien spaceships that have come to Earth and occasionally abducted people, perhaps to study us, perhaps to help us out. Astrobiologists and SETI scientists are quick to dismiss such opinions, even while adhering to their own strong faith in our ability to establish radio contact with like-minded aliens. On both sides of this divide, many people nourish the fantasy that advanced extraterrestrials with superior wisdom will lead us beyond the threat of high-tech self-destruction to a safe and wondrous future.

Ironically, widespread belief in UFO aliens may be partly responsible for the strong public support enjoyed by NASA's astrobiology program. Certainly our regular stream of "Life on Mars" headlines has lent comfort to the UFO believers. People think there's life out there, and they want us to find it. Although the science/antiscience divide is wide, both sides believe in aliens. Perhaps, deep down, our reasons for believing are not so different.

PART II

Science

5 | The Greatest Story Ever Told

I believe a leaf of grass is no less than the journey-work of the stars.

—WALT WHITMAN, *Song of Myself*

To punish me for my contempt for authority, fate made me an authority myself.

—ALBERT EINSTEIN

May you build a ladder to the stars. May you climb on every rung.

—BOB DYLAN, "FOREVER YOUNG"

NEED TO KNOW

To assess our universe's potential to create other life and intelligence, we need a framework for understanding our own arrival on Earth. What is this place and how did we get here? We need to know, so we construct cosmologies. A multitude of fine origin tales suggest themselves, but they can't all be true. How do we choose? We can test our answers against the nature of nature itself.

The tale of Cosmic Evolution is our new origin story. Over the past four hundred years, we've found some answers, at least partial and provisional ones, to many of our oldest questions: "How did the world begin, and when?" "How does life work?" "Who were the first humans?" "What are the heavens made of, and how are they related to us?" Once these may have seemed to be disconnected, independent mysteries. Now we've found that the answers can all be woven together into one continuous chronicle: our journey from universal birth through galaxies, stars and planets, elements, molecules, and life, into mind, and then, perhaps, onward to something beyond.

This story has not been handed down, but dug up and cobbled together. It is full of holes, inconsistencies, and paradox. An inexhaustible stream of new evidence ensures that it is a living myth, a script always in rewrite. Paradoxically, this provisional and changing nature of the story of Cosmic Evolution is what gives it its great credibility. We see it improving before our very eyes, admitting error and making corrections. It's getting so much better all the time.

What do I mean when I say this is "our" origin story? Just whom am I referring to? A bunch of males who are not only dead but white? Is this story accepted and embraced by everyone? Of course not. But our knowledge of Cosmic Evolution is not in conflict with the core beliefs of most of the religions, and it certainly isn't necessary to discard or discredit older origin stories to embrace this new one. Even if you go to church, temple, or ashram for the singing and the dancing (that's the part I like), for the comfort of spiritual community, or to receive ancient wisdom, you probably accept that science has clued us in to some big truths about our origins that the writers of our ancient texts could not have known. Except for some Rastas I used to play with in a reggae band, and some Jehovah's Witnesses who've knocked on my door, I haven't met many people who take a seven-day Genesis literally.

Clearly, if our society endorses any "official" story of genesis, it is the scientific account. It is the most generally accepted origin scenario in our culture, and the one we teach in public schools, when we bother to teach about origins at all. Strange, isn't it, that science has become the keeper of official wisdom? Since Galileo's time, science has grown from an upstart, radical fringe to a dominant worldview. To punish us for our contempt for authority, fate made us secular priests with pocket protectors.

Yet, the story of Cosmic Evolution itself is not well-known or widely celebrated. This new creation story is typically regarded as an impressive work that is quite incomprehensible to nonexperts. Most people do not look to this story for vital beliefs and satisfying answers to deep questions. Rather, they think of it as science. Many people associate science with a dry recitation of facts or torture by algebra while waiting for the clock to signal freedom from incarceration. So here we've found what seem to be convincing clues to our real origins, a story about a point of nothingness exploding and evolving over billions of years to generate our entire universe of galaxies and flowers, and many regard the tale as Dullsville. Maybe we're not telling it in the right way.

Part of the problem is that an ethos running deep within our profession suggests that science, to be objective, must be devoid of passion. The conduct of research does require a dispassionate attitude. When we *do* science, we have to practice nonattachment. To keep ourselves honest, we have to pretend not to care and always be willing to accept nature's verdict on our precious theories.

But we've spent too much time in the lab, and something has gone wrong. This necessary emotional distance has led to a Spock-like detachment in the way we share science with the rest of the world. It's all too easy to lapse into a science-class drone when telling the story of Cosmic Evolution—the same, almost obligatory dry tone in which it was taught to us. When we discover something fantastic about the age of the Earth, the history of life, or our true location among the stars and galaxies, we should be shouting it out with glee, not droning on with only dim life signs. Why aren't we singing the song of the galaxies on television and doing the DNA dance in every town square?

Certainly Cosmic Evolution is the story I was brought up to believe, but I've also studied enough of the evidence that now *I think* my continuing belief represents more than the tendency to retain the views of one's parents. It wasn't exactly *Fiddler on the Roof* in our household, but in my upbringing as a secular humanist Jewish American, I did learn some of the traditional stories. Each year, we held a Passover seder with a gathering of relatives and friends and recited the Haggadah, the tale of the escape of the Jewish people from slavery, as the kids and some adults fidgeted in eager anticipation of matzo ball soup. I suppose the association with good food is part of the secret of the success of this ritual—whose story has survived intact for thousands of years, much of that time primarily as oral history. Maybe what the story of Cosmic Evolution needs is a similar association with a family gathering and an enticing hot meal.

Though I was deprived of (or saved from) spending my Sundays learning the tale of Genesis, I picked up the basics later on. Genesis, like many other prescientific takes on our origins, contains a kind of temporal anthropocentrism: not only are humans the focus and the crown of creation, but the universe itself is only 5,775 years old. Thus the entire scope of all existence is held to be not much longer than the time during which people have been walking the Earth (five days longer according to Torah).* In the absence of evidence to the contrary, why assume that the world is older than we are? But we now know that our entire written history is just a one-sixteenth note in a long cosmic symphony of many movements.

Try to stretch your mind around the sweep of time envisioned in this new tale: ten thousand times a thousand times a thousand years. Just as the Copernican revolution and its aftermath put us in our place in space, teaching us that we are very small and not centrally located, Cosmic Evolution shows our time to be tiny. Humans have been around for perhaps a few million years, a hundredth of a percent of all existence, so far. Next to nothing.

Feeling diminished? Not to worry. This story invites us to identify with something much larger: our biosphere—the totality of life on Earth. "We," Earth's biosphere, have been around for a thousand times longer, for several billion years. This is a significant fraction of cosmic history, perhaps a third of it. So we are not so small, in time.

In one sense, this tale cannot compete with the older origin stories because there are no people in it. Most chapters have no characters at all, just particles and forces following blind compulsions of physical law, and from this perspective it's only the story of an elaborate machine. Not that exciting, really. But through all its stages, the universe has been molting, preparing to awaken, becoming us and everything else.

The presence of human minds, then, permeates this story, along with the minds of any counterparts who may be watching the same drama unfold from distant stellar perches. We've been in it all along. Cosmic Evolution is not "merely" science. Longing to know our origins is a spiritual itch that this story can help us scratch.

WHAT'S THE STORY?

The word *evolution* in a scientific context usually conjures up the procession of life-forms parading across Earth's deep timescape. You know: amoebas, algae, trilobites, apes, rock 'n' roll drummers, humans

---

*A notable exception is Hindu cosmology, which describes a universe much older even than that described by Cosmic Evolution.

and dolphins. But this chain of events on Earth is embedded within a much larger sequence of transformations that first prepared the ground and planted the seeds for local biological evolution. Cosmic Evolution follows the universe through its own progression of forms. We can trace a train of causality in which, following physical law, each developmental stage of the universe creates the conditions leading inevitably to the next.

In every life there are pivotal moments: your first step, first words, first kiss, the birth of a child, the death of a loved one. In human history there have been similar moments that have left us permanently changed: starting to walk upright, the invention of agriculture, the printing press, the electric guitar, humans walking on the Moon. The universe, too, has passed through critical experiences that divide its history into distinct phases. Here I'll briefly sketch out the major evolutionary transitions that the universe, that we, have passed through. It's a story with chapters we know, even if the verses have not all been worked out.

As the good witch said, it's always best to start at the beginning. At first, we were very small. And hot, too. According to a literal interpretation of big-bang genesis, at the very beginning the entire universe was all piled on top of itself in a single point of zero size with no surroundings—a point of infinite density and infinite temperature from which sprang forth all matter and energy.*

In that ultradense primitive universe, there were no atoms, no structures of any kind. It was just too hot. Forming atoms or molecules in such an environment would be as hopeless as trying to execute an elaborate tango in a mosh pit full of aggressive slam dancers. Try as they might to hold together and execute their steps, the tango dancers would instantly be pulled apart, merging with the joyously colliding crowd. In the beginning, this was the only dance in town. It was way too hot to tango.

But not for long. Soon, things had chilled enough for matter to make something of itself. After about .00001 seconds the first simple structures formed. The temperature had dropped to 10 trillion degrees, just

---

*That's the "standard theory." A more recent variant of this holds that it was never zero size and infinite in density and temperature. Rather, according to "superstring cosmology" the universe started out as a little nugget .0000000000000000000000000000000000001 meters across with a temperature of 100,000,000,000,000,000,000,000,000,000,000,000 degrees. The beginning still needs work.

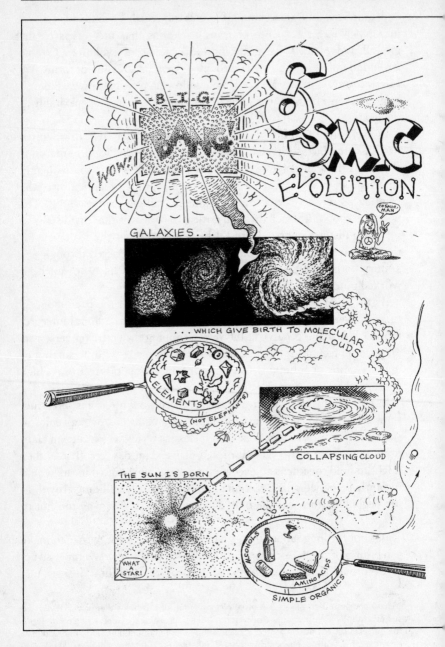

(Borin Van Loon)

cool enough for protons and neutrons to assemble out of the initial hot soup of quarks. Thus began a long series of ever more intricate partner dances, the latest of which is, at least around these parts, complex, communicating sentient life.

A gradual increase in complexity over time is one of the major themes tying together the tale of Cosmic Evolution. Matter is able to assemble itself into more and more elaborate forms as the universe expands and cools off. As the general heat-induced jostling subsides, the self-attracting forces of matter are better able to resist the tumult, and ever larger and more delicate creations materialize.

After a few seconds, when the universe had cooled to about 1 billion degrees, protons and neutrons were able, by the attraction of atomic forces, to gather themselves into hydrogen and helium—the first atoms. Except, they were not complete atoms yet because the electrons were running wild. In that heat, no nucleus could hold them, so the entire universe was plasma—which we see today in a glowing star or a flash of lightning.

Plasma ruled the universe for ages. Fast forward another three hundred thousand years, when things had cooled down to a few thousand degrees. This slowed the electrons down just enough so the nuclei could reach out and hold them against the thermal storm. Atoms at last! The plasma gave way to a hot gas of hydrogen and helium. This was a significant moment in our cosmic history. Light cannot travel through a thicket of free electrons, but now the universe as a whole suddenly became transparent.

And there was light. The universe was white-hot. As it continued to grow and cool, the ubiquitous rays of light expanded along with it, and their wavelengths stretched out to ever redder and cooler colors. Over billions of years, this background radiation has marched down the spectrum from the visible range through infrared, then radio, down to cold rays of microwave. The "cosmic microwave background" we see coming from all directions today is the chilled-out, stretched-out afterglow of that moment, only three hundred thousand years into Cosmic Evolution, when light was set free by the formation of atoms.

Even as the microscopic structure of matter was slowly evolving from chaos, climbing a ladder that had passed through quarks, protons and neutrons, and simple atoms, the structure of the cosmos at the very largest scales was also evolving. The universe was getting ready to build galaxies, the environments in which the rise of matter could continue, where atoms could achieve the next great leaps of complexity.

Although, as we have seen, in the earliest state of the universe no stable constructions of any kind could form, all was not completely uniform. From the beginning, random ripples pulsed in the formless fireball of the primeval universe. In the rapid expansion that ensued, these infinitesimal clumps were frozen into the structure of the universe and they grew along with it. Eventually they became giant clouds and tendrils of gas millions of light-years across.

Then gravity took over. Establishing another theme that we will see repeated, gravity tends to accentuate large-scale differences in density. It's the anti–Robin Hood of the universe, increasing disparities in the distribution of mass, stealing from the poor and giving to the rich. Any clump of matter has self-gravity. In the absence of other forces, a gas cloud will contract, falling in on itself and becoming more dense, robbing matter from its less dense surroundings. Gravitational contraction amplified the primordial density differences, and these condensing clouds of gas became the spiraling galaxies. Thus the tiniest things imaginable, little bumps smaller than anything we can even detect with instruments, grew to become galaxies—the sprawling, spinning arenas where generations of stars are born, live, and die. On the face of it, this is harder to swallow than any biblical, magical, or mythical tale. However, the evidence is compelling. Our Milky Way and all the other whirling galaxies, the orbiting clusters of dozens of galaxies and the vast superclusters containing thousands of these—the largest structures in the universe—all started as infinitesimal random fluctuations when we were very small.

Galaxies formed fast. Since light moves at a finite speed, telescopes are also time machines. With the Hubble Space Telescope, we can see young galaxies already formed some 12 billion years ago.

As far as we can tell, the universe is only 13 or 14 billion years old. There is some slop in these numbers, and they will continue to change as the theorists and observers wrestle on into the twenty-first century. But right now, it looks as if the galaxies are nearly as old as the universe itself.

Unlike stars, which are still being born all the time, the galaxies are all the same age, and still in their first and only generation. Though they've collided, merged, fragmented, and morphed into new shapes and colors, no new ones have been born since the cosmos was very young. This is our universe's one and only crop of galaxies. They'll have to do, unless someday we, or someone, figure out how to make new ones. And they are all slowly aging.

GALACTIC ECOLOGY

In the beginning, the Milky Way and all other galaxies were made only of the hydrogen and helium created in the big bang.* Picture the periodic table, fixture of every science classroom, in which the 103 elements are arrayed in columns according to their chemical behavioral types. Now, imagine a chemistry class during the youth of our galaxy. The chart on the wall would have been small, just two boxes, one each for hydrogen and helium, the two simplest atoms. The only chemical reactions possible were the coupling and uncoupling of hydrogen atoms. And you think science classes today are dull. But, not to worry, such a class would have been quite impossible. Without more elements to work with, there would have been nothing with which to build the classroom walls or print the chart. Without a richer palette of atoms with which to build complexity, it's hard to imagine life of any kind evolving, let alone scientists and bored chemistry students. Fortunately, the galaxies evolve chemically as they age, slowly gaining more heavy elements. Thus they increase their potential for interesting chemistry, which, if local experience is any guide, means greater potential for life.

Where does this increasingly diverse array of chemical elements come from? From the metabolism and decay of stars. For galaxies are the fields where stars are born, live, and die, and stars are element factories where nuclear fusion creates all the chemical richness of the modern universe.

Stars result from the collapse of giant, diffuse clouds of gas that float among the galactic spiral arms. That collapse, initiated by random breezes that blow through the galaxy, soon reaches a point where, once again, gravity takes over, accelerating the contraction. From the pressure of the collapse, it becomes ever hotter in the center of the cloud. Eventually the gas at the center is squeezed so tightly and the collisions are so violent that individual atoms can no longer hold each other off. They start to merge in nuclear fusion reactions. At that moment of

---

*Purists will note that a sprinkling of lithium was also made in the bang, but no manic-depressives would evolve to take advantage of this for billions of years.

nuclear ignition, a star is born. In the hot cores of stars, matter continues its climb toward more complex structure. While it liberates the energy that makes the stars shine, nuclear fusion also builds new and heavier elements. Hydrogen is fused into helium, helium to carbon, and so on, filling out the periodic table of the elements.

During its lifetime, each star maintains a balance between the inward force of gravity and the outward pressure of heat and radiation liberated by the nuclear reactor at its core. When a star reaches old age, having spent much of its hydrogen fuel, it can no longer hold this balance. For the most massive stars, the end is dramatic and quick, a cataclysmic supernova explosion in which a dying star can briefly outshine its entire galaxy. In the intense flash of a supernova, pressures and temperatures dwarf even those found inside a normal, healthy star, and still more new elements are made. The nickel in Earth's metallic core, the silver hanging from your ears, and all the gold in Fort Knox were made in the creative flashes of stellar demise.

In the calamitous explosions of stellar self-cremation, supernovae spread their remains far and wide, fertilizing the galaxy. Their ashes make their way into the great clouds of gas and dust that form raw material for new stars. Thus enriched, subsequent generations of stars, and their accompanying planets, start off with a greater endowment of the heavy elements. The greater the diversity of elements, the more the potential for building planets like Earth, made of iron, silicon, and oxygen. This also increases the possibility of complex chemistry, including carbon chemistry, the starting material for our kind of life. Our galaxy is a compost heap of elements, growing ever more fecund with the life and death of every star. Each successive stellar generation is born with the potential for more diverse planetary evolution and more complex chemistry. Our own Sun is probably a third-generation star.

A galactic ecology is at work here. Just as a forest ecosystem has new plants and trees growing from the decayed remains of their predecessors, in a galaxy new stars are continually born of the gas and dust spread by their dying ancestors. Our galaxy is still young—a vast hydrogen sea of possibilities.

ANCESTRAL CLOUD

If we're got it right, parts of our origin story must be truly universal. Our version of events will be somewhat familiar to the children of other stars sprung from the same field of galaxies when we finally meet them and compare notes. We don't exactly know where the universal part of Cosmic Evolution ends and the "local flavor" begins. We're still trying to figure out how normal or rare are the events that have transformed planet number three from a rock into a reef. Yet, we do know where we got started as a distinct entity, separate from the universe at large, with our own local history and, now, culture. Our differentiation from everything else, our individuation, began some 4.6 billion years ago, when the universe was about two-thirds its current age, with the collapse of our ancestral cloud.

Long before we became creatures living within a planetary biosphere, and just before we became a preplanetary disk of debris circling a young Sun, we were a molecular cloud floating in the arms of the Milky Way. Then, everything in our entire solar system was smoothly blended, drifting together in this diffuse cloud of gas and dust. You, me, the Elephant Man, the Dalai Lama, the neighbor's barking dog, the flower shop down the street, the Great Wall of China, the core of the Earth, the Sun, and the planet Neptune: we were all one. Of course we still are, but back then it would have been obvious even without the aid of meditation, psychedelics, or quantum mechanics, as we were all ground up and comingled, all one and the same cloud.

In the journey toward complex structure that matter had been following since the bang, the next big step was chemistry. For interesting molecules to form, first we needed the more diverse array of atoms forged in the nuclear fires of living and dying stars. Then we needed the cool, nurturing environs of a molecular cloud, where solitary atoms could safely assemble. Later still, beneath the sheltering sky of at least one planet, the chemistry that began in the cloud would come to life.

Our cloud was a potent brew of primordial hydrogen and helium heavily seasoned with the remains of dead stars: nitrogen, carbon, oxygen, and a smattering of other tasty elements. Ultraviolet light from nearby stars played with these atoms, sometimes knocking off their outer electrons. An atom with missing electrons is in desperate need of others to complete itself. Wounded atoms will band together, sharing and trading electrons, combining and rearranging themselves into molecules. Where carbon was involved, these elements formed complex organic Tinkertoy structures that blew about in the cosmic breezes.

The birth of our solar system was helped along by the ashes and the splashes of the Sun's dying ancestors. In addition to seeding our cloud with heavy elements, the final spasms of spent stars sent shock waves through the cloud, squeezing it in some places and thinning it in others. The temperature in the cloud varied, and it was more dense in the cold spots. One of the cold dense spots, nudged by the hot breath of a nearby supernova, started the fateful collapse that led to our solar system.

Once it got started, the collapse happened fast. For the first million years it was in free fall, accelerating until enough gas pressure built up in the center to resist gravity and slow things down. As it shrank, it started to develop a distinct shape, morphing from a roughly spherical cloud into a flattening disk.

Again with the disks! Why always a disk? Some basic physics tells us why a collapsing cloud becomes a spinning disk. Like a break-dancer spinning on her head, pulling in her knees, our cloud spun faster and faster as it contracted. It had to. It's a law: the conservation of angular momentum. The more a spinning object contracts, the faster it spins. Like other physical laws, there's no penalty for breaking it—you simply can't.*

The spinning cloud began to flatten out into a disk, for the same reason that a spinning ball of dough tossed into the air by a pizza chef stretches out into a pie shape. The centrifugal force caused by the spinning motion flattens any fluid object—whether made of soft dough or diffuse interstellar gas and dust.†

Finally, when things got hot enough at the center of the collapsing cloud, nuclear fusion began and the Sun turned on, heating and lighting the surrounding preplanetary disk.

MAKING WORLDS

At this point the picture should sound familiar: this is the flattened, spinning disk described in the eighteenth century by Laplace with his nebular hypothesis. In the intervening centuries, and especially in four decades of space exploration, we've learned a lot more about how a disk became a system of worlds.

---

*In fact there are huge rewards if you do succeed in breaking one of these laws. A Nobel Prize and maybe even an appearance on *Letterman*.
†You can feel this same force by simply twirling your body. Your arms take flight, pulled outward by the same force that pulled the twirling preplanetary cloud into a disk.

The first step was a blizzard of snow made of metal, rock, and ice. When the collapse slowed, the disk began to cool off. Solid material condensed out of the cooling gas, just as snow condenses out of a cooling cloud of water vapor. In this case the snow was made of a wide range of materials, which varied with the changing temperatures at different distances from the Sun. In the hot regions near the Sun it snowed flakes of metal and rock. Farther out, around the present orbit of Jupiter, it was cold enough for ices to form: both the familiar snowflakes of water ice that adorn winter on Earth and more exotic snows of frozen methane and ammonia.

When these snowflakes collided, they stuck together, forming larger aggregates—dust bunnies of metal and rock near the Sun, and icy snowballs farther out. These continued to collide and stick, forming larger clumps, beginning the process of growth by *accretion* that eventually created the planets. Between the time when all was gas and dust and the final formation of a small number of large planets, our solar system was a raging storm of billions of tiny planets orbiting the Sun—the *planetesimals*.

As planetesimals grew to a kilometer in size, they began to hold together by their own self-gravity. They also started to yank one another around, gravitationally perturbing each other's orbits. The near misses, which were more common than actual collisions, scattered orbits far and wide, churning up the disk and facilitating further collisions. The few planetesimals that, by chance, grew the fastest became the embryos of planets. Their gravitational hunger only increased as they grew, sucking in and consuming the lesser planetesimals.

The last stages of this growth were violent and chaotic. As the planets approached their final sizes, giant also-rans, the contenders that could have been planets, came hurtling down to Earth (and Mercury, Venus, etc.) at speeds of tens of thousands of miles per hour. These final giant impactors left a trail of destruction throughout the solar system, stripping Mercury of its outer rocky mantle, leaving Venus spinning backward, and knocking Uranus on its side. And in an event as propitious for us as it was random, a Mars-size protoplanet smacked into the young, still-forming Earth, splashing a massive ring of vaporized rock into Earth orbit, which quickly condensed to make our singular, giant Moon.

This entire growth process, from snowflakes to planets, was largely complete in 100 million years, although the tail end, a rain of leftover debris from planetary construction, continued to terrorize the planets for several hundred million more.

The end result of all of this snowing, growing, clumping, bumping, and grinding was our solar system as we see it today. The initial segregation of material by temperature, which made metal and rock near the Sun, and ice farther out, has been preserved. Four small planets of rock and metal (which we call the terrestrial, or Earth-like planets) orbit close to the Sun, and four planets dominated by gas and ice (which we call the Jovian, or Jupiter-like planets) live farther out where cooler temperatures prevail. And then there's tiny Pluto, the largest denizen of the frozen farther reaches of our system, where the icy planetesimals that never accreted into larger planets still trace out their long, slow orbits.

How universal is the planetary part of our story? Everything we've been able to glean about how the planets formed here leads us to believe that when the Sun was born, the Earth and other planets were the inevitable aftermath. So, it seems, planets like ours should be a natural by-product of starbirth, at least for stars that formed under the same conditions as the Sun. But how rare, or commonplace, were these circumstances of birth? In a later chapter I'll report on recent discoveries that are finally shedding some light on that all-important question.

What we do know is that at least once the expanding, cooling, coalescing, condensing, accreting mess of the early universe was formed into a sheltering tide pool of a planet, sufficiently protected from the pounding of the cosmic surf to harbor the delicate chemical structures needed for the next steps in the long climb of matter into life.

THIS IMMORTAL COIL

Now we enter the planetary phase of the cosmic fable, where matter made the most astounding leap yet. The surface of a planet can be a good place for elements and simple molecules to get together, try new variations on their structural themes, and make ever more complex molecules. Especially if, as was this particular planet, the third stone from a third-generation star, it is blessed with a sprinkling of holy water rich in carbon, nitrogen, oxygen, sulfur, and phosphorus—the "biogenic elements." It also helps if, when the music stops after the random accretionary dance, your planet winds up at a healthy distance from the irradiating glow of its newborn star. If a planet is not so hot that complex chemicals are ripped apart by the frenzied molecular collisions that constitute heat, and not so cold that chemistry itself grinds to a halt in the atomic lethargy of frigid matter, then things could get interesting. Better yet, if you orbit at a distance where, with the help of some "greenhouse warming," surface temperatures allow for plentiful liquid water, you probably can't go wrong.

The young Earth had no shortage of simple organic molecules, the building blocks of terrestrial life. They may have been made here on Earth out of water and methane, using energy from lightning, ultraviolet sunlight, submarine volcanoes, or the explosive shock waves of falling comets.* Or they could have come from outer space. Heaps of organic-rich interstellar dust made their way to Earth, frozen in icy comets.

The young Earth was awash in organics. In the early, warm oceans, our organic molecular forebears played at combining and recombining in myriad forms. It was a chemical Kama Sutra where molecular adventurers got busy, trying anything at least once. The result was a period of "chemical evolution," a primitive molecular survival of the fittest where stability and longevity were the only criteria for success. Molecules that found longer-lasting, more stable configurations were naturally selected. Such hardy structures, favored by the gods of trial and error, became concentrated in the primordial sea.

And then it happened. This magic moment. Carbon chemicals, innocently cavorting in water, hit upon something new and extraordinary: structures that could make copies of themselves. Eureka! In the 10-billion-year history of the settling down of matter, and the bubbling up of order, a powerful new force was unleashed: heredity—the preservation of useful information. Those structures that could copy themselves most accurately and efficiently multiplied like rabbits in a world without foxes. Self-replicating chemicals rapidly populated the young oceans.

We don't know what the first self-copying chemical mechanism was, but we do know which one stuck, because it's still here, working its magic: DNA. In a sense, there has only ever been one strand of DNA, dividing and multiplying on Earth for nearly 4 billion years. DNA never dies, it only improves. Evolutionary tributaries dry up and disappear, but the mainstream of DNA-driven life keeps on growing. Barring

---

*The Miller-Urey experiment, described in chapter 3, demonstrated that these molecules are easy to make in conditions simulating those of the young Earth.

disaster, it will flow onward until, 5 billion years hence, our dying Sun becomes a red giant star and incinerates the Earth. This helical miracle may even outlive its star of birth, if we leave and take it with us.

DNA is a molecule that may have achieved immortality. And we, along with all other terrestrial life, are its by-products, reproduction mechanism, and life support system. We are the ships that DNA has devised to navigate the world. But when the machines get smart enough, they can take over the factory. With the first faint, fragile glimmering of intelligence on Earth, we can see the potential for a new, as yet unwritten, chapter of Cosmic Evolution, the self-aware phase in which matter wakes up, takes a look back at where it's been, and consciously decides how to proceed.

WHAT'S THE POINT?

This new origin story need not replace all others, but it does augment them. I do not mean to be cavalier in comparing the story of Cosmic Evolution to the world's great religious texts. Obviously, this story does not address all the needs met by our older origin stories. It doesn't answer the question "Why?" about anything. It provides us with no rules on how we should live, at least not in any obvious way. I do not subscribe to *scientism,* the view that science is all you need. If pressed, I'd say that John Lennon's view that love is all you need is closer to the truth. But a combination of the two is even closer.

So then, what's the point? How does it help us if it cannot provide us with a purpose or tell us how we ought to live? Why science's attitude of superiority? Well, the attitude is not justified. But there is a point. Several, actually. Science does not tell us why we are here, but it has taught us a lot about where "here" is. Science limits the range of the possible and, in so doing, points us toward what is true. And I think that a careful reading of this story does convey some powerful messages that I, personally, am not afraid to call religious.

You, me, everything you can see except the stars, and a great deal that you cannot, were all once mixed together in one giant, diffuse cloud. Even the stars themselves, and all the distant galaxies imaged by the Hubble, we and they were all one, even earlier, when the universe was a hot little ball of fire. Once a singularity, always a singularity. That Zen master hot dog vendor doesn't have to make you one with everything. You already are.

Cosmic Evolution carries a message of complete and profound unity, which I think can be read as a reason to care deeply for all things, especially for the living Earth and its creatures, the most highly evolved local products of matter's slow climb from formlessness. Against this backdrop, the anthropogenic mass extinction that we are currently inflicting upon Earth seems a desecration of cosmic proportions. Maybe there is a moral to this story.

FIERCE MAGIC

Last night, I was up late writing this chapter, trying to distill the story of Cosmic Evolution into a brief narrative. I had to take years of detailed study, throw out most of the details, and try to extract the core events, the highlight reel of the universe. With visions of galaxies dancing in my head, I drifted off to sleep and had a vivid dream. An earnest woman who might have been Native American—she looked like the poet Joy Harjo—was telling me about something important that had just happened to her, something that made her realize how incredibly fleeting our lives are. She implored me to remember to treasure each day.

I don't know whether this dream was related more to the cosmic timelines ripping through my mind or simply to the emotionally mixed experience of shutting myself off from my people and my planet for some of my precious days to get some writing done. But, I awoke that morning with these Joy Harjo lines in my head:

I can hear the sizzle of newborn stars, and know anything of meaning, of the fierce magic emerging here. I am witness to flexible eternity, the evolving past, and I know we will live forever, as dust or breath in the face of stars, in the shifting pattern of winds.

This poem expresses another lesson of Cosmic Evolution. That we are immortal. There is life after death in a sense that is completely real and does not require blind faith. You don't have to see yourself as separate from everything else. We have been here, gathering ourself for at least 10 billion years. This is just the beginning. We'll live on for tens of billions more.

We can't study life as a cosmic phenomenon without studying the universe as a whole. The seeds of life were planted in the early primeval fireball, and the roots of life grow in the very fabric of space, time, energy, and matter. What we don't yet know is how to place ourselves in this story. Are we peripheral or accidental, or somehow integral and central to the main theme? The meaning will not be clear until we find out if there are other souls circling distant stars, others who have emerged from this same cosmos into an awareness of their own.

Either way, this is our story. It grounds our existence in the unfolding of the universe. We carry reminders of it in every cell, every atom, of our bodies. We have learned that our ancestors are stars. And we are water—enclosed sacs of Earth's salty oceans that have conspired with carbon molecules to achieve mobility and a fledgling form of consciousness. Looking out, and looking back, we see that we are the eyes of the world and the soul of the galaxy. We just don't know whether we are the only ones.

Each of us, growing up, reaches a point at which we become curious about our origins. We ask our parents challenging questions about where babies come from and which came first, chickens or eggs. Here on this rocky speck of living stellar afterbirth the climb of matter has somehow arrived at what we proudly call intelligence. Here the universe has grown up to an age where it wants some answers about its own provenance. Here on Earth, the cosmos has awakened from a 12-billion-year dream. It seems that our consciousness, in inchoate form, was here all along, waiting for the right conditions to precipitate out of inanimate matter. Elsewhere, is it slumbering still, or were we among the late sleepers?

Why here, why now? Once life got started, it quickly embedded itself in the workings of the planet, altering the atmosphere, the oceans, the rocks, and the soil. As life has adapted to the changing environments of the Earth, Earth has been remade continually by life. So ancient is this partnership that it is impossible to know what Earth would have become if it had not become alive some time around its 500 millionth birthday. What happened on Earth to give rise to one species that makes music, Mandelbrot mandalas, and Mars probes? How were we transformed from a newly formed, warm, wet, organic-rich planet into a profuse, prolific biosphere that has now started looking back out at the planets, stars, and galaxies to piece together this story? To better understand the context of our awakening to consciousness, and to assess its significance for life on other worlds, we must closely examine the history of Earth and its life.

Earth Birth

If seeds in the black Earth can turn into such beautiful roses, what might not the heart of man become in its long journey towards the stars?

—G. K. CHESTERTON

Did I know you?
Did I know you even then?
Before the clocks kept time
Before the world was made
—U2, "WILD HONEY"

INSIDE OUT

Who are we to say that our Earth is such a special place? Like parents certain that their baby is the cutest ever born, of course we think our planet is the chosen one. Nevertheless, in all modesty, you have to admit: Earth stands out. Ours is the only planet around these parts able to tell its own story.

All other local planets appear to be dumb as rocks. Somehow, our world has sprouted eyes, hands, minds, and mouths, not to mention microscopes, telescopes, and microchips. Since we seem to have, for now, become Earth's only voice, our reconstructions of geological history amount to an autobiography of the planet. Many pages are missing, scattered by the winds and crumbling with age. The story is written in the rocks, fossils, and whiffs of ancient air. As we learn to read it, our world begins to tell its tale.

We planetary scientists are used to taking a global view. We've had no choice, really. Holism is easiest when you know next to nothing, and we've learned about the other planets from the outside in. At first we knew them only as bright lights wandering the sky. Telescopes transformed them into dusky disks with faint, suggestive markings. Spacecraft have turned them into real places with landscapes of rocks, dunes, riverbeds, craters, mountains, and ice. But each is still a tiny dot in the sky that you can cover with your little finger.

When planetologists look at life on Earth, we regard our own world as we might examine another planet in searching for signs of life. From space we don't see organisms, cells, tissues, and chromosomes. We see global patterns of vegetation and habitation, and the telltale traces life has left in the atmosphere and oceans. We look at the totality of life to see what distinguishes our planet from nonliving worlds. The planetary perspective allows us to see large-scale features and trends that we might miss if we stayed mired in the details.

Because we spend our time unraveling planetary histories that have unfolded over billions of years, ours is a time-lapse view of terrestrial evolution. In this fast-forward movie of Earth, the coming and going of individual species, mountain ranges, oceans, and ice ages are accelerated to a blur. We see only those major transformations that would be visible from far outside, catching the eyes (or the antennae) of interested aliens with long attention spans.

What follows in the next few chapters is a brief account of some of the more memorable, formative events in the maturing of Earth. My purpose is to make sure that, when we turn outward to consider life in the rest of the universe, at least our guesswork will be educated. Like all biographies, this will be highly selective. In particular, I'll focus on developments that seem to have been pivotal for the continued flowering of life and eventual awakening of consciousness here, and that might be part of the story on other worlds as well.

We won't know anything with certainty about the evolution of life elsewhere until we find some, or it finds us. From a cosmically enlightened perspective that we do not currently enjoy, the path of life on Earth may appear either typical or unique. Either way, it is worthy of careful study, if only because it is the path that we ourselves have taken from the inanimate trajectories of elementary matter, through a nested series of communal bodies with ever subtler reflexes of survival, on into the first sputterings of conscious awareness.

BORN IN STEAM

As the newborn Earth grew by the assimilation of lesser worlds, her gravitational reach extended farther into space. Stray planetesimals or comets wandering nearby were sucked in, feeding the rocky little bundle of joy, increasing her appetite for more. As the planet's strength and hunger expanded, approaching bodies were accelerated to ever faster velocities and incoming rocks increased their punch. The harder they came, the harder they fell. Mountains of rock, metal, and ice came crashing down, each one raising a massive spray of vaporized rock and leaving a round pool of incandescent magma.

Though Earth was dotted with newly formed craters, each briefly heated to thousands of degrees, on the whole the planet stayed quite frigid. In its earliest stages, the growing planet had no atmosphere at all, so these puddles of molten lava were touched directly by the frozen void of space. Some heat was buried underground, but the surface cooled quickly, radiating away heat just as fast as the impactors could bring it in.

All that changed when Earth began to cloak herself in steam, which was liberated from the falling rocks themselves. Virtually all of the boulders that assembled to make the Earth contained some water, locked inside the crystal lattices of minerals. If you hit such a rock hard enough, the water is knocked loose from its mineral cages. Once Earth grew beyond a certain size, roughly half her final diameter, every new rock fell too fast to hold its water inside. Now each impact spat out an angry puff of vaporized water and carbon dioxide, which, bound by gravity, began to accumulate around the growing sphere. This "impact-generated atmosphere" was Earth's first air. Soon our planet was blanketed in steam.

Water vapor is a greenhouse gas, meaning it absorbs infrared radiation and helps a planet hold on to its heat. Our early steam atmosphere formed an insulating blanket, swaddling the infant Earth. Giant bodies from space continued to fall at cosmic speeds, but now the heat from their impacts was trapped. The young Earth, insulated by its new steam atmosphere, became absurdly hot, melting surface rocks, which liberated still more steam. Our planet was covered by a global ocean of molten rock and enveloped by a dense, sweltering atmosphere.

Sometime during those early, steamy days, before oceans appeared and life began, the young planet suffered one truly Earth-shaking impact that, traumatic as it was, would later prove essential in building its unique character. In the chaos of planetary growth, when the solar system was crowded with planetesimals on crazy, intersecting orbits, another planet, about the size of Mars, plowed into Earth. The glancing blow blasted off a big chunk of Earth, forming an orbiting ring of vaporized rock that later coalesced into our Moon. This apocalyptic event melted our planet in its entirety and had major effects on its later atmospheric and geological evolution. Ever since, the Moon has influenced Earth in numerous ways, slowing its rotation, raising tides in its oceans, steadying its spin axis and climate, and inspiring its poets and lovers.

So Earth was born a burning sauna with the thermostat jammed into the red zone—a tumultuous, noisy place with mountains continually falling from space at supersonic speeds, spraying showers of molten metal and rock, which then rained back down through the suffocating sky.

Finally, about 100 million years after the beginning of planet formation, the primordial pounding from space began to subside, depriving the steam atmosphere of its sustaining energy source. The surface began to cool, and the steam to condense.

It rained. For a thousand years it rained, filling Earth's oceans for the first time.

Peace on Earth at last, or so it would seem.

However, when Earth formed out of smaller bodies, the finish was not tight, but scattershot, like an unrehearsed or wasted band trying to end a song. The gathering of Earth tapered off gradually and irregularly. Toward the end, after the first rains came, there were quiet spells during which the planet was spared the trauma of large impacts. The carbon-rich ocean settled down to begin the fragile dance of organic evolution. Repeatedly, these placid periods were rudely punctuated by late hits from space that shattered the calm and plunged the world back into chaos. The last few giant impacts took a huge toll. An impactor more than two hundred miles in diameter falls to Earth with enough energy to boil off the entire ocean and heat the planet's surface to thousands of degrees. There were probably four or five of these impacts between 4.4 and 3.8 billion years ago, each returning Earth to steam, endangering any complex organics that had, during the interregnums, made tentative steps toward life.

Several times the world was lulled into an apparent cease-fire, and began the chemical climb toward life, only to be thrust back into hell. Oceans formed and were blasted back into steam, which then condensed and fell again in thousand-year rains. No one knows how far chemical evolution progressed in the many calm periods between. Life may have gotten started more than once, only to be destroyed by massive impacts. If life did start many times, it died out every time but one. This we know, because we are such close kin to all life on Earth. The biochemical evidence shows clearly that all Earth's diverse organisms are related and stem from a single origin. We are family. Bugs, slugs, and us—we all share a single origin that occurred sometime around 4 billion years ago, soon after the last of the mighty, ocean-boiling impacts.

Do you ever wonder what might have happened if you had met your true love too soon, or too late? Though the thought is disconcerting, sometimes the most important things in life come down to luck and timing. The same is true of the life of our planet. Since life may have started several times, only to be snuffed out by massive impacts, we may owe some deeply ingrained features of our basic biochemical machinery to the random timing of the last giant collisions. Maybe some of life's earlier doomed experiments were quite promising. If fate had allowed one of these other beginnings to inherit the Earth, then evolution on Earth could have proceeded along a very different path. Perhaps large animals and intelligence would have come along after only 1 billion years, instead of 4 billion. By now we could have been off roaming the galaxy, with Earth a fond but distant memory. Or, perhaps life would have evolved much more slowly, and we would never have made it beyond the bacterial stage in the entire 10-billion-year life of our Sun.

HOLDING WATER

There is still some mystery about the source of Earth's life-giving oceans—the ultimate headwaters for our rivers and seas. Where in the solar system was the water before it landed here? Was it locked up in the rocky bodies, orbiting in Earth's vicinity, that smashed together to form the bulk of the planet, escaping as steam when the world got large? Did it arrive frozen in comets, plunging headlong from the cold fringes of outer planetary space, each exploding upon crashdown like a snowball shot into a blast furnace? Or did it waft gently down to Earth, stowed away on tiny grains of interplanetary dust?

These questions about planetary water have been an obsession of mine since grad school: where it comes from, where it goes, what kind of planets end up with it, and what it does to climate, geology, and biology. Why do we care where it came from, since we know we got it? Because water is key to so much of what makes Earth Earth. Water erodes the mountains, keeps plate tectonics* chugging smoothly along, and governs global climate. Most important, water is essential for life as we know it. The same triangular molecule—a tricycle with one big oxygen and two little hydrogen wheels—also plays host for the 3-D carbon dance party that carries our genetic memory down through time, composes our bodies and, perhaps, forms our every thought. Understanding how the planets of our solar system have gained and lost water can help us nail down the conditions that other planets need to come alive.

After studying this problem for a couple of decades, I've concluded that getting enough water was not the problem. Water was everywhere in the early solar system. The tricky part for any young planet is hanging on to it. The more we learn about planet formation, the more it seems that the planets must have all lost large amounts of water during their earliest phases. Especially when most of the water was exposed in a hot, extended steam atmosphere, it would have been vulnerable to getting stripped away to space in several different ways.

The Sun, in its wild youth, radiated furiously at energetic far-ultraviolet wavelengths, heating Earth's upper atmosphere and causing a steady breeze of hydrogen to blow back into the interplanetary void. No hydrogen, no water. The late phases of the Earth's assembly must also have taken their toll. Explosive impacts can blow bits of a planet's atmosphere right out into space. When Earth was a steaming lava ball, its water supply was susceptible to these losses.

Later, after the rains, the remaining water was in much less danger of being lost to space. Once it was condensed into surface oceans, Earth could keep its precious water closer to its chest. Good thing, or our life-giving oceans could have gone the way of those of Venus or Mars, a topic I'll soon get back to.

---

*Earth's outer, solid "Lithosphere" is broken into about a dozen plates which drift around, driven by internal heat, making mountains, causing earthquakes, and recycling the planet's surface. We call this activity *plate tectonics*.

SETTING THE SCENE

Wherever the water came from, Earth was awash in it from the start. As the sterilizing steam rained down for the last time, the curtain rose on a new drama. What was it like here when our planet first became infused with something that could be called life?

The young Earth was heavily cratered and largely covered in oceans. The circular rims of the largest craters formed rugged mountain arcs jutting sharply from the sea. Still reeling from the mighty collision that had made the Moon, Earth spun rapidly, completing a day and night in about five hours.* A dim Sun and a nearby, looming Moon bolted across the sky. The atmosphere that remained after the steam condensed was largely composed of carbon dioxide.

Baby Earth needed all the help she could get to keep warm, as several factors conspired to threaten a deep freeze. In its early days, the Sun was much dimmer than it is now. Throughout its life, our star has gradually been brightening as the hydrogen in its core burns slowly to helium, increasing the Sun's density, and requiring a hotter nuclear flame to fight off gravity. If the modern Sun is a hundred-watt bulb, at the time of Earth's formation it was only about a seventy. This wimpy Sun could not have kept Earth very warm without help. If the greenhouse effect had been as feeble then as it is now, the oceans would have completely frozen over. Earth would have been the solar system's largest skating rink, and it is questionable whether life of the kind we know would have formed.

The earliest known rocks, some 3.8 billion years old, are constructed from deposits of water-borne sediments, betraying the presence of a liquid-water cycle well before that date. Clearly, something was keeping the young Earth warm against the weak Sun and the frigid vacuum of space.

Given that our neighboring planets Venus and Mars have atmospheres of nearly pure carbon dioxide, the consensus these days is that the early Earth also had a thick atmosphere consisting mostly of CO_2. As you well know from presidential debates and cereal boxes, CO_2 is a greenhouse gas that warms a planet. Back then, the dreaded greenhouse effect was a good thing† that efficiently trapped the weak sunlight, keeping Earth cozy and warm. Global warming to the rescue.

---

*The moon itself, dragging on Earth through the tides, has gradually slowed us to our current twenty-four-hour day.

†It still is, actually. Without it we wouldn't be here. But anthropogenic global warming could quickly cause too much of a good thing.

A weak young Sun was not the only problem for a newborn Earth struggling against the elements to stay warm. Making matters worse were the thick clouds of obscuring dust constantly lofted high into the air by the continuing bombardment. Though the early, fearsome ocean-vaporizers, the two-hundred-mile-wide monsters, were finally gone from the inner solar system, a steady rain of lesser impacts continued to pepper the planets. Many of these were big enough to cause severe environmental changes.

An object only a few miles across falling from space will raise enough dust to darken the skies globally for a couple of years. When this happens, most incoming sunlight is absorbed by dust in the upper atmosphere and the surface grows dark and cold. Even now, such events still happen every 10 to 100 million years. The last big one was the "K/T impact" 65 million years ago, which did in the dinosaurs.* Long ago, when the solar system was still sweeping up the mess from planet formation, impactors came much more frequently. Between 4.3 and 4.1 billion years ago, several objects this large were hitting Earth every century, causing enormous temperature oscillations at Earth's surface. A thick cloud of light-absorbing dust intermittently shrouded our planet.

How can we be so sure about all this? After all, we're talking about a time that is, as I've already admitted, older than any preserved surface on the planet, and older than the oldest Earth rock ever found. Aren't we just guessing? Nope. Fortunately a well-preserved record extends back to this time. You can see it with your own eyes on any moonlit night.

Our nearest neighbor has not had nearly as interesting or eventful a life as has Earth. While Earth's surface has continually been remade by mountain building, weather, and life, destroying all traces of the earliest rocks and landscapes, the Moon is dead. Geologically, meteorologically, and biologically, it's dead. It just sits there, passively taking whatever space tosses its way, never washing its face with rain or regenerating its skin with plate tectonics. The moon's entire surface is much, much older than any place on Earth, and it has been getting shot up with craters for billions of years. Consequently, its pockmarked face preserves a record of the intense bombardment that hammered both Earth and Moon when remnants of the preplanetary swarm still men-

---

*The "Cretaceous/Tertiary impact event" which ended the Cretaceous geological age and ushered in the Tertiary.

aced the inner solar system. The Moon serves as a cosmic rain gauge, recording the environment of near-Earth space back to circa 4.1 billion years ago. (Before that, the cratering itself was so vigorous that it continually obliterated all traces of earlier surfaces, so we have no direct record of the bombardment rate for the first half billion years of our planet's existence.)

The pockmarked face of the Moon tells us, without a doubt, that conditions on Earth's surface were dominated by the effects of explosive impacts right up to the time when life here first got started. The entire globe oscillated between periods of freezing dark gloom and hotter spells when the skies cleared and the surface was bathed in intense, deadly solar ultraviolet irradiation (the protective ozone layer would not be invented for billions of years).

These surface conditions were not healthy for children or other living things. For this reason, we think that life may have originated deep underground, or at the bottom of the oceans, places that provided natural fallout shelter from the cosmic bombs still wreaking havoc at the surface. A currently popular location for life's origins is at hydrothermal submarine vents on the ocean floor. Plenty of chemical energy was supplied by the hot, mineral-rich waters pouring out of these vents, and the deep ocean was relatively immune to the extreme environmental hazards plaguing the surface at the time when life seems to have gotten its start. As the impact storm raged above, the first glimmerings of life on Earth may have been safe and warm below the storm in an octopus's garden beneath the waves.*

---

*If life can get started at the bottom of a planet's ocean, caring little about hazards plaguing the surface environment, this could have interesting implications for life beyond the Earth. Keep this in mind when, in a few chapters, we return to the question of life on Europa, Jupiter's oceanic moon.

7 | Life Itself

What do they call it . . . the primordial soup? The glop? That heartbreaking second when it all got together, the sugars and the acids and the ultraviolets, and the next thing you knew there were tangerines and string quartets.

—EDWARD ALBEE, *Seascape*

All is but a woven web of guesses.
—XENOPHANES

WHAT IT IS

Speaking of life, I suppose I've danced around the question long enough. What is it? I was hoping you wouldn't ask.

Next question?

Okay, how about this? *Life is something that eats, grows, reproduces, and evolves.*

Except for when it isn't. There is no airtight definition. Paradoxical exceptions are easy to find. What about a mule, or a cat that has been "fixed"? They can't reproduce, so does that mean they aren't alive? Certainly their individual cells are performing the functions listed above. So, if you are sterile, your cells are alive but you are not? Perhaps you are alive if you are made of living subunits. If that's the case, where do we draw the line? Is the Earth alive? How about the universe?

Further, individuals do not evolve (not in the biological sense). Yet no being alive today could ever have existed without a long evolutionary

saga written in the lives and deaths of countless generations of ancestors. It seems that aspects of the above definition do not apply to individual life-forms, but work well for the larger continuum of life. In fact,
the existence of a living organism requires the existence of a larger biosphere. If there is no such thing as an individual organism independent
of a biosphere, then perhaps we don't need a definition of life that
works for individuals.

If we just say that life is something that eats, excretes, and makes
more of itself, then you could say that a forest fire is alive. Perhaps it is,
a little bit. But, can something be a little bit alive, or would that be like
being a little bit pregnant? Intuitively, it seems like an all-or-nothing
deal. Alive or dead, with no in-between.

Try this: *Life is a self-perpetuating, self-contained chemical phenomenon that extracts or manufactures high-energy nutrients from its environment, excretes waste material of lower chemical energy, and surfs
the energy difference between food and shit to go on living.* Life is a
breakfast cereal, a board game, a very long sentence, a bitch and then
you die. I'll let you in on a dirty little secret: We don't really know what
life is. We may as well try and catch the wind as pin life down with a
tidy definition.

Even if we found a decent definition that worked for all life on Earth,
we wouldn't know if it applied anywhere else. We have no outside perspective. The fact that we have only one form of life to study was not
obvious when biology was a new science. After all, trillions of diverse
creatures are on Earth to compare and contrast. Now we know that
they—and we—are all branches of one sprawling evolutionary shrub
with a single root. Our limited and parochial knowledge of the nature
of life makes any confident statement about life elsewhere an affront to
the scientific method. Nevertheless, we can't help it, because we so desperately want to know about life in the universe. We will study Earth
life with the finest-toothed combs we can find, drawing great and universal significance from what may be random or unique events.

Of course, we can always use our definition itself to limit what it is
we are looking for, declaring that any extraterrestrial phenomenon that
does not conform is, by definition, not alive. A better approach is to
accept the ambiguity. Though we cannot precisely define life, we can
describe many of its properties and make reasonable guesses about
which ones are universal. It may be that life, like true love, is impossible
to define, but you know it when you see it. And perhaps finding

extraterrestrial life will be more like falling in love than confirming a specific hypothesis. When it happens, we'll know.

HOW IT STARTED

Now that I've established that we don't know what life is, I'll continue to describe where we think it came from. After the rains, the first oceans were laced with amino acids and other goodies. A seething brew of organic goo began to whoop it up. Carbon chemicals combined in new ways, evolving without memory or intention, but with plenty of time. It was a self-organizing organic orgy, each molecule getting it on with all comers, unafraid of the consequences of complete promiscuity. For some it proved fatal, but a lucky few wound up in long-term arrangements of greater stability. Once some of these learned to start copying themselves, there was no turning back.

The first self-replicating molecules didn't have to be very good at it. Any random assemblage that could make even imperfect copies of itself found its chemical type increasing in number. Structures with self-replication proclivities became more abundant, interacting and combining to form new molecules, with novel properties and behavior. Some of these new models were even better at self-replicating. You see how this could quickly get out of hand. In the right kind of environment, with a ready supply of organics and without catastrophic interruption, what was there to stop it? Maybe this ocean just had to come alive.

We believe that it did. The story goes as follows: *Chemical evolution led inexorably to self-replicating molecules, which in turn evolved into the first primitive cells. Through Darwinian selection, these cells evolved into modern organisms.*

This statement seems so reasonable and consistent with what we know about the natural world that we scientists accept it as true. The problem is, it's difficult to prove. No one has succeeded in creating life from nonlife in the laboratory. It would be hard to repeat the experiment the way nature did it originally, because that probably took millions of years. Even the most patient scientists or the most aimless grad students don't have that kind of time on their hands. In our labs we try various tricks to speed things up: concentrating the most promising chemicals and adjusting temperature, acidity, or other conditions to encourage evolutionary activity. We've come up with many promising and suggestive results, chemical brews that point down the path toward self-replication. But nothing we've created has crawled out of a flask and introduced itself, or even met the obvious minimal requirements for a living organism.*

Though we don't want to admit it, our belief that chemical evolution can lead to life is still an article of faith. Let's call it informed faith, to give justice to the great strides that we've made in finding the potential pathways of life's origins. The creationism-versus-evolution debate has unfortunately pushed science into a defensive corner from which we exude overconfidence, pretending to have certainty in places where we really have only reasonable inference. Instead of saying, in effect, "We have proof whereas you only have faith," we could, more honestly, say, "At least our faith is testable in principle, and wherever tested has been borne out by observation."

It is certainly not immediately obvious that the beauty and complexity of life on Earth all came about through billions of years of random variation and selection. Our prescientific forebears can be forgiven for their intuitive inference that such a wonderful design requires a superhuman designer. Science has given us reason to doubt this need, but science has also revealed the design to be far more intricate, complex, and finely tuned than anyone imagined hundreds of years ago. Modern thinkers, too, are reasonable to doubt that natural selection could come up with all this. If you have never, ever, doubted it, then you've never really thought about it, only accepted the ideology and authority of your teachers. Within each living cell, from paramecia to paramedics, is a chemical factory far more complex and elegant in design than the most sophisticated chemical plant ever built by humans.

HOW IT WORKS

Without a chemistry book and a chemist, how does nature know how to construct these intricate factories? What keeps them running? Proteins. What makes the proteins? DNA.

Life on Earth is largely a game played between two types of macromolecules (giant molecules), proteins and nucleic acids (DNA and RNA are nucleic acids). Each is a long, thin, tangled chain of thousands of

---

*If we succeed in creating life from scratch in the lab, we may still not know how it actually happened historically. But our belief that it did happen will gain several notches in credibility.

nearly identical subunits. Because they are made up of an enormous number of these smaller units, they are analogous to sentences made up of many letters. In the run-on sentences of protein, the letters are amino acids. Proteins are the basic structural materials of living organisms. They are what you are made of—except for some parts such as fat, bones, and teeth, but even these are constructed under the close supervision of proteins.

Even more important, proteins control the chemical machinery of life. Every chemical reaction in every cell of your body—all your life's work—is mediated by proteins acting as *catalysts*. The colloquial use of this term is closely analogous to its meaning in chemistry. Someone who is known as a catalyst makes connections, brings people together. A chemical catalyst grabs this molecule over here and that one over there and says, "Why don't you two get together? Let's make something happen."

Proteins are organic catalysts with an incredible ability to recognize other molecules, pull them together, and moderate their interactions. They regulate all the chemical reactions that, collectively, we call life. How do they do this? It all has to do with the unique 3-D shape of each protein.

A typical protein is made of thousands of amino acids. An amino acid looks like this:*

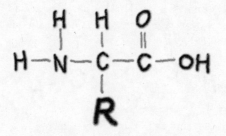

They are all exactly the same, except for the side group, here labeled R, which is different in each one. This amino acid, the simplest, called glycine, has a single hydrogen atom for its side group:

---

*H, N, and C are atoms of hydrogen, nitrogen, and carbon, respectively. Notice that each carbon atom insists on forming exactly four bonds with other atoms. Thus the C that is "double-bonded" to an oxygen.

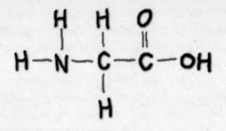

Here's one, called tryptophan, that's among the most complex:

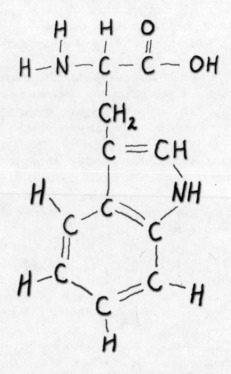

When thousands of amino acids are strung together in a particular order, you've got a protein. It is the specific sequence of amino acids that gives each protein its unique abilities. Here's how: When you string a large number of amino acids together to make a protein, the different side groups (the R's), all dangling off the main string, interact with one another. Some of them are strongly attracted. Other pairs find each other repulsive and can't get enough distance between them. These forces of attraction and repulsion cause the protein chain to fold up into a complex twisted shape.

Imagine a long rope with a large number of cats tied to it at regular intervals. Some of the cats would try to get as far from each other as possible, whereas others would seek each other out to fight, groom, or play. Now, picture this happening in a fluid tank where they all have kitty Aqua-Lungs, or in the weightlessness of an orbital cat house where all the cats can move about in three dimensions. The final shape of the rope would be quite twisted because of the complexities of feline social life. Proteins become twisted and folded because of the social interactions among all the side groups of their amino acids. Each protein folds in its own way, and the final shape is precisely determined by the specific amino acid sequence. It is the 3-D, folded shapes of these molecules that give them their amazing ability to "recognize" and bind to other molecules. Each protein has evolved so that its amino acid sequence causes it to fold up into just the right shape to precisely fit specific molecules and encourage them to react in ways needed to keep our cells running.

How do the amino acids know what order they should assemble themselves in to make a protein fold in just the right way to work its magic? That's where the other group of macromolecules, nucleic acids, come in. DNA is a nucleic acid.* It stores and passes down the information on how to string together amino acids in the right way to make the proteins needed by living cells.

The famous double helix is made of two strands of DNA, each built up from a long string of subunits called nucleotides. Just as amino acids are the individual repeating "letters" in a protein molecule, the nucleotides are the letters in DNA. There are only four nucleotides in DNA: adenine, thymine, guanine,† and cytosine. These four structures function as the four letters in the genetic code, and we abbreviate them with the letters A, T, G, and C. The information content of the genetic code is entirely contained in the ordering of the A, T, G, and C nucleotides along strands of DNA.

---

*So-called because Friedrich Miescher, the German chemist who first isolated them in 1869 (while experimenting with pus!), didn't know what they were but suspected they came from the nuclei of cells.

†So-called because it was first isolated in bird shit or *guano*. I wonder how many more disgusting bodily fluids I can work into the footnotes of this chapter. . . .

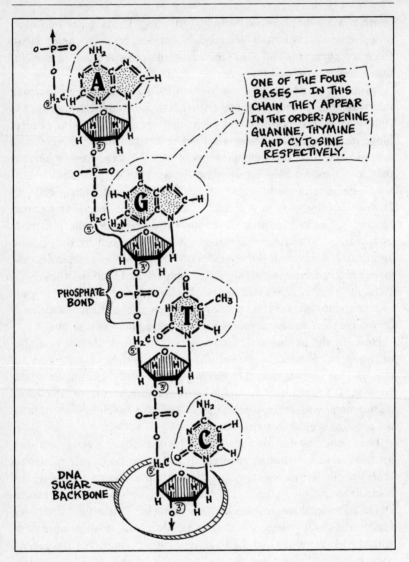

ONE OF THE FOUR BASES — IN THIS CHAIN THEY APPEAR IN THE ORDER: ADENINE, GUANINE, THYMINE AND CYTOSINE RESPECTIVELY.

PHOSPHATE BOND

DNA SUGAR BACKBONE

(Borin Van Loon)

Now for the vital connection between nucleic acids and proteins. Contained in the sequence of nucleotides composing a strand of DNA are coded instructions for stringing together a list of amino acids in the right sequence for making specific proteins. That's all it is. Nothing more and nothing less. The entire genetic code—the information on how to make all of you out of a single fertilized cell—is a set of instructions for making a large batch of proteins that will fold into the shapes needed to get the job done.

Each "word" in the genetic code is three letters long. That is, every three nucleotides in a strand of DNA codes for one amino acid in a protein. For example, the sequence GGT stands for the amino acid glycine, and CAA stands for glutamine. So a sequence of DNA reading CAACAAGGT contains the instructions "Add two glutamines and then a glycine." A DNA code for making an average protein contains thousands of these little words. Inside each cell is chemical machinery that can read the DNA chain, putting together amino acids in the specified order. The resulting proteins then promptly fold into their 3-D shapes and make the "desired" chemistry happen in your cells. If "love is just a four-letter word," life is just a long series of three-letter words.

THE TWIST

That's quite a stunt for dumb old nature to pull off—encoding the 3-D shapes of our all-purpose molecules (proteins) within the linear sequence of a different molecule (DNA). Why didn't we think of that? But, wait, it gets better. DNA molecules can perform another amazing, essential trick: they can make identical copies of themselves. That gives us heredity, without which we would still be nothing more than a skanky brew of chemicals sloshing around in the ponds of a dead rock.

In DNA's double helix, each coded strand lives in a twisted pair with another. This allows each molecule to contain a template for its own reconstruction. The two strands of the double helix are identical except for the letters of the code, which form a complementary message. Bonds form between the nucleotides on each strand, joining the two together like the rungs of a twisted ladder. Each nucleotide reaches across and bonds to one on the sister strand. Because of their shapes, they are choosy about whom they will bond with. G and C bond only with each other. Likewise, A and T are a faithful, exclusive pair. So, the sequence of the nucleotides on one strand is exactly specified by the sequence on the other. Each contains a complete description of the other's structure.

When they're in the mood to replicate, the DNA molecules, with some protein midwives to help them unzip, sequentially break the bonds forming the rungs of the ladder, leaving two naked strands dangling free. The nucleotides strung along the two resulting individual DNA strings, suddenly finding themselves single again, are quick to bond to any attractive nucleotides floating by. In this way, each of the two separated strands immediately builds itself a new partner. But G will only bond to C, and A will only bond to T. The result? When each nucleotide along these chains hooks up with its desired counterpart, the two new double chains are in every way identical to the parent double helix. Each of these will then build the same proteins as its parent did. Really, each one *is* its parent. The parent molecule never died, but simply replicated. In this sense there is only one molecule of DNA on Earth. The ones dividing right now in your toes and in the grass beneath them are all pieces of the original founder. And so it divides, never forgetting, forever and ever, amen.

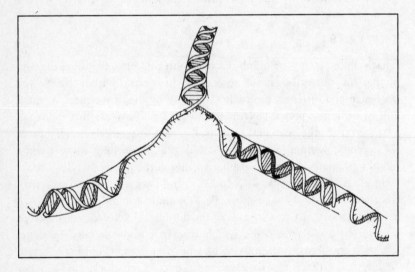

(Borin Van Loon)

To recap: What keeps the complex chemical factories in the cells of all Earth life running? Proteins. What makes the proteins? DNA. How does DNA copy itself? Proteins. A very clever design, indeed, yet it seems to have arrived here through evolution by natural selection.

When I learned the details of this surprisingly complex machinery for the first time (and here I've only scratched the surface), I felt that intellectual honesty required me to rethink my opinion of evolution. Doesn't it strain credulity to think that the intricate, streamlined, fantastically clever, and totally uniform building code found in all life, even its simplest known forms, could ever come into existence through such an aimless process? I've heard this many times, and I've thought it myself: that our planet and the life on it are too beautiful and "well designed" to have just happened by accident.

Ultimately, I did not change my mind and reject Darwinian evolution. I realized that the character of evolutionary change is such that a highly evolved system will, after hundreds of millions of generations of trial and error, be so optimized for survival that it will seem, upon first examination, to have been designed by an imaginative, clever, and ingenious mind.

OCCAM'S RAZOR

We cannot prove that no other force, such as divine or alien intervention, has directed evolution. But the scientific attitude, which I find very appealing, is to reject any hypothetical, hidden mechanisms when known mechanisms are adequate to do the job. We have a name for this attitude. We call it Occam's razor after the fourteenth-century monk William of Occam, who said, "It is vain to do with more what can be done with fewer." We take this to mean, "Why assume that things are complicated if a simple theory can explain all of the observations?" The razor is a tool we use to cut the crap from theories that seem too contrived to describe the apparent simplicity of nature.

We assume that the universe is simple until proven complicated. Why invoke forces, mechanisms, creatures, or gods that are not really necessary to explain what we observe? For scientists, such arguments are almost as good as proof. Science is driven by a belief that there are simple laws, which we can discover, that govern the behavior of much of the universe. If we can conceive of multiple explanations for a given observation, the simpler explanation is more likely to be true.

Why do we believe this? Science is supposed to have no dogmas, to be ready to question and discard every idea if the evidence does not support it. Is there any logical, a priori reason to believe that the universe should be explicable with simple laws? Or is this merely received knowledge, an article of (gasp, shudder) faith?

I think that it is more an aesthetic principle: Wouldn't it be nice and elegant if the universe turned out to follow simple laws that we can figure out? The original Copernican revolutionaries advocated a Sun-centered solar system largely out of an aesthetic preference for this scheme, compared to the more complex and cumbersome arrangements required to keep everything spinning around the Earth.

Or, perhaps we believe in it because it works. You can think of Occam's razor as a hypothesis about the natural world, a position to take for the sake of argument, an idea to try on the universe to see if it fits. After all, you've got to start by assuming something to get anywhere in science. Then your results serve to test your initial assumptions, and you either confirm or reject them as you proceed.

In the case of Occam's razor, the "results" are the hundreds of years of progress made with science. Science, guided by the search for simplicity, has uncovered many deep patterns and hidden connections in our universe. All of the inventions that work, and the predictions that come to pass, help to confirm the original working hypothesis. Science, operating under the doctrine of simplicity, clearly works. That means it's a good assumption. A keeper. I would describe Occam's razor as a hypothesis, based on an aesthetic intuition, that has proven to be "true" in the sense that it is quite fruitful.

I find the logic and the evidence of evolution to be completely convincing. A deep look at the world, digging into the rocks and dirt, shows a record of change and adaptation. The mechanisms described by Darwin, tweaked with 150 years of subsequent insights, marvelously equip us to understand this process. Given variation, death, and heredity, there is no escaping that evolution will happen. Fossils and numerous other clues show clearly that it has. For the scientific mind, guided by Occam's razor, there is no reason to invoke any other force in evolution, and the case is closed.

MICROCOSMIC GODS

All living cells, from the bacteria lounging in your gut to the neurons humming in your brain, depend on an intricate pas de deux between nucleic acids and proteins. Neither can be made without the assistance of the other. The DNA code cannot be read without an elaborate protein transcription machine. The machine cannot be built without the code. Once it's up and running, this system works wonderfully. But it presents us with a serious chicken-and-egg problem. Which came first, proteins or DNA? And how did such a tangled web *evolve?*

In a more general sense, the problem of getting from organics to organisms is still unsolved. There is a large gulf between the most complex self-replicating molecules that we can easily imagine arising from chemical evolution and the incredibly elaborate chemical machinery common to all cells. There is still vast, unmapped territory bordered by the familiar lands of chemistry on the one side and biology on the other. Many scientists are seeking to retrace the route that nature evidently found across this terra incognita. In this quest we have found many promising leads that point from chemistry toward biology and, on the other side, bits of biochemistry that seem to hint at nonliving precursors.

One of the cool things about having Carl Sagan as a friend and mentor was that he was constantly feeding me reading tips. Mildly disapproving of my hard-core science fiction habit, he tried, during my teen years, to steer me toward "good" science fiction. Once, Carl gave me a short story that he described as one of his favorites: "Microcosmic God" by Theodore Sturgeon.* The story, written in 1941, concerns an iconoclastic biochemist, James Kidder, who worked on an island off New England. Fascinated by the mystery of life's origin, he endeavors to create life in his laboratory: "When the cloudy, viscous semifluid on the watch glass began to move of itself he knew he was on the right track. When it began to seek food on its own he began to be excited. When it divided and, in a few hours, redivided, and each part grew and divided again, he was triumphant, for he had created life."

He not only succeeds in creating primitive organisms, but also learns to accelerate metabolism, so that his creatures pass through many generations in a single hour.† The evolutionary process speeds up to a dizzying pace. As weeks and months go by, he observes his creations passing through many of the phases that took billions of years for nature to achieve on Earth. Things get interesting when they develop intellectual and technological abilities vastly exceeding those of human beings. Fortunately for Mr. Kidder, his "Neoterics" have always worshiped him as their god, which, practically speaking, he is. They have no such respect for the rest of humanity, however. Let's just say that the military is called in, but beyond that I won't spoil the ending.

---

*Sturgeon was one of Carl's favorite SF writers. He also wrote several *Star Trek* episodes. Which is funny, because Carl hated *Star Trek*. The week of the *Viking 2* landing on Mars, his son Dorion and I got him to sit through "The Menagerie," the pilot episode of the original *Star Trek*. Carl admitted that it was much better than he'd expected.
†Part of the secret comes from chemicals he extracts from *Cannabis indica!*

A few months after reading this story, I had a summer job as an undergraduate intern working in Carl's Planetary Simulation Lab at Cornell. Most of that summer (1978) I spent working closely with Reid Thompson, Carl's grad student and a brilliant chemist. Reid was an animated and patient teacher with a wrestler's build and a thick beard. Blessed with a Kentucky accent and a rambling Southern sense of humor, he had a passion for fast American cars with powerful engines, and he possessed an encyclopedic knowledge of many subjects, including organic chemistry and trees. As we walked to lunch every day, he would tell me about the trees lining the pastoral walkways of Cornell— their species, ancestry, seasonal growth patterns, sexual preferences, whatever. On the way back from lunch, he would quiz me.

There were many other enticements that summer: the Ithaca music scene, the local skinny-dipping pond, a girl named Katie, and the fireflies dancing around the cemetery at night. I was eighteen and living in a group apartment in College Town that never had anything but beer in the fridge, so some mornings I would be moving a little slow. But Reid was tolerant, even a bit of a mischievous rebel himself, and the work was engaging enough to compete for my attention with raging hormones and experiential curiosity. Spending days in the lab with Reid was always a good time.

Recall the groundbreaking experiments of Miller and Urey in 1953, which made the first baby steps down the route from chemistry to biology by showing that amino acids are easily made in conditions simulating the early Earth. That summer we were doing a series of experiments that were the evolved descendants of the Miller-Urey experiment, trying to induce the first steps of organic evolution in a range of conditions simulating the environments of other planets. We set up an impressive array of glassware, including a maze of coiled tubes for heat exchange, "cold fingers" encased in liquid nitrogen for trapping condensed gases, valves, flasks, and chambers in which we could subject gas or liquid mixtures to various provocations: heat, cold, ultraviolet radiation, or electric shock. The whole sprawling contraption was held together with clamps extended from a scaffolding of metal rods. It was the prototypical mad scientist's lab, complete with foul-colored bubbling mixtures and electric discharge chambers. We did not wear white lab coats, even though we would have been screwed if government auditors had shown up.

Carl was fascinated by the results of the Miller-Urey experiments (Urey was one of Carl's scientific mentors) and the further questions they implied: What is the range of environments in which biologically promising organics can be made from simple, ubiquitous chemicals? Could it happen in the clouds of Jupiter? On Saturn's gas-shrouded moon Titan? In the surface layers of carbon-rich icy moons?

Not at all an exact science, this lab work proceeds by hunches and trial and error. At the end of an experimental run, you are left with your precious chemical product: a sealed flask with some unknown gases or a residue of yellowish or brown crud. You analyze this gunk with a range of high-tech instruments, and depending on what you find, you then start over with a modification of the original experiment.

We've learned from such experiments that it is surprisingly easy to make the organic preludes to life in various environments that exist in the solar system and elsewhere. Take a source of carbon, the fourth most common element in the universe (after hydrogen, helium, and oxygen), add some hydrogen, nitrogen, oxygen, sulfur, and phosphorous, tap into a sufficient source of energy, and you almost inevitably get amino acids and other simple but vital organic precursors. One thing the experiments of Reid Thompson and his colleagues showed was that the necessary first steps toward organic life should occur commonly throughout our universe. Reid died in 1996 of cancer, which seems to be an occupational hazard of experimental chemists (although in his case I have no idea if there was any connection). I'll always be grateful for the tutelage I received from him among the glass tubes and tall trees of Cornell.

In the fall of 2000, twenty-two years after that summer at Cornell, I served on NASA's review panel evaluating new proposals for funding in exobiology. This particular panel meeting was in some ways not unlike a cult indoctrination. There we were, locked in a small, air-conditioned building surrounded on all sides by the deadly cultural desert of southern Houston, thirteen scientists in close quarters, reviewing ninety-one research proposals in four days. We were immersed in the material constantly from sunup to well after sundown, barely leaving to sleep and eat. By the third or fourth day of sleep deprivation, this kind of experience would start to take on a surreal air even if you weren't reading about fishing for alien squid beneath the ice of Jupiter's moon Europa.

Though it is customary to complain about being fingered for a NASA review panel (no, you don't get paid), it can be fascinating and fun.

There is that peculiarly giddy sense of camaraderie that comes only from focused group concentration over long hours—it reminded me of cramming for finals with fellow students or all-night recording sessions with various forgotten bands. Forced to endure such conditions, nerds can get pretty silly by the end of the broadcast day.

Like a voyeuristic stroll down a dark city street, peeking into random rooms and lives, an assignment to a review panel gives you a great glimpse into random labs and minds. For me, serving on this panel provided a wonderful opportunity to see what kind of research into life elsewhere NASA was funding, and what was being proposed but not funded.

Because astrobiology, or exobiology as it is still called in this particular NASA program, is an attempt at forging a metafield from many disciplines, I was closeted not only with fellow planetary scientists but also with biologists, geologists, chemists, and others. The attempts at cross-disciplinary communication, some more successful than others, were enlightening. One proposal came from a cosmologist who said he could explain some important mysteries of biology using simple structural principles adapted from cosmology. The physical scientists (astronomers, planetologists, chemists) looked at it and thought, "Cool! Why not?" We thought it was insightful, innovative, and deserving of the highest rating. The biologists on our panel looked at it and thought, "How dare he? Who the hell is this guy? This makes no sense." They regarded it as naive, foolish, grandiose, and entirely undeserving of support.* This extreme case of interdisciplinary cognitive dissonance reminded me of the challenges facing us as we search the universe for something we all desperately seek but can't exactly define.

I was particularly interested to see what kinds of experiments chemists are now doing in origin-of-life studies. We didn't see any proposals that claimed, "We will be as Microcosmic Gods and create life itself from inanimate matter!" But, in fact, many groups are chipping away at that problem from several different angles, mapping the uncharted pathways of organic matter, seeking life's primordial route out of the chemical wilderness.

My overall impression was that the field had not advanced greatly since the late 1970s when Reid and I had fooled around with organic

---

*Unfortunately, for legal and ethical reasons, I can't be more specific.

synthesis at Cornell. What Miller and Urey had done in their lab and Sagan had done in his lab with help from young lackeys like me, several groups were still attempting: making organics from mixtures of simple chemicals, egging them on with various energy sources, and analyzing the results. It's still as much art as science. We've mapped out many possible parts of the path from chemistry to biology, but the overall route is still far from clear. Our experiments in prebiotic chemistry are still more like medieval alchemy than we would like to admit. You add a little of this, take out a little of that, and see what you get. It's more like cooking than quantum mechanics. We are like those earliest biomolecules, casting about in a sea of enticing chemicals, hoping to find some magic.

THE MISSING LINK: RNA WORLD?

What was the evolutionary step, or series of steps, between simple self-replicating molecules and the elaborate reproductive machinery common to all living cells on Earth? How did organics beget organisms? This, not some shadowy ape-man, is the real "missing link" in evolution.

The link may have been found in the form of remarkable chemicals called ribozymes, the discovery of which bagged a Nobel Prize for chemists Thomas Cech and Sidney Altman in 1989. Ribozymes are molecules of RNA that, in addition to encoding information like other nucleic acids (including DNA), can also behave like proteins. That is, they can catalyze and orchestrate chemical reactions between other organic molecules. This is huge, because in all life today that catalyzing role is played by proteins. If a nucleic acid such as RNA can do it, there could have been life without proteins.

Another way of putting it is that DNA, strictly speaking, cannot self-replicate. It needs its dance partner, protein, to propagate itself. If molecules of RNA can somehow catalyze their own duplication, then they would be true self-replicators.

A popular idea now is that life emerged through an early stage where RNA was both the keeper of the genetic code and the catalytic enabler of replication. At one point on its travels from chemistry to life, Earth may have been an "RNA world," awash in self-replicating ribozymes.

Is RNA world really the missing link between complex chemistry and sophisticated cell? There are still a few loose rungs on the ladder of

Cosmic Evolution. A description of RNA world is valuable as a "proof of concept." Like building the *Kon-Tiki** and sailing it across the Pacific, the RNA-world theory removes some of the historical mystery by showing that the task could have been done with the tools at hand. It bolsters the faith of those of us who think we can see an emerging path winding from chemistry to life.

---

*A balsa wood replica of a prehistoric South American raft, which was sailed from Peru to Polynesia in 1947, in an attempt to prove that Polynesia was first settled from South America.

8 | Childhood

Originally you were clay. From being mineral, you became vegetable. From vegetable, you became animal, and from animal, man. During these periods man did not know where he was going, but he was being taken on a long journey nonetheless. And you have to go through a hundred different worlds yet. There are a thousand forms of mind.

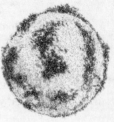

—RUMI

The difference between an amoeba and a human is one step.

—KARL POPPER

GOING CELLULAR

If we started out as self-replicating strands of naked RNA, we did not stay that way for long. Soon the RNAs (or whatever the first genetic stuff was) started assembling coteries of attendant molecules to help them survive and reproduce. Catalysts, such as RNA, can selectively surround themselves with other molecules, assembling some and attracting others, choosing their company to control their immediate chemical environments. Self-replicators that surrounded themselves with the right molecules would come to dominate the mix. The "right molecules" would be those that helped them to replicate.

A class of organic compounds called lipids spontaneously form spherical structures in water. These thin linear molecules are water-loving (hydrophilic) on one end, and water-fearing (hydrophobic) on the other.

(Borin Van Loon)

Put a bunch of them in water and they quickly form spheres, called vesicles, by "circling the wagons," forming a spherical wall against the surrounding water, with the hydrophobic ends huddled together inside.

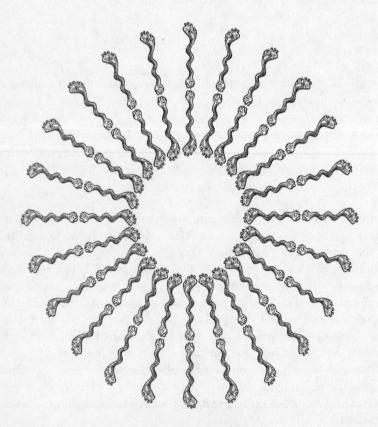

These spontaneously forming structures are suspiciously similar to the membranes surrounding the cells of modern organisms, which are also dominated by lipids. Any primitive genetic material that could make or gather lipids would soon find itself protected from the environment by a surrounding vesicle. This is probably how the first cells formed.

Becoming cells was a giant step in our evolution. Perhaps this was the moment of the actual origin of life. Before this, life (if you could call it that) was a sea of chemicals with no clear boundary between one individual and the next, no separation between living creatures and their surroundings, and thus no real organisms.*

With the advent of membranes to divide interior from exterior, life gained the potential to regulate its internal environments. Here on Earth, life is chemical. The stability of those chemicals, and the rates of their reactions (our metabolism), require specific temperature, acidity, and concentrations of many trace chemicals. Organisms that could control any of these aspects of their internal environment had a huge advantage over others that could not.

This was the microscopic equivalent of learning to live in caves or to build huts to survive changing climate. When cells evolved the means to regulate their interiors, they were not helpless against the randomly changing chemistry of their watery home. After this, the chemistry of life was forever more immune to harmful or fatal changes in the weather. The internal environments of cells became stabilized by various chemical and physical feedbacks that evolved to increasing sophistication. Competition for survival was no longer between naked self-replicating molecules, but between evolving cells.

COME TOGETHER

However it happened, wherever it has happened, the birth of life is big news in Cosmic Evolution, right up there with the formation of quarks, atoms, molecules, stars, and planets. Smaller, previously assembled bits of the universe had repeatedly banded together to make big, new

---

*On the other hand, maybe the first life-form was actually this undifferentiated ocean of organic replication, a global organism that lived before differentiating into cells.

things. But something changed with life. Before this point, matter self-assembled only by answering the call of various forces of attraction. There was never any blueprint. The advent of heredity marked a huge shift. The cosmos, having stumbled upon chemical memory, would never be the same. It could now start learning from its mistakes. Life evolves by throwing possible organisms into the world, seeing what survives, and passing the successful designs on to the future. Life remembers.

An average humanoid handful of times, the universe has undergone major changes in the way it pulls itself together, differentiates, organizes, evolves, awakens, and reflects. Heredity was one of these giant steps, allowing entirely new possibilities of change, facilitating a metamorphosis in the roles of matter and energy. I think that the advent of conscious awareness* is probably the next change on this scale. But, how did we get from there to here?

Once the twisted ladder of DNA was firmly planted in the primordial ooze, it kept building on itself. Improving and diversifying its reproductive tricks, it built more complex cells, then multicellular bodies with specialized organs, nervous systems, sensory structures, hair, scales, feathers, and flowers. From DNA's perspective, all Earth's creatures are stationary or mobile reproduction units, even those weird new brainy, bipedal ones that make fire, print books, and build rockets. All this in the service of the master molecules. We are DNA's spaceships preparing for launch and trying not to get ourselves killed.

There are many ways to describe the history of life on Earth. We could think of it as a sequence of geological dioramas—you know, Jurassic©, Cambrian, Rastafarian, Bohemian—with depictions of the flora and fauna populating each. But the procession of species making cameos in life's rich pageant is largely irrelevant to the questions of life on a cosmic scale. Chance seems to be behind so much evolutionary innovation. The intricate, colorful details of life's transmutations are fascinating to study and beautiful to behold, but they are transient ripples in the sinuous stream of evolution. The profundity is not in the details but in the breakthroughs, the moments when the evolutionary river spills its banks and surges in new directions.

The truly cosmic changes, in this sense, are those involving major

---

*Something at least as difficult to define as life. I prefer to define it as something humans have not yet fully attained, but I'll get back to that.

transformations in basic biochemistry, in energy source, in reproductive and hereditary mechanisms, and in the level of complexity and organization of life that allowed new abilities to emerge.

The first of these transcendent changes was the formation of the cell. Echoing the pattern of Cosmic Evolution, many of the key developments of cellular life on Earth involved the formation of larger assemblages from smaller, simpler subunits. The first cells were *prokaryotes* (pro carry oats). They resembled bacteria today in that they possessed no internal organization. These earliest cells were just rudimentary watery sacs with all of life's chemicals mingling together, a well-mixed cocktail of nucleic acids, proteins, and other organic molecules.

The next important transition came when cells began to differentiate internally, acquiring subunits that were specialized for certain functions. All cells in our bodies, and those of all creatures except bacteria, are *eukaryotes* (you carry oats). These more complex cells have their genetic material separated into an internal nucleus, distinct from the rest of the cell. Inside eukaryotic cells, various tiny subunits, or organelles, developed to play specific roles. Small units called ribosomes became the sites where proteins were assembled from DNA instructions, with RNA acting as intermediary messengers. Other specialized structures brought new abilities. Organelles called chloroplasts and mitochondria took on the functions of photosynthesis and oxygen respiration respectively.

How did these organelles come to inhabit every living eukaryotic cell (every cell of every animal, plant, or fungus that ever lived)? These were early instances of evolution by *symbiosis,* in which once-separate organisms begin living in close association and then, somehow, merge into larger unified organisms. The energy-transforming subunits of animal and plant cells (mitochondria and chloroplasts) have their own evolutionary history. Their ancestors were once separate bacteria, living independent lives. Somehow these little guys gave up their individuality to become the energy-processing parts of the collective eukaryotic cell. The joining of formerly separate organisms to form new superorganisms with expanded abilities is a persistent theme in evolution. Much of the evolution of complexity in Earth life comes from the symbiotic joining together of simpler organisms.

Certain steps in evolution have a Borg-like quality. For those of you who have just woken up from a decades-long coma or are for some other reason unfamiliar with *Star Trek,* the Borg is a fearsome entity that evolves by assimilating other species, incorporating their technology and culture into the Borg Collective. The price for becoming part of the ever-growing perfection of the Borg is that you give up your individuality. On Earth, complex cells were created by the assimilation of once separate, simpler life-forms whose abilities were added to those of the collective. Resistance was futile. We are the Borg.

Symbiosis is a biological form of "enlightened self-interest." Survival of the fittest still applies, but often the most fit are those who can form strategic alliances with others. Extreme cooperation creates a new identity born of the cooperative as a whole. A new kind of individual emerges when the group starts to reproduce as a single entity.

In evolution, the continual, random meandering of forms and species is largely fueled by competition. But the real progress, the major innovations, often involve new forms of cooperation between formerly separate creatures.*

The next great leap forward in the coming together of life on Earth occurred when large numbers of eukaryotic cells joined together to form multicellular individuals, or metazoa (big life). Even when individual cells grew much more sophisticated, they were still just cells. Cells can't do certain things. If you want to fly, see, grow skin and bones, dance the merengue, or calculate triple integrals, you are going to want to go multicellular.

On the face of it, it is mysterious that single-celled creatures, each acting on its own imperative to survive and reproduce, would give up their individuality to become part of a new, larger individual. The advantages to you—the metazoans—are clear enough.† But what's in it for your cells? Why should they give up their sovereignty to join the United Cells of You? From an evolutionary perspective, the most precious thing that any creature has is the ability to reproduce. Somehow, in the metazoan contract, individual cells ceded the power of reproduction to the centralized cells residing in specialized reproductive organs.

What the other cells got in return was the chance to be in on the ground floor of something new and really big, with opportunities for advancement and innovative ways to survive. Multicellularity allowed

---

*Some biologists and philosophers are fond of saying that there is no progress in evolution. On a certain level, this is obviously true. On a more profound and interesting level, it is obviously wrong.
†If there are any microbes reading this book, I apologize for my presumptuousness and insensitivity.

whole new levels of differentiation and organization. Multitudes of specialized cells developed into novel tissues and organs. Animals, plants, and fungi: we are all consortia of huge numbers of eukaryotic cells, which are themselves forged from ancient bacterial alliances.

WHY THE WAIT?

Did life need to be multicellular? My perspective is warped, since I'm writing this book with fingers for typing, eyes for reading, and a rudimentary brain for musing. I like being multicellular, and I'm quite fond of people, cats, redwoods, sea turtles, and prickly pear cacti. From where I sit, it's all the freaky multicellulars that make the beauty of the world.

If microbes could talk, maybe they'd tell a different story. But they can't. Then again, in a sense, microbes *can* talk—in the same sense that human beings can build space stations and chunnels. Overbudget and behind schedule? No. What I mean is that an individual human can't build a space station, but complex organized groups of us can. Microbes manage to talk by getting together to make us, by becoming multicellular.

When we look over the history of life on Earth, a gnarly question leaps out at us: If multicellularity is so cool, why did it take so long? There is a gap of roughly 3 BILLION YEARS between the time when cells were born and the time when they figured out how to join together in large numbers to make animals. Three billion years is a long wait by any standard—human, geological, even cosmological. It is more than half the age of Earth and probably about a quarter the age of the universe. A 3-billion-year gap is no temporal chump change.

I'll admit I find this disturbing. If life always self-organizes into more complex entities, why did it get stuck? What kept us for so long at the stage where individual cells were the greatest show on Earth? We do not know the answer, but we have no shortage of explanations. Maybe cells first had to develop some special capabilities that took a long time to evolve. Or maybe this leap required changes in Earth's environment that happened very slowly.*

---

*Note that this is not really an either/or situation: changes in the physical state of Earth and changes in biological evolution are not independent. Extreme proponents of the Gaia hypothesis believe that no separation at all exists between the two.

Multicellularity is a neat trick, but it's not all that easy to pull off. Before going multicellular, a few fundamental problems have to be worked out. The development of a complex plant or animal always starts from one single, fertilized cell made from the fusion of sperm and egg. This *zygote* starts dividing like mad, producing the millions of cells needed to make a large organism. The strange beauty of this development is that somehow each cell "knows" what kind of tissue it is to become part of—a slice of liver, the tip of your nose, a piece of your heart, or a piece of your mind. How do they know?

Some scientists used to think (quite reasonably) that the instructions were split up so that your toes got only the genes to make toes and your nose had only nasal genes. Nope. It turns out that the entire genome, the whole instruction manual, is present in every cell. This is why embryonic stem cells have such great versatility. Each dividing cell of a developing embryo has the potential to generate any body part. There is some system—which hasn't completely been worked out yet—for letting cells know which kind of tissue they are to become. Without such a master control system, multicellularity would not be possible.

This problem is solved through a multilevel genetic control system. It is as if there is another genome within the genome of each cell that somehow learns what kind of cell it is to be and activates certain genes (calls them to action) and suppresses others (asks them to sit this one out). This centralized control is quite elaborate, yet it is essential to have something like this in place if multicellularity is going to work. Difficult things do take time to evolve, but can that explain a 3-billion-year gap?

Alternatively, we can ask what changed on Earth during this interval that may have made the leap possible or even inevitable. One thing that changed was the mixture of gases composing our atmosphere. Oxygen, good old O_2, which was apparently absent from Earth's earliest air, gradually increased in concentration until it became quite abundant, second only to nitrogen. The most common explanation for the timing of the multicellular leap is that it had to wait until the concentration of atmospheric oxygen rose high enough to make efficient respiration possible. When you blow lightly on a flickering flame, it flares up. Similarly, the more oxygen there is around, the more organic fuel life can burn in the slow flames of metabolism. Perhaps the increased energy source of an oxygen-rich atmosphere was needed to power the larger bodies of the metazoa.

Where did all this bountiful O_2 come from? Life made it as a way to avoid starvation. The history of oxygen on Earth is one of life coevolving with the planet.

LET THEM EAT LIGHT

When life first formed, finding a meal was not a problem since the early Earth was loaded with organic molecules. But this paradisiacal situation, with tasty molecules everywhere for the taking, could not last. As the food supply became depleted, a new long-term, reliable food source was crucial for the continued survival of life on Earth.

Fortunately for us, before it was too late, life evolved the trick of photosynthesis—using the energy of sunlight to make organic food. This had to happen fast. Otherwise, when all the leftover crumbs were gone, the party would have been over. Life would have perished, unless it found a way to mooch off the geothermal energy coming out of submarine vents or other exotic sources. But Earth life is almost entirely solar-driven. Had life not developed photosynthesis early on, our planet would now be unrecognizable, and there might well be no one around this part of the galaxy capable of recognizing anything.

Photosynthesis is so pervasive and essential to life on Earth that it is not inaccurate to describe the biosphere, as did Russian biogeochemist Vladimir Vernadsky in 1926, as a continuous, thin film enveloping the planet, within which sunlight drives matter through incessant transformations. In other words, life is something the Sun does to Earth. Earth life is the way (or at least one way) that our star has found to express its biological potential. We are the life of the Sun.

Photosynthesizing microbes began ripping the H out of H_2O, combining it with carbon to make organic food, and spitting out O_2 as waste. Even so, at first oxygen did not build up in the air. Everybody loves oxygen, and many other elements were in line for its favors. The young Earth had lots of iron, which is an oxygen hog if ever there was one. For hundreds of millions of years, most of the photosynthetically released oxygen went into oxidizing Earth's iron, which was constantly showing up in volcanic flows fresh from the interior, always demanding oxygen. There was also methane (CH_4) from volcanoes, methane from bacteria, ammonia (NH_3) from lightning storms, and other hydrogen-rich gases, all composed of elements that eagerly forgot their other dance partners the moment oxygen entered the room.

There was always more than enough water to go around, and plenty of sunlight. Life went on using both and leaving oxygen behind. Eventually, the world's thirst for oxygen was sated, or at least the demand leveled off. Finally, free oxygen started ever so slowly to build up in the atmosphere.

Hallelujah!

Except that this was a terrible disaster.

Oxygen is dangerous. Because of the promiscuous reactivity of oxygen, no organic molecule is safe. Rust never sleeps. Have a little fire, scarecrow. Oxygen was poisonous to organic life. The buildup of oxygen was Earth's first global environmental crisis, and life brought it on itself. At first, it seemed as though the careless photosynthesizing microbes had really screwed themselves, with their shameless oxygen emissions causing global change that threatened their own extinction.

LIFE EXPLODES

Catastrophes, viewed from a different angle, are often opportunities. It's true that oxygen reacts ferociously with organic molecules. It's also true that these reactions release a lot of energy. Uncontrolled, this energy will burn you up. But life, turning adversity into advantage, found a way to harness fire: respiration, which uses controlled oxidation to constantly charge its batteries. Aerobic life was born.

When respiration started, oxygen was still only a minor trace gas in our atmosphere. Over billions of years, it gradually built up until it reached its present level about 1 billion years ago. Since then, it has had minor ups and downs but has not strayed too far from 20 percent of the air we breathe.

The buildup of oxygen in Earth's atmosphere had another tremendously fortunate and enabling side effect: it made the ozone layer. Once there was enough oxygen in the air, ultraviolet light started splitting O_2 and recombining it in various ways in the upper atmosphere. A byproduct was the production of O_3 (ozone). Ozone happens to absorb the same wavelengths of solar ultraviolet light that are fatal to our kind of life because they destroy complex organic molecules. Before ozone, life mostly hid under the surface layers of the ocean, where water absorbed harmful UV. Now it was safe to colonize the land.

Like an athlete trained at high altitude coming down off the mountain, life reveled in the ever richer air and revved up its metabolic engines. But if more O_2 allowed the possibility of bigger living creatures, you would never have known it for most of Earth's history. Respiration was around for billions of years before multicellularity. Life did not get gradually larger as the oxygen level rose, but stayed unicellular. Maybe this was because it could not make the move until a certain threshold was crossed—a minimum oxygen richness for big life.

When multicellularity did arrive, however, its entrance was not subtle. After waiting for so long, it burst onto the scene dramatically. Six hundred million years ago, animals materialized, and soon they were everywhere, appearing in a multitude of forms. We call this event the Cambrian explosion.

It was as if some repressed creative force in nature was finally set free. The evolutionary creation of major animal body types is not something that began in the Cambrian and has continued to the present day. Rather, these templates seem to have mostly formed all at once. Species come and go, riffing endlessly on the grand structural themes that were all established at the time of the Cambrian explosion.

GETTING SMART

Once we became multicellular, entirely new possibilities for organization and specialization opened to us. Legions of cells could now be called upon to dedicate their lives to specific structures and tasks. Skeletal, circulatory, and digestive systems provided the infrastructure for large bodies. Muscles and limbs sprouted for swimming, running, climbing, and flying.

It doesn't do any good to have such nifty toys if you can't control them, though. A nervous system was needed. The ability to sense the environment and respond to it would be a definite plus. From such humble needs were born refined senses and mechanisms for sending signals throughout the body. Now you're receiving information and coordinating reactions and movements, so you need some kind of central processing site. If you only had a brain.

Once you've got that, then you've got the rudiments of a cognitive system. You're on your way now, kid. You've got a leg up if you can respond to your environment in flexible ways. This creates a pull for more complex nervous systems with increasing power to sense, manipulate, navigate, anticipate, and remember the world.

Then, 600 million years after the Cambrian explosion, something new happened. Matter woke up in human form. Was it a gradual focusing or a sharp blink of self-recognition? We don't remember. I doubt we're fully awake now. Intelligence, consciousness, self-awareness, the divine spark of Jah—call it what you will.* We don't know exactly what crucial ingredient was added, or what threshold value was surpassed. One species started talking, building tools, and creating new societies, not just out of instinct but through culture, ritual, and design.

A number of changes happened quickly as our hominid ancestors sprinted into the new, big-head niche. We developed opposable thumbs for grasping. An upright stance freed those hands to make tools and throw spears. Language gave us the ability to communicate complex ideas within groups. A rapid increase in brain size, doubling in about a million years, facilitated all of the above, but made childbirth difficult and dangerous. Evolution responded by pushing birth back earlier in development, so that our expanding heads could make it through the birth canal. As a result, humans, compared to most other animals, are born unformed and helpless. This required prolonged and attentive infant care, which increased social cohesion.

Recall that previous great leaps of evolution involved new associations between preexisting simpler organisms. Similarly, the origin of human beings is inseparable from the origin of human societies. The beginnings of language, the rapid growth in neural capacity, the formation of social groups with division of labor and the ability to plan and learn collectively—all seem so tightly linked that it may be meaningless to ask what caused what. These abilities seem to have bootstrapped one another into existence. With language came the advent of a powerful new form of heredity. We are cultural animals, and we pass on information through word of mouth and artifacts: artworks, songs, rituals, rhythms, stories, and now books, films, and disc drives. Our minds expanded beyond our bodies and our thoughts came to survive the death of individuals. A new kind of multi-organismic structure—the society—was born.

Less than one hundred thousand years ago a small band of us, possibly no more than fifty brave souls, left our native Africa and began spreading

---

*I'm not saying that other animals don't have some degree of it. Some elephants can play the marimba pretty darn well.

around the Earth.* These pioneers are your ancestors, unless you are of pure African descent, in which case their less restless cousins are your ancestors. Perhaps only fifty thousand years ago, we started using syntactical, symbolic language, which allowed us to communicate abstract concepts and poetry and planted the seed for the language of mathematics. The more we can understand the tangled causal relationships that catalyzed the jump to human awareness, the more informed will be our speculations on the likelihood of similar evolutionary events on other worlds.

THE PSYCHOZOIC AGE

However it happened, it is clear that in just a few million years, in barely the blink of a cosmic eye, one lineage of primates went through an intense metamorphosis, and Earth acquired thought and self-awareness. Several scientists and philosophers have recognized this as a profound moment of transformation in Earth history. Pierre Teilhard de Chardin, the Jesuit paleontologist/philosopher described it in 1955 as the beginning of a new geological age, the "psychozoic era." He described the web of interacting thoughts, communications, and artifacts rapidly covering the Earth as a new terrestrial sphere, the *noosphere* (new-oh-sphere), which emerged out of the biosphere as the biosphere had emerged out of the rocky lithosphere.

Although I cannot follow Tielhard all the way to his Christian conclusions, I find his vision of the human place in Cosmic Evolution to be prescient and inspiring. I am with him when he says, "With hominization, in spite of the insignificance of the anatomical leap, we have the beginning of a new age. The earth gets a new skin. Better still, it finds its soul."

Am I giving us too much credit here, going on about the arrival of humanity as though we were the second coming of sliced bread? Does this view of planetary history place humans at the apex of creation? Isn't this picture of evolution a hopelessly self-serving glorification of the human race?

Not necessarily. Read on, and you will find that I do not see us as the apex of evolution. I believe that humans are the resident example of something extremely significant on a cosmic scale. We may not even be

---

*Some scientists believe, on the basis of DNA studies, that it was less than fifty people.

a particularly promising example. But that "something" is essential in Cosmic Evolution. Without it, what's the point? I believe we are part of the very beginning, locally, of a phenomenon integral to the conscious awakening of the universe. Whether we will be part of later stages remains to be seen.

Conveniently, this realization gives us a new way to be human-centered, even though science has robbed us of an Earth-centered cosmos and a purposeful human creation. Are we conflicted, hapless humans really the vanguard of a new, conscious phase of Cosmic Evolution? Deserve it or not, we are.

Lest our already dangerously bloated braincases become even more swollen with this thought, keep in mind that of the species that ushered in the age of multicellular life in the Cambrian explosion, all are long gone and forgotten to everyone but curators of paleontology. But multicellularity lives on. The phenomenon is more important than the ephemeral particulars of who was first. Humanity does represent something the likes of which Earth has never seen, but the jury is still out on our legacy and longevity as a species.

It seems as though, on principle, some people want to deny any evolutionary importance to humans. I see this as a reaction to the historical tendency to assume that humans were the purpose of creation, the center of the universe and the pinnacle of life. We now know better than to believe any of these egotistical fantasies, but the reflexive reaction against these views can be as illogical as the narrow thinking that prompted them.

Sure, humans have historically overemphasized our role in things and foolishly imagined that we were "conquering nature" and running the world, where at best we've been temporarily mucking it up. Sure, we need the microbes more than they need us. If they disappeared from the face of the Earth, we would perish. If we disappeared, most of them would not notice. But microbes, without us, are not capable of noticing anything. Through us the microbes have found a mind, a voice. It is foolish to ignore that human evolution represents a new phase in the life of Gaia,* a phase with enormous potential and peril. It's merely a matter of perspective whether you regard Earth as planet of the microbes or planet of the apes. Each brings essential qualities to our

---

*The *Gaia Hypothesis,* a scientific portrait of a living Earth, has implications for life on other planets that will be discussed in chapter 17.

home world. I believe, as Teilhard argued, that mankind is not just another species, but an important evolutionary phenomenon—the precursor of the psychozoic age.

LOVERS OF EXTREMES

Now, having found mind and the beginnings of technological prowess, Earth's biosphere is starting to take a look around, exploring the neighbor planets for the first time, and searching for life in the bright and dark corners of the solar system. Much of this search takes the form of self-examination, seeing the home world with new eyes.

I've been describing the history of life as a series of leaps to greater complexity and organization culminating, at least so far, in the psychozoic age. While all this has been going on, over the 4-billion-year life of Earth's biosphere, other evolutionary dramas have been playing out. Life has expanded on its bag of chemical tricks to facilitate survival in a bewildering array of environments.

In recent years, we've discovered life in the strangest of places: in unlikely corners of our planet where no one had thought to search because they seemed so obviously uninhabitable. We've found bacteria thriving in acid so strong that it would dissolve your skin instantly, and creatures soaking contentedly in superheated thermal springs above two hundred degrees. Some of these hyperthermophiles, or extreme-heat-loving organisms, *require* temperatures above the normal boiling point of water to survive. At the opposite extreme are those that survive in intense cold. In frigid arctic tundras that appear lifeless, we've found colonies of bacteria hiding out inside frozen rocks. We've even found organisms that can survive after being frozen for weeks in liquid nitrogen!

The green plant *Welwitschia mirabilis* can survive for thousands of years in places with only one centimeter of rainfall per year. The Dead Sea, it turns out, is alive with salt-loving bacteria and algae. In 1997, Japanese scientists discovered a species of marine worm living in an ocean trench twenty-one thousand feet beneath the sea at a crushing pressure 650 times that of sea level. Bacteria have survived for 3 million years in Siberian permafrost at fifteen degrees below zero with no sunlight, air, or food. They don't do very much down there but survive simply by waiting, for eons if necessary, until the ground thaws and they can resume living at a healthier clip.

Large, diverse communities of previously unknown organisms crowd the hot, nutrient-rich waters surrounding "black smokers," volcanic vents on the bottom of the sea. The denizens of these recently discovered ecosystems include sulfur-eating shrimp and giant tube worms up to ten feet long. As weird and unearthly as these deep-ocean communities seem to us, many scientists are starting to think that our most distant ancestors came from just such a place.

There are even bacteria living a mile underground and eating nothing but basaltic rock and water.* In fact, it now seems possible that *most* life on our planet is in the "deep Earth biosphere," a realm extending miles underground whose existence we never before suspected. This would be the biological equivalent of "dark matter" in that the majority of life even on our own planet could as yet be unknown to us. We've been sharing a planet with these unlikely creatures for billions of years, but who knew?

Our own planet is crawling with "aliens." We continue to find extremophiles (lovers of extremes) that break our conceptual barriers of life's range in temperature, pH, diet, and pressure. They show us that life is even more robust, adaptable, and resourceful than we imagined, and this encourages us to think that it will find ways to persist in diverse and extreme environments on other planets.

In fact, life may not even need a planet. When the *Apollo 12* astronauts retrieved pieces of the old *Surveyor 3* spacecraft, which had been sitting idle in a lunar crater fully exposed to the harsh radiation and vacuum of space, investigators back on Earth were shocked to find viable *Streptococcus* bacteria that had survived a three-year stay on the Moon. Who is to say that living creatures cannot survive longer spells in outer space?

This possibility was amplified by another recent discovery: bacteria such as *Deinococcus radiodurans* that live happily inside nuclear reactors, flawlessly reassembling their damaged genomes from hundreds of fragments, despite radiation doses a thousand times stronger than those that would kill a human. Suddenly, the merciless radiation of space seems less of a barrier to survival than it once did. Scientists have also revived bacterial spores that had been encased in amber for at least 25

---

*Fontenelle anticipated this in the seventeenth century when he wrote about "innumerable small worms, living in imperceptible gaps and feeding themselves by gnawing on the substance of the stone."

million years. Such findings have revived talk, or at least whispers, of "panspermia," the idea that Earth life may not have originated on Earth.

PANSPERMIA RECONSIDERED

Though Earth's creatures live in a hugely diverse range of environments and power ourselves with an impressive variety of chemical energy sources, we all use the same basic chemical operating system. There is no "think different." It's a complete monopoly.* What is the meaning of this?

A remarkable aspect of Earth's story is that the planet became *inhabited* as soon as it was *habitable*. Once the sterilizing impacts died down, Earth sprang to life—in less than a couple hundred million years, and maybe much faster. Then, a short while after life on Earth started, it had already constructed the fantastically intricate, interlaced chemical engines of DNA replication, protein transcription, and a great many more standardized cycles and systems that would seem so elegantly conceived if only they seemed conceived. Life did not take billions of years of evolution to perfect these fine and fertile, far-flung factories. Not only the origin but most of the essential chemical evolution of life on Earth happened before you could say "evolutionary divergence."

Does this mean that, given half a chance, a biosphere with such refined chemical machinery is easy to make? A widely accepted interpretation is that on a planet like ours chemical evolution leads rapidly and inexorably to life. This inference is a mainstay of scientific belief in a universe with many inhabited worlds.

Although it is not popular to do so, we could also look at this auspicious timing as supportive of the theory of panspermia, elaborated by Arrhenius in 1903—the idea that Earth life was seeded from elsewhere. If space contains spores wafting and waiting for watery worlds, life would be expected to spring up instantly as soon as it found a stable ocean. The strongest objection is that we cannot imagine spores that would survive the drift between star systems. This is a valid argument against panspermia, but not a damning one. Time and again, evolution has proven more clever than we are at solving survival puzzles. I wonder if we really don't like the idea because it gives us the creeps.

---

*Hey, could I be channeling Bill Gates?

A far-out variant is "directed panspermia"—the idea that some creatures from elsewhere in the galaxy have played cosmic Johnny Appleseed, deliberately spreading life through the universe. That could explain the sudden appearance of chemically sophisticated life. Could the first living cells on Earth have been designed by a species with very advanced biotechnology?*

If this strikes you as pure science fiction, consider the biotechnology progress (and I use the word loosely) of the last decade. Let's assume for a second that we don't soon commit high-tech suicide with the genetically modified tomato that ate Manhattan. Can you imagine the biotechnology and space technology we might possess in just one hundred years, in one thousand years? It doesn't stretch the imagination much to think that we could soon be able to do some directed pansperming ourselves. So why does it seem like such a fantastic notion that someone else may have done so?

We hate the idea of directed panspermia because it demotes us, like the Neoterics in "Microcosmic God," to being the product of someone else's ingenuity. It mocks our current biology and origin-of-life studies, to say nothing of our egos and established sense of place in the world. Our rational minds whir into action, finding logical objections to safeguard us from the disturbing thought. How ironic if science should have slain all obvious creator gods only to find that some godlike aliens had created us in their image many billions of years ago.

This model could actually meld the supposedly inimical theories of intelligent design creationism and Darwinian evolution. Life evolved elsewhere by natural selection to the point where it could develop advanced biotechnology. It then proceeded to spread intelligently designed seeds throughout the universe. These took root on Earth, where life evolved by natural selection to the point where it is now again starting to muck around with intelligent design of organisms.

This is not a favored explanation, because scientists don't like these kinds of explanations. Some people like them very much. But to us, it violates Occam's razor. It is logically almost the same as invoking a biblical miracle, only aliens, unlike gods, are scientifically acceptable creatures.

A frequent objection to panspermia is that it does not solve the origin-of-life problem but simply pushes it out into space. Life still had

---

*What if we find out that some multigalactical corporation owns the patent on DNA-protein biochemistry and they come back through our system demanding licensing fees?

to get started somewhere. Obviously, this is a lame argument against panspermia.

A much less radical variant that's come into vogue in the last few years is "impact panspermia" or "transpermia." In the mid-1980s we realized that some of the meteorites in our collections are actually pieces blasted off of the planet Mars. There are several lines of evidence for this, but the clincher is that some of these rocks have bubbles of trapped air that exactly match the measured composition of the Martian atmosphere.

Ever since we learned that rocks from Mars have landed on Earth, we've wondered whether life could stow away and ride between worlds. Were meteorites the bees that cross-pollinated the flowering planets, carrying the seeds of life across the vacuity of interplanetary space? Especially when the solar system was young and unruly, with a nonstop demolition derby of planetesimals constantly spraying the planets with each other's shrapnel, any critters that could survive the launch and the journey would have had no problem hitching a ride to another world. So, we cannot regard the early environments of the planets as isolated systems, quarantined by space. Maybe, in their early days, the planets were all sneezing on one another, sharing germs, with meteorites as the vector. If any of them became infected with life, the others would quickly catch it, too. It is conceivable that our earliest ancestors lived on Venus or Mars and that we are all descendants of interplanetary immigrants—or should I say resident aliens?

According to this picture, we started out as a multiplanet biosphere (a biopolysphere?). Gaia was born as planetary "Siamese triplets" joined by meteorites. Then, when the rates of interplanetary transfer declined, the three became isolated, and Earth was more or less on its own. I will discuss the fate of the other two orphaned biospheres in ensuing chapters. Transpermia would have slowed down before it effectively stopped. In an intermediate stage, perhaps long-lost Martian and Venusian microbial cousins occasionally arrived on the shores of Earth, after millions of years of isolation, stirring the genetic pot in interesting ways.

Now, fueled by the discovery of possible fossils in one of our Mars rocks, we are experiencing a revival of the nineteenth-century idea that life might have arrived on Earth in falling meteorites. We are attacking the problem from several angles with twenty-first-century science. Could anything have survived the impact that launched these rocks into space? We test for that with impact simulations using high-speed guns. Could they have survived the journey through space and their fiery arrival on Earth? With sophisticated computer models of orbital dynamics, we follow the paths of impact shrapnel from Mars (and Earth and Venus) to see how likely they are to reach another planet and how long the journey would take. Experiments in space and in terrestrial laboratories allow us to study the survival of organisms subjected to the hardships of space travel.

Microbes are much tougher than the rest of us. The damn things are hard to kill. Thank Gaia they're mostly on our side. They've survived experiments with intense radiation, heat, cold, vacuum, shock pressures, and that ear-crushing tape of bad rock music that the U.S. Marines used to flush out Manuel Noriega. So far no showstoppers have been found to affect the survival of microbes in space.

Even given all of these intriguing alternative possibilities for life's origins, my instincts are in accord with the majority scientific opinion. I think that planets, Earth included, probably make their own life. Why? Because they can. The steps to life on Earth seem promising without any external help or prompting. As for the possibility that life here was deliberately seeded, all we can do is smile up at the stars, shrug, and keep looking for the missing links in our cosmic genealogy, keep searching heaven and earth for our roots.

9 | So What?

There is that within us which believes us worthy of the stars.
—SALMAN RUSHDIE, *The Ground Beneath Her Feet*

Intelligent life understands the void must be filled.
—CONVERSATION BETWEEN TWO VENUSIANS IN *The Quiet Invasion* BY SARAH ZETTEL

BETWEEN THE LINES

We could look at Earth life as a trivial by-product of the immense cycles of galactic ecology, something odd that happened on one small planet, in a random spiral of a largish but unremarkable galaxy. We are that, but we are something much more. Earlier, I described Earth's biosphere as the Sun's way of creating life. Well, our star is a product of the Milky Way galaxy, which coalesced from the cooling ashes of the big bang. If the universe is a self-tilling orchard slowly cultivating life and mind, then we, Earth's awakening biosphere, are a bright flowering of the vine. We strongly suspect that there are others.

It all really boils down to two questions:

How often do new biospheres get started on planets? How likely is life, once started, to develop intelligence? What might intelligence become?

Okay, three questions, then. We'll deal with the third one later.

Can we really say anything in answer to these, based on writing and reviewing our biosphere's autobiography? We've put together a reasonably credible story for the formation of our solar system, for the birth and evolution of the Earth and its biosphere. The hard part is figuring out how universal this story might be. We study it intensely, trying to glean whatever meaning we can from the clues at hand. But there is a danger in this close reading—a temptation to see more significance in the details than may actually be there. We think we're deciphering an instruction manual for making habitable worlds, but maybe we're just learning the colorful story of one eccentric place. On questions of extraterrestrial life, we make the most of the evidence we have, because we can't stand not knowing. But we may just be reading tea leaves.

So, what lessons can we take away from our autobiography that might help us gain perspective on the biographies of other distant biospheres? Can we find any messages with significance beyond this orbit?

First of all, there is evolution itself. Any place where sufficiently flexible and variable structures (organic or otherwise) are given the time and freedom to play at combination and recombination, natural selection will grow beyond a simple contest of stability and longevity to self-replication, making the jump to hereditary memory. Logic dictates that any chemical system that can do it, will. Once you have self-replication that is imperfect (so there will be variation), then some varieties will be better at reproducing themselves, and evolution must occur. On this level, evolution is not a radical proposition. The interesting question then becomes, how far can it get in different environments?

Life on Earth is all made of cells. Will life elsewhere have "cells"? Yes, I think so, because the separation of interior from exterior environment seems like a certain requirement. Some sort of minimal organizational unit with an outer boundary separating organisms from their surroundings seems called for. Once such units exist on any planet, then competition and Darwinian evolution will lead to refinement of the cellular machinery.

Every survival trick in the book, and then some, will be tried sooner or later. Some of these "cells," or cell-like thingies, will do better by teaming up. Symbiosis will lead to the formation of larger, more sophisticated assemblages. The resulting compound structures will have subunits optimized for specific functions. To maintain the individuality of the collective, they may need something like a nucleus, which stores the genetic material and handles reproduction. Nothing says they have to look like our cells or be made out of the same stuff, but these overall structural principles seem like universals.

What else happened here that seems essential? Photosynthesis. We jumped one crucial hurdle when our little green brethren learned, before it was too late, to feed off the Sun. We do not know if this was a lucky break, or if the ability to manufacture food from sunlight is something that always evolves easily in the right conditions.

Photosynthesis seems somehow ingrained in the idea of a living universe. I mean, what's the first thing that comes to mind when you think of the universe?

Quick, what is it?

Stars, right?

We live in a universe of stars. They're the most energetic, stable power sources in the cosmos, and for life not to use them would be like a hungry pack of dogs ignoring a pile of juicy T-bones. Starlight is there for the taking. Photosynthesis is probably not the only way to power a biosphere. Even on Earth, some species ride the chemical surf of hydrothermal vents or otherwise draw power from the planet itself. However, even these creatures are not completely "off the grid." All reside in a solar-driven biosphere. Over the long haul, all live off the Sun in various indirect ways.

To support life without photosynthesis, a planet would need some long-term source of chemical energy either from geothermal sources (or whatever-planet-you're-from thermal sources) or other kinds of radiation from space. There are many types of energy pouring through space, most of which we make no use of. What we call light, radiation in the visible range, is a tiny portion of all available electromagnetic radiation, and not the part that carries the most energy. More energetic radiation can kill. But that's just us. I'll bet life has evolved elsewhere that can use ultraviolet light for serious power, rather than cowering from it in fear as we do. High-energy cosmic rays could be another awesome source. We think of these things as lethal, and we are grateful for the sky that shelters us from them. Someplace else, creatures may have learned to embrace these huge sources of energy, though they would have to be made out of different stuff than we. Complex organic molecules are ripped to shreds by these hard radiations.

Even nuclear energy may be exploited by some kind of biology elsewhere. One good trick is to find ways to indirectly use energy sources that would burn you if you got too close. Life on Earth keeps its nuclear fusion reactor at a safe distance, 93 million miles away. Similarly, other energy flows that we think of as dangerous may be used indirectly in other environments by "clever" (i.e., well-adapted) organisms.

The most obvious alternative to solar power in our solar system is tidal power in a tidal system. Some moons of Jupiter (and of other giant, outer planets) are left cold by the Sun but are heated by tides. Jupiter's fearsome gravity raises tides that make the Bay of Fundy look like a birdbath, tides so strong they squish and warm the insides of the grateful moons. Creatures living in these moons would pray to Jupiter first, and the Sun would be a lesser god. Solar energy might never occur to them, and the idea of life on a tidally challenged world such as Earth could seem laughable.

Although we don't exactly know what life is, I think we can safely say that some long-term, stable energy source will always be required. Life needs energy, and early on it will have to successfully plug into a star or some other live wire if it is to survive over planetary timescales.

What other general principles can we abstract from the particulars of our own story? It seems likely that the nature of life will always be to multiply exponentially. Therefore, it stands to reason that if life survives for long, it will become a planetary entity, a biosphere, exchanging matter and energy with its environment on a global scale. Sooner or later, life will always cause major chemical transformations on its planet, and these changes will create new challenges to survival. At these junctures, life will either die out or learn to love the new reality. The best example here is oxygen. We are all extremophiles living in an oxygenated world that is hostile to organic life. But that which doesn't kill us only makes us stronger. Oxygen could have done us in, but instead we learned to breathe. We evolved the capacity to use the highly reactive nature of oxygen to charge our chemical batteries like never before.

One critter's poison is another one's fuel. Life is opportunistic and resourceful, and it's a fine line between a dangerous toxin and a bountiful energy source. Life, through adaptation, can cross this line. For this reason, I think we have to be extremely cautious in declaring planetary environments with an excess of "dangerous" energy or "caustic" chemicals to be uninhabitable. The very property that make these conditions dangerous to us, the powerful ability to induce chemical reactions, could turn them into a gravy train for some clever thing.

The only places that are surely uninhabitable are those where absolutely nothing is happening. I'm not talking about Pauls Valley, Oklahoma. I mean places where really nothing is going on, where there is no chemistry, no flow of matter or energy, where everything is just sitting there at equilibrium, at rest. Such places are safely declared dead.

SOMETHING OLD, SOMETHING NEW

What about the recent discoveries of extremophiles living on Earth? How do they really change the equation for ET life?

Recently in a used-book store a few blocks from the Natural History Museum in Manhattan, I found, tucked away on a high, dusty shelf accessible only with a rickety ladder, an original edition of *Biography of the Earth* written in 1941 by George Gamow, the Nobel Prize–winning physicist and prolific author. The price printed on the cover is thirty-five cents, but it cost me two bucks (seventeen cents in 1941 dollars). As I read it, the brittle pages have been disintegrating in my hands, leaving a trail of little yellow crumbs wherever I go.

Gamow begins a section entitled "Conditions for Life on the Planets" as follows (I need to hurry before the page crumbles completely):

> When we discuss the possibility of life on other planets, we come to a delicate point, for we do not actually know what life is or what forms of life different from those on the Earth are possible. Life in any form is doubtless totally impossible at the temperature of molten rock, or at absolute zero, at which all materials become quite rigid, but these are extremely wide limits. If we restrict ourselves to the ordinary forms of life found on the Earth, we can narrow these limits roughly to the temperature range in which water, the most essential constituent of organic structure, remains liquid. Some bacteria, of course, can stand boiling water with impunity for a time, while polar bears and Eskimos live in regions of perpetual frost.

These 1941 ideas about the limits of life would not seem out of place in a modern astrobiology journal. We still recognize, as Gamow did in the forties (and others suggested in the seventeenth century), liquid water to be the magic elixir for our kind of life. In some ways, the message of extremophiles is "no surprises." All life we know, no matter how freaky in other respects, is still based on organic molecules dissolved in water, and we all use the same basic cellular machinery.

Extremophiles haven't fundamentally changed the way we think about strategies to look for life, but they've bolstered the optimism with which we search. Right now anywhere with liquid water is considered a possible habitat, and this guides our quest.

What we have learned from the extremophiles is that life is much more adept than we ever thought at adapting to use different chemical sources of energy. The microbial world, in particular, has found ways to gain energy by trading electrons between a surprising variety of molecules that trickle through the recesses and backwaters of our world. We've found creatures exploiting the energy released in reactions between different combinations of hydrogen, methane, iron, sulfur, manganese, and nitrogen. The big surprise in the world of the extremophiles is really this wide range of chemical fuels.

Each one of these power sources required evolutionary innovations. That this has occurred many times on Earth probably tells us something profound about life. Once started, life can make huge leaps, fundamentally changing its energy-extracting metabolism while conserving its systems of heredity and self-construction. Like modular homes that can be run on AC electricity or go off the grid and use wind, solar, hydro, or geothermal, living cells on Earth have learned to run on an astonishingly diverse range of chemical sources. If need be, we can become whole new kinds of beasts feasting on energy sources poisonous to our ancestors or distant relatives.

Extremophiles teach us that on a planet with rampant disequilibrium,* life will find multiple chemical systems to feed off. It must be "easy" from an evolutionary perspective to switch to new chemical food sources. This tells us that once life gets started on a planet with deep and long-lived energy sources, it can survive drastic environmental changes.

Extremophiles, like the rest of us, are carbon-based. They all use DNA and proteins. The lack of other chemical architectures even in a biosphere that has found diverse energy pathways probably means something. It could mean that carbon is the only way to live in this universe—at least on a planet like Earth. Or, it could just mean that once a certain basic architecture takes hold on a planet, others don't stand a chance.[†]

These are the kinds of clues that we sift through in search of general principles of cosmic biology. Yet, no matter how extreme some organisms may seem to us, we have studied only one form of life. It is only

---

*One with active chemical flows, where different chemicals coexist that will spontaneously react when mixed.
[†]The Microsoft effect again.

through exploration that we will learn anything about ET life—whether it is we or they who are the successful explorers.

WHO WANTS TO KNOW?

As it has on Earth, life elsewhere is likely to mimic the overall trend of Cosmic Evolution, with aggregates of smaller bodies coming together to form successively larger and more complex objects and organisms. Multicellularity seems obvious and inevitable. So, we are rankled that cells took billions of years to make this step on Earth. Life seems to have passed through some kind of bottleneck before it could proceed from microbial forms to larger, more complex organisms with highly differentiated bodies and breakable hearts. Can we draw sweeping conclusions from this historical observation? Is this good evidence of a universal obstruction on the road to complex life, or could it be just an extended rut that life accidentally stumbled into here that has no bearing on the probability of complex life beyond Earth? It has often been conjectured that this history might mean that planets with simple life are common in the universe, but planets with more complex life are quite rare.

For those of us who worry about the odds of complex life existing elsewhere, that such complexity came so late in the game is one of the most widely interpreted facts, perhaps the most overinterpreted fact, of Earth history. Many learned opinions regarding the prevalence of complex and intelligent life in the universe rest largely on interpretations of this timing. "If complex life was a likely or inevitable product of biological evolution, it would have happened much sooner in Earth's evolution. Therefore it must be exceedingly rare in the universe and perhaps exists only on Earth." The great conviction with which this opinion is often expressed might be justified if we understood why this step took so long.

The final biographical fact that demands an interpretation is this: the development of intelligence happened so late in the game. Even though complex, multicellular animals have been around for 600 million years, we "got smart" only in the last few million years—just a nanotick of the cosmic clock. SETI theorists debate whether this fact is profound or mundane. Is intelligence, because it took so long to arrive here, an unlikely adaptation that might never occur on many other worlds with life?

When we ask ourselves this question, the first reply we should consider is "Who wants to know?" We have to watch carefully for any bias that may creep in because of our unusual place in history. Immediately after any major evolutionary shift, the shift will seem very recent. This statement is only saved from being a meaningless tautology by the word *seem*. The first autobiography written by any planet will always say, "Intelligence did not evolve until very recently—only in the last tiny sliver of time did we become able to tell this story." This will always seem a striking fact whether, on the planet in question, intelligence arose after 100 million, 1 billion, or 10 billion years. The first ones to discover cosmic time will always say, "The universe is so old, and we are so young!"

Yet, it's not "all relative." There is an absolute clock of sorts against which we can measure the arrival of intelligence on Earth and ask, "Fast or slow?" We can measure our progress against the lifetime of the Sun. If intelligence is always slow to develop, *compared to the life expectancy of the star powering its biosphere,* then intelligence could be a rare thing. No matter what planet you live on, your days in the sun are fleeting. It could be that most biospheres do not awaken to consciousness before the lights dim and the show is over. In this context, it is somewhat disconcerting that it took roughly half the expected life of the Sun (5 billion years out of an expected 10 billion of sunshine) for us to get to this point. If it had taken only twice as long, it would never have happened. Therefore, it is sometimes said, we are damn lucky to be here at all and may be the only ones around for thousands of light-years. Do the prospects for consciousness in our universe come down to a race between stellar evolution and the biological evolution of intelligence?

Some evolutionary biologists have argued that intelligence in the universe must be rare because there are no other examples on Earth. We are it. Dolphins and chimps notwithstanding, no one else has our cultural, communicative, and technological abilities. Among all the millions of species that have come and gone, and those that are around today, intelligence has evolved only once. If it was very useful or likely, then many species would have it. Therefore, most planets will not develop intelligence.

Am I the only one who finds this argument incredibly lame? The fact is, we haven't been here long. We speak these opinions as if we were the omniscient observers, looking over the whole stretch of Earth history and drawing final, sweeping conclusions. We discuss evolution as if it were a done deal, for which we were providing the wrap-up commentary, rather than an ongoing, unfolding process that we are bound up in. We are more like the first sunflower shooting up in a patch of thousands, opening our petals, and confidently declaring, "There are no other sunflowers and therefore sunflowers are extremely unlikely." The first intelligence will always be the only intelligence. Only if it is really not too swift will it conclude from this that it represents something that must be highly improbable.

RARE EARTH?

How unusual is the evolution of advanced life? Two books published in the last few years show the full range of answers that we can logically reach: *Rare Earth* by Peter Ward and Donald Brownlee, and *Probability One* by Amir Aczel. Each is written by credible scientists using rigorous arguments backed by mathematical equations. Yet, like the Epicureans and Aristotelians of ancient Greece, these thinkers reach opposite conclusions, as can be seen when we juxtapose the subtitles of the two works: "Why Complex Life Is Uncommon in the Universe" *(Rare Earth)* and "Why There Must Be Intelligent Life in the Universe" *(Probability One)*. Each is written as a polemic, putting opposite spin on the same set of observations to make antithetical points.

The focus of *Rare Earth,* as the title implies, is the many ways in which Earth is an unusual planet. These, it is argued, have uniquely qualified our planet to develop complex life. According to this view, a great many ducks had to line up in a most fortuitous way for the evolution that led up to us. If just one of many factors had been different, our planet would, at best, be a world of microbial slime. Since the odds are against such a harmonic convergence, most or all other planets must lack complex life.

Look at the myriad ways in which Earth is exactly suited for us: It is the right distance from the Sun, neither too hot nor too cold. It is just the right size to hang on to its atmosphere and oceans and retain enough internal heat to drive plate tectonics. Earth has enough water to maintain life, but not so much that the continents are entirely submerged. We are blessed with a perfect atmosphere to support the kind of life that has evolved here.

Several random historical events seem to have worked strongly in our favor. We were fortunate to acquire our large Moon in an early freak accident. The Moon helps stabilize our orbital tilt, keeping climate steady.

Not only is Earth exceptionally lucky (or rare) in all these ways, but we also grew up in a decent neighborhood. The Sun is the right kind of star for us—stable, long-lived—a good provider. Our solar system has orderly, stable circular orbits so Earth can count on avoiding nasty, chaotic orbital shifts that would doom us. Our good neighbors have pitched in to help us: Mars may have seeded Earth with life if we couldn't make it on our own. Jupiter, the big bouncer of the planetary club, guards the inner solar system. It is just the right size and distance from us to toss out most threatening asteroids and comets with its mighty gravitational pull. We even orbit a star that is ideally located within the galactic disk. Closer to the center we'd be threatened by radiation from other stars, especially dying, exploding ones. Closer to the edge, stars may not have enough heavy elements to make planets and living creatures.

Reading through this (nonexhaustive) list, it is easy to get the impression that only on a planet just like ours, in a planetary system just like ours, orbiting a star just like ours, located at just our position in a galaxy like ours, could such a wondrous thing as us come to be. There is something almost biblical about the way in which Earth seems to be so uniquely and improbably suited for our comfort.*

But, so what? Planets, like people, are all pretty strange when you really get to know 'em. I'm sure any slightly thoughtful creatures on Titan believe that a methane atmosphere is essential for complex life. The Plutonians are convinced that no life is possible above –200°F.

I do believe that Earth is a phenomenally rare and special place. Bless its round little head. But I don't agree with the conclusions the authors of *Rare Earth* draw from this fact. They make several crucial errors. One is the failure to fully recognize the role of life in *creating* the Earth's unusual character. The weirdness of Earth is at least as much a product of life as it is a precondition for life. *Rare Earth* ignores Gaia— by which, in this context, I mean all of the myriad, blatant, subtle, and interwoven ways in which life has made Earth what it is today. Is life a passive, helpless victim of cosmic forces? A passenger strapped in the

___
*In fact, *Rare Earth* has been accused of having a hidden creationist agenda (in David Darling's book *Life Everywhere*).

backseat while Earth drives us through the universe? Hell, no. Life is a major player in Earth's story. What seem, after the fact, to be lucky breaks in the creation of a planet perfect for us are actually a testament to the resourcefulness with which life has exploited geological developments and shaped Earth to its needs.

This misunderstanding of the role of life in manifesting Earth's idyllic state is just one problem with the "Rare Earth Hypothesis." The biggest error is the conclusion that "this is the best of all possible worlds." Why should we think that conditions on our planet are optimized for development of complex life? This leads me to propose the Rare Wookie Hypothesis. There is no other cat just like our Wookie.* An incredible set of circumstances had to occur for Wookie to become the best of all possible cats. A chance encounter between his parents in a Curtis Park alley, a roll of the genetic dice that endowed him with thick, dark gray fur and a tail dipped in white, a narrow escape from the dog that mauled his sister, a rescue, the party where I met him. What if the dog had gotten him instead? What if I had not gone to the party? What if Tory had never found him hiding in the *Encyclopaedia Britannica*? The chances of all this happening ever again are vanishingly small. Clearly we and Wookie are extremely lucky to have each other. This leads inescapably to the Rare Wookie Hypothesis: there are no other cats—at least none as perfect as ours.

The fallacy here is thinking there is only one way that complex life could have happened and only one set of circumstances that could facilitate it. Heartbreaking as it is to think so, if we didn't have Wookie, we might have some other fine kitten. If Earth had come out differently, it might still be a pretty decent planet. Hell, it might even be better. The Rare Earth thesis implicitly assumes that our planet is perfect—that any deviations from our evolutionary path would have had negative consequences for life. But we do not know that this is the best of all worlds. We don't even know if it's a particularly good one. Sure it seems great to us, but we've evolved for billions of years to take advantage of all of Earth's quirks and idiosyncrasies. Looking back at the many random events that led to a world where we feel perfectly at home, of course we say, "How fortunate for us." But we are the way we are because we've rolled with Earth's punches. Life and Earth have shaped each other.

---

*Wookie's hip-hop name used to be Fluff Daddy, but he recently changed it to Flea Diddy.

If saguaro cacti were intelligent enough to theorize about life else-where, but not smart enough to think about observational bias, they would conclude that most of Earth is uninhabitable since most of it is not desert. (But they would be comforted to see that more of it is becoming habitable all the time.)

While we're searching for other Earths, we should also be asking what a *more* habitable planet would look like. More habitable than Earth? Dare I say it? Won't lightning strike me down? What could possibly make a planet better than our perfect Earth?

People are always giving the Earth a hard time about its slow development: "Look at that stupid world!" they say. "They didn't go multicellular until their star was half used up. No wonder they're still in the primitive, factional, warring phase."

But, picture a planet that is similar to the Earth in many ways except that it formed originally with much less iron in the mix (perhaps orbiting one of those stars farther out in the galaxy having less of the heavy elements with which to make planets). Remember that on Earth iron hogged all the oxygen for the longest time and kept it out of the atmosphere, stunting our growth for billions of years. On this iron-depleted world, all other things being equal, oxygen would build up much faster in the air. If that's what complex life needs, on that planet they'd have been finishing up grad school while we were still microbes in diapers.

Another variable is the way planetary size affects atmospheric evolution. We do know that all planets are constantly losing hydrogen. The more hydrogen departs, the more oxygen is running free, not locked up in H_2O. Over long periods, this hydrogenous exodus helps planets become oxidized. Now, the smaller a planet is, the faster it loses hydrogen. So, other things being equal, smaller Earth-like worlds might become oxidized sooner. If high atmospheric oxygen content is the key to animal complexity, then Earth may not be the fastest out of the gate by a long shot.*

You could come up with countless other possible improvements, I'm sure. The examples I've just given assume a biosphere with essentially the same needs as ours, but an improved ability to meet them on time and on budget. What about life that thrives on very different condi-

---

*But, if a planet is too small, it might lose all of its atmosphere to space. That's what happened to the Moon and Mercury. Mars lost most of its atmosphere early on, rusted over, seized up, and, I suspect, dropped dead.

tions? Remember the extremophiles. Evolution is clever, or can pass itself off as clever through sheer persistence. Life is adaptable and resourceful. Those scary regions nearer the center of the galaxy are rich in alternative energy supplies. Who knows what grows there?

I'm not complaining about Earth. This is my favorite world by far. I'm perfectly comfortable here and for the most part I'm glad things have worked out the way they have. I'm just saying let's think twice before assuming that this is the only or best world that some kind of cogitatin' creatures could call home.

We're just getting to the point where we can start to make decent predictions about the likely evolutionary paths of other worlds. This is an area of active research, funded by NASA's exobiology and astrobiology programs, where meaningful new results should be forthcoming. Is Earth a late bloomer, a precocious planet, or typical? We don't know yet, but stay tuned.

In the next few chapters I'll look in more detail at what we actually know about other planets in our own solar system and the rest of the universe. As we survey the lives of planets, think about what a planet really needs for life to arrive and thrive. Which qualities of Earth are simply local oddities to which life has adapted? Surely fate has dealt the Earth some of the cards a living planet truly needs. But surely also the wonderful adaptability of life, and its ability to shape its own environments, combined with our lack of imagination and perspective, could make a random draw seem like a rare winning hand.

I've had discussions with Peter Ward and Don Brownlee, the authors of *Rare Earth*, and heard them speak. I don't think they are as certain about their pessimistic conclusions as you might think from reading their book.* They do provide a useful antidote to the wishful thinking about extraterrestrial life that dominates current thought. Many books reflect the current mainstream scientific belief in "the extraterrestrial life hypothesis"—a confident belief in what used to be called a "plurality of inhabited worlds." In *Probability 1* Amir Aczel reaches this conclusion using arguments based on probability theory. The bottom line, he believes, is that other intelligent life must be out there. There are simply so many stars that it doesn't matter if planets like Earth or

---

*My trickster wife has altered the dust jacket of my copy to say "Medium-Rare Earth," which may actually reflect the authors' views more accurately.

places with complex life are highly unusual. Astronomers have been making this case for centuries.

We won't find another biosphere exactly like Earth's, just as there'll never be another cat like our Wookie. But there are lots of cool cats and, if the logic holds, many places with complex life and intelligence. As Aczel writes, "If you give something enough of a chance to happen, it eventually will."

TEA AND SYMPATHY

What kind of universe do we live in? My sympathies lie more with the logic of plenitude than rarity. As the poster in Agent Mulder's office says, "I want to believe." However, given what we actually know about the universe, it is entirely possible that there is no other life. Even the Rare Earthers believe that simple, microbial life is probably spread throughout the universe. It's the jump to complex life that they feel may be unique to Earth. Given Earth's history, where life stayed simple for so long, we must at least accept this possibility.

For me though, that's not enough. I want company. A universe thoroughly infested with bacteria but devoid of thought is not one in which we can make any acquaintances. If microbes are widely spread among worlds, but more complex life is nowhere else to be found, then the answer to "Are we alone?" is "Yes." What we really want to know is if there is anyone else to talk to.

Now that we've found science, we fancy ourselves masters of the universe. Like a terrible two-year-old, we won't take "no" from the cosmos. Tell us, dammit! Cosmic Evolution is a pretty good story and getting better all the time. But now we want someone to compare notes with. So strong is our desire for certainty in the face of this grand mystery that we cannot resist the urge to vastly overinterpret every bit of potential evidence.

We are like desperate lovers waiting by the phone, fretting about the silence, then hanging on every nuance of every word. Our hunger leads us to find significance whether or not it is there. Like relatives at a séance straining to hear their dear departed knock on the table, we cannot bear the hush from the great beyond. We find signals in the noise of existence and read into them the conclusions that we want to reach.

In a recent issue of *Skeptical Enquirer,* a pseudoscience was described as a belief system that "begins with the desired answer and works backward to the evidence." Does astrobiology escape this description? Sometimes yes and sometimes no. Science is supposed to serve the truth, without prejudice or bias. If we are going to aspire to this ideal, then we must sometimes be able to say "we have no idea" about an important topic.

We obsess over the meaning of every twist in our narrative, but are we just studying tea leaves that settled at random? For now, we have to live with the uncertainty of not knowing. We can hold fast to unsupported opinions if it makes us feel better, but who do we think we're fooling?

While we wait, look, and listen, we can keep revising and studying our autobiography. Sustained study, contemplation, and revision of our story will help us build our intuition about the ways that living planets can evolve, and will sharpen our perceptions and our ideas as we search the cosmos for other life stories.

As the American astronomer Harlow Shapley (1885–1972) liked to say, "Mankind is made of star stuff." For 12 billion years, we've been riding the cascading energy flows of gravity and starlight, haltingly pulling ourselves together into conscious beings. I like to think we'll share these realizations with creatures on other worlds, born of the same expanding, cooling, coalescence, who ponder their own histories and wonder whether or not *we* exist.

The Lives of Planets

The secret of good science fiction is that the author should be free to invent anything he or she can think of, providing no one can prove that it's wrong. No one can prove that intelligent creatures are not swimming in the planet-wide ocean of Jupiter. No one will be able to prove it until we send spacecraft into that vast unknown sea.

—BEN BOVA

Why does every fucking poem mention the ocean?
—ERICA JONG

THANK YOU, LUCKY STARR

I have a confession to make. Part of the reason why I'm in this planetary game stems back to a strange and powerful vision I "received" back in the fifth grade, when I happened upon an interplanetary time vortex transmitting extraordinary extraterrestrial visions from the 1950s. This information was communicated to me across time from the brain of Isaac Asimov, using a primitive device called a "typewriter." His Lucky Starr series, full of obsolete but utterly captivating ideas from pre–space age astronomy, provided my earliest vivid awareness that planets are places. It was also my first serious literary addiction.

These interplanetary adventures featured the exploits of David "Lucky" Starr, space ranger. Lucky, working for the benevolent and powerful Council of Science, was a detective, scientist, and action hero. His sidekick, Bigman, was a tough little Martian. Though they wound up in some tight spots, Lucky and Bigman always managed to blind the crooks with science and reason, ultimately insuring that peace would guide the planets. I eagerly devoured Lucky's battles with saboteurs under the big Sun of Mercury, terrorists on Mars, pirates out among the asteroids, interstellar invaders in the rings of Saturn, and mind-controlling aliens beneath the oceans of Venus.

I was turned on to the Lucky Starr books by my grandmother Sally G, a.k.a. Grammy, an old friend of Asimov's. They had worked together at Boston University where she was a reference librarian and he a biochemistry professor secretly spinning science fiction yarns. For a time he hid these from the academic thought police by publishing them under the pseudonym Paul French. Grammy was quite fond of him and took great pride in his success. For decades after my Lucky Starr fixation, she continued to give me a copy of every new novel Asimov wrote up to his death in 1992—which added up to quite a shelfful. Shortly before she passed away in 1995, her ninety-five-year-old body running out of gas, but her mind still sharp as a tack, Grammy gave me her worn copy of *The Stars, Like Dust* (1951), Asimov's second novel, which bears the inscription "To Old Lady Green Spoon. Love, Isaac."

My parents and the Asimovs ran with the same crowd, so I had the good fortune to meet him. Isaac was a kind of adult that kids love: he poked fun at everything in a conspiratorial way that made you feel like you were in on the joke. Asimov stayed on the scene as a writing machine for over five decades, tirelessly using science to enlighten and entertain the masses. He was a devout positivist, a true believer in the power of rationality and science to lift humanity out of the gutter, and his fictional heroes were always champions of science. In this he felt part of a movement, a ringleader of the closest thing to a religion I grew up with—rational humanism. The optimism inherent in planetary exploration, and the possible futures it could help create, seemed an integral expression of these ideals. For me, it all started with Lucky Starr.

For Lucky's adventures, all written between 1952 and 1958, the arena was the entire solar system on the eve of real planetary exploration. Science fiction writers and fans speak reverently of a golden age that began in the thirties and ended roughly (and not coincidentally) when the space age began. Asimov's planet-conjuring was informed by, and helped to inform, the latest ideas of astronomy. He was one of the architects of the planets of the golden age.

Some of these golden age planets seemed so complete and compelling that it has been hard to leave them behind. Back then it was widely accepted that Mercury spun on its axis perfectly in time with its orbit, always keeping the same face to the Sun, as the Moon does to the Earth. This made for a planet one-half of which was always broiling day and the other perpetual frozen night. In between was the twilight zone, a frequent fictional setting for indigenous life or human settlements. Now we know that Mercury doesn't really have one side locked to the Sun in this way. All sides are gradually rotisseried.

Even tougher to lose were the deep, warm oceans of Venus, where Lucky battled monstrous jellyfish and telepathic frogs. When *Mariner* 2—the first human-made spacecraft to successfully visit another planet—flew past Venus in 1962, just a few years after Lucky's exploits there were written, the fabled, fertile oceans had vanished without a trace, leaving a parched, red-hot surface. When *Venera 9* landed in 1975, there were no jungles or swamps, just rolling volcanic rock and dust. Science can be cruel to our fantasies.

As we rocketed farther into space, we gained our first close-up peek at each planet. In dramatic, often anxious, moments of revelation, one by one the beloved worlds of the golden age became antique visions, imaginary yet familiar places solidly grounded in obsolete science.

We are just now finally learning about the existence of other planets, scores of them, orbiting other suns. It will be some time before our knowledge of any of these new worlds approaches even our 1950s-era knowledge of our own solar system. In the meantime, science fiction writers ought to have a field day connecting the dots and filling in the blanks.

CONNECTING THE DOTS

Before this age of exploration, we were not entirely clueless about the planets. We had blurry photographs and measurements of temperature and atmospheric thickness. But our acute desire to know the other worlds has sometimes led us to overinterpret our pictures, constructing images of planets that, in some cases, seem as though they should really exist. Perhaps they do, elsewhere. If you believe in the principal of plenitude, then somewhere there is a Mercury with one side locked toward its sun and a bio-ring girdling its temperate longitudes, a Venus with warm seas, and a Mars cut with canals and adorned with shrubbery. Someday, some bored million-year-old cyborg will build this golden age solar system out of fallow molecular clouds for a science fair project.

When we started going to other worlds, we got a reality check on hundreds of years of science and fantasy. We learned that the most rigorous of scientists and the most imaginative of storytellers had consistently tended to picture the other planets as much more like Earth than they really are.

Planetary qualities have been assumed Earth-like unless proven otherwise. When observations force us to accept a more deadly or alien environment, we still find visions of ancient, destroyed Earths buried beneath wasted landscapes, or future Earths waiting latent beneath the skin of dead or dying worlds. Home is where the heart is, and wherever we travel, Earth shines back at us, the beginning and the end of all our journeys, the standard by which worlds are judged.

This tendency to view the planets as minor variations of our home world, rather than accepting them on their own terms, has always been especially pronounced for those two that flank us in the solar system. Venus and Mars, the nearest worlds to home, are our closest siblings both as the crow flies (assuming the crow has a pressure suit and a rocket pack) and in physical resemblance.

Believing is seeing. We often saw what was familiar and comforting and missed that which was truly alien. For centuries before the space age astronomers knew that Venus had bright clouds and a thick atmosphere. These facts encouraged scientific fantasies of a warm, water world—a twin Earth. Through telescopes our aqueous eyes saw aqueous clouds. When we finally sent our robots, we were shocked to find clouds of strong acid shrouding a scorched earth where no water could condense.

What we knew of Mars also once seemed encouraging. We've known for centuries that Mars really does have a "normal" length of day, rotating once in just over twenty-four hours. This astonishing coincidence encouraged us to think of Mars as a modified Earth. It was long assumed that the bright polar caps easily seen through telescopes were made of frozen (you guessed it) water, seasonally melting and refreezing. Biologist Joshua Lederberg, writing in *Science* in 1960, expressed the mainstream view when he wrote, "The most plausible explanation of the astronomical data is that Mars is a life-bearing planet." Now we know that the ice caps are largely frozen carbon dioxide (dry ice) and not a drop of water can persist in the frigid, wispy Martian air. On both Venus and Mars we saw white and inferred green.

BEGINNING OF THE END

Real comparative planetology first became possible on December 14, 1962, when *Mariner 2* whipped past Venus and radioed home disquieting news about the surprisingly sweltering and utterly dry surface. The *New York Times* ran an editorial entitled "Venus Says 'No' ":

> The finding of extraterrestrial life in some form similar to that on earth, even at the lowliest stage, would lend support to the widespread belief—rooted deeply in the aspirations of mankind—that life as we know it is not unique to this insignificant corner of the universe, but exists in many other systems similar to ours throughout the universe. Indeed, there has been speculation among scientists, philosophers and poets that some of these systems have reached a stage of evolution much superior to ours. The message from Venus now reduces hope of finding evidence in support of this speculation to one half, so far as our solar system is concerned.
>
> Mars now remains our only hope of turning this universal dream into reality, and the evidence so far is not very encouraging. The message from Venus may mark the beginning of the end of mankind's great romantic dreams.

And so an era of romanticism ended, and an age of exploration began. It was hard to give up on our relatively unconstrained fantasies, but the exciting prospect of voyages to unknown lands and an unending stream of new discoveries helped ease the pain.

FIELD OF DREAMS

Each new planetary mission reached farther, replacing vague expectations with clear glimpses of a reality that had been waiting patiently since the world was made. For my generation, the planets have, one by one, gradually been transformed from unknown arenas for speculation and fiction to places we have visited, photographed, scratched, and sniffed. Sometimes I have to remind myself that this is not a normal part of growing up.

The first generation of planetologists, the people who designed the initial exploratory missions to the planets and who also trained my generation, were not themselves trained in planetary science. They had to invent it. Our impending ability to travel to other worlds, and the need to interpret what we would find there, had forced the question "Who knows how to study planets?" The answer was astronomers, geologists, meteorologists, physicists, and chemists. All those who had been describing the Earth and the heavens as seen through their own disciplinary lenses would now have to work together to figure out the planets.

This was a great challenge because each field has its own subculture and language. The sad truth is, it is not just the public, the "layman," who can't understand what scientists are saying. Quite often, we cannot understand each other. Sometimes, however, a task arises that demands that we leap out of our disciplinary trenches, into a no-man's-land where the view is wider. In the early sixties, the imminent stream of new information flowing toward Earth from the other planets created a sudden need for a new, multidisciplinary framework. All the blind men, accustomed to studying the terrestrial elephant from their own perspectives, would have to get together and make sense of the new beasts in the planetary zoo.

Out of this need, a synthetic culture of planetary science was somehow made. Today, many academic departments call themselves Planetary Sciences, and their faculty ranks are filled with a growing number of people with planetary science degrees. A steady stream of newly minted Ph.D.s are being churned out, waiting to inherit their mentors' offices. We have become a field.*

This young science is the best example I know of a truly interdisciplinary field, a distinction that carries special challenges and joys. I loved going to grad school in a department where students both scramble over rock outcrops on geology field trips and spend nights on the mountain in telescope domes. Meetings are always an adventure, with topics ranging from the atmosphere of Pluto to the craters of Mercury. Who wants to be a specialist? At a planetary conference you have to describe your work using language that can be followed by a general

---

*Also sometimes called planetology or comparative planetology.

science audience. Since we are always having to explain our jargon to each other, I think that planetary scientists are on average better communicators and educators than scientists from most fields.

Planetary science is a field born from our realized dreams of exploring new worlds. In just a few decades, the planets have magically metamorphosed from distant objects glimpsed through telescopes into places we've actually visited. We now know enough about our sibling worlds to do meaningful comparisons of planetary functioning and evolution. We are placing the Earth and its life in a cosmic context that Copernicus and Galileo could scarcely have imagined. We can begin to assess the potential of our universe to create other habitats for life.

COMPARATIVE PLANETOLOGY 101: THE RULES

Is our planet normal or some kind of freak of nature? We desperately need context for the Earth story. The first thing to do is visit the neighbors and see what they're like. So what do we know about the lives of the planets? What made them what they are today? Were their fates preordained, bred in the bone? Or was it experience and happenstance that gave each its own unique character? Through comparative planetology we're seeking an understanding of the similarities and differences among worlds.

What patterns emerge from close study of the busy mess of the solar system? We want to explain why planets are the way they are, based on "first principles": on simple rules and conditions of birth. At first glance, this deterministic goal is not unlike the goal of astrology—to predict your personality from planetary positions at the time of your birth.

I doubt that the other babies born at the Newton/Wellesley Hospital on winter solstice 1959 all grew up to be just like me. But can a planet's position at birth determine its own personality and fate? It's easier to see how there might be a connection. In fact we do find that two variables, each set at birth, have a huge influence on planetary destiny. These two qualities are *size* and *location* (distance from its star).

Size controls both gravity and internal heat. The larger a planet is, the more easily it can hang on to both the matter and energy it was born with.

The story of planets is largely the story of their "thermal evolution": how they acquire, store, and lose heat. All planets emerge hot from the oven of accretion. From the start, they are balls of excess energy left over from the violent collisions that assembled them. It's all downhill from there. Like steaming, freshly poured bowls of porridge they slowly cooled off at different rates, depending on their size. This is true about anything that is heating or cooling—smaller things exchange heat with their surroundings faster than big things. On a cold winter walk from the café back to my office, a small coffee cools off faster than a grande.

The same is true for steaming young planets: larger servings cool more slowly.* A larger planet shields its own interior from the cold of space and so hangs on to its "energy of accretion," its initial heat of birth, longer than a small one. The bigger the planet, the longer it stays hot inside.

The interior of the Earth is a giant heat-engine. As the heat locked inside finds its way out through convective churning, it creates new crust, sucks old crust back into the cauldron, and pulls the continents around, building mountains and driving earthquakes and volcanoes.† The level and type of geologic activity on Earth is a direct manifestation of the amount of heat bubbling up from the interior. You might expect a smaller planet with a lesser flow of heat to be a less happening place overall. Indeed, that is what we have found.

As we look around the solar system, we see a clear relationship between planetary size and surface age. The bigger worlds are hot and vigorous inside, and this is reflected in surfaces that are more recently active. Thus larger planets have younger surfaces.

The Moon is relatively tiny compared with Earth (2,200 miles in diameter versus 7,900 for Earth), and not surprisingly, it is cold inside and has been for billions of years. No plate tectonics or active volcanoes on the Moon.

Mercury has a diameter of 3,000 miles, larger than the Moon but still quite small as planets go. In appearance it is quite lunarlike—an ancient, cracked, cratered orb. If it ever had vigorous surface activity like Earth's it was only for a brief time in its molten infancy.

How do we know? You can tell the age of a planetary surface by counting craters: older, inactive surfaces are more pockmarked by the stray falling rocks of space. A surface full of craters is the signature of a long-dead world. Conversely, if you see no craters, that means the surface is

---

*This is also the reason why babies need more protection from the cold than linebackers.
†This is the interconnected global system we call plate tectonics.

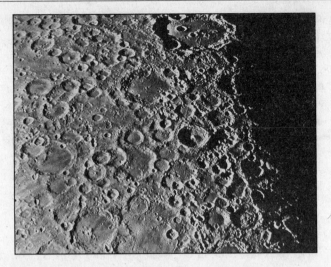

The heavily cratered highlands of Earth's Moon. Stone cold, and old.
(Lunar and Planetary Laboratory, University of Arizona)

young, and some kind of recent activity (volcanoes, erosion, earthquakes, etc.) has refreshed its appearance and wiped away the craters.

Mars is bigger than the Moon (diameter, 4,200 miles) but still much smaller than Earth. Mars shows signs of a much longer geological life than the Moon. In keeping with its larger size, after birth it remained hot for much longer. Mars, however, is long past its prime. Whatever vigorous geological activity it had is now a distant memory. The signs of this illustrious past are scattered about the surface, being slowly dented with craters, buried in dust, and scoured by the winds.

Venus and Earth are remarkably similar in size. Venus, at 7,500 miles in diameter, weighs in as the slightly smaller twin. Until recently we didn't have much of a clue about the age of Venus's surface. Now, with *Magellan* images, we have counted every crater on Venus and learned that the average age of the surface is less than 1 billion years old, making it, next to Earth, the youngest place around. Nowhere on Venus do you see signs of the ancient heavy bombardment that saturated large areas of Mercury, the Moon, and Mars with craters. Venus alone stands with Earth as having erased all signs of this traumatic past with a long life of more varied, more recent experience.

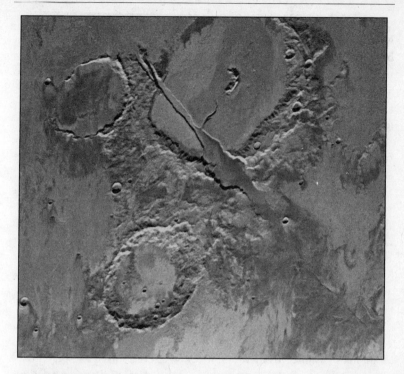

These ancient craters on Mars look very different from those on Mercury and the Moon. The obvious signs of erosion reveal a less ancient, more geologically active world. (NASA)

When you ask of a planet, "How hot is it there?" it's not the interior heat that concerns you, but the temperature you would experience standing on the surface—the climate. Planets need atmospheres to keep warm, and bigger planets generally have thicker atmospheres for two reasons. One is a direct result of the geologic forces I've been discussing. Geological activity doesn't just push rocks around and make mountains. It also gushes gases into the air. The more volcanically active a planet, the faster it supplies itself with new air.

The other way that large planets maintain atmosphere is through sheer brute force. Big worlds possess stronger gravity, which helps them hang on to their atmospheres over the long haul. The gradual loss of gas to space will always doom small planets to an airless existence.

For both these reasons (larger planets have more active volcanism, which supplies new air, smaller planets lose air more easily) we expect bigger worlds to have thicker atmospheres. What we've seen on our planet treks largely conforms to this. Mercury and the Moon have no atmosphere. Mars has a wimpy one—only about one-hundredth as thick as Earth's, so that the surface pressure on Mars is equal to that at an altitude of about 130,000 feet on Earth.

Going just by this logic, however, we would expect the atmosphere on Venus to be slightly thinner than Earth's due to its slightly smaller size. That is most definitely not what we find there. Instead, Venus has the thickest, heaviest atmosphere of any rocky planet around. This shows us that there is more to planetary character than just the size-dependent effects of gravity and internal cooling. Clearly, with the thick atmosphere of Venus we see the influence of something other than size coming into play.

To make sense of this difference we need to consider the role of *location*—the second major factor controlling planetary characteristics.

LOCATION, LOCATION, LOCATION

A planet's distance from the Sun (or whatever star it happens to be near) plays a key role in its birth, life, and death. Like people milling about a campfire on a winter night, planets closer to the Sun are hotter and farther ones are colder. The average surface temperatures on Venus, Earth, and Mars, in degrees Fahrenheit, are 864, 59, and −67. No surprises at first glance: temperature falls with distance from the Sun. However, when you look at the data in more detail, it is not quite that simple. The difference in temperature between Venus and Earth is much greater than you can explain by their differing positions with respect to the solar heat lamp. Something else is going on. It is the dense air of Venus that keeps its surface so intensely hot. But it takes more than just a massive atmosphere to keep a planet warm. The air has to be made of the right mix of gases.

An airless planet absorbs the warmth of sunlight and sends it right back into space as infrared (IR) radiation. Planets with atmospheres try the same trick, but something gets in the way. Certain atmospheric gases act as selective filters that, like bouncers at a snotty club, let the svelte little visible rays of sunlight pass right on by, but block the heftier

IR rays from passing through. These bouncers are the unfairly maligned, infamous greenhouse gases.

What makes one molecule, such as CO_2, a potent greenhouse gas, while another, such as O_2, is not? Certain molecules find light at infrared (IR) frequencies irresistible, like music with a certain beat that compels you to dance but leaves others cold. These molecules dance so vigorously that the energy of the radiation is completely absorbed, transformed into molecular vibrations. Other molecules, in the presence of the same infrared music, just sit there.

It's all in the size and structure: how big and loose a molecule is determines how well it absorbs IR. A molecule with only two atoms trying to absorb IR is like a rhythmically challenged person trying to dance to reggae music. If you can't shake your hips and swing your arms, then you might as well have a straitjacket on. These uptight little molecules have nothing to move, no "degrees of freedom" as we say in the biz. So, for example, little diatomic (two-atom) molecules such as oxygen and nitrogen (O_2 and N_2) cannot absorb infrared. They may want to dance but they're not hip to the right vibrations, and the IR passes them by like the little stiffs they are. Slightly larger molecules, with three or more atoms, such as carbon dioxide, sulfur dioxide, methane, and water (CO_2, SO_2, CH_4 and H_2O), have a lot more stuff to shake, and this makes them all strong IR absorbers. Pass some IR through air that's thick with these willing dancers and it's easy skankin': they just bounce, bounce, bounce, until the IR wave (or particle or whatever the hell it is) is history and its energy is lost to the vibrations.

We have another word for all that bouncing about of molecules. We call it heat. The more of the big, flippy-floppy molecules in an atmosphere, the more they intercept the IR trying to escape into space. In this way, the triatomic molecules (plus fat, wobbly methane) catch IR and harvest its heat for the planet. Those gases that dominate Earth's atmosphere right now, oxygen and nitrogen, do not dance. With only one bond each, O=O and N≡N are stiffer than boards and wouldn't notice an infrared photon if it walked up and bonked 'em on the head.

The atmospheres of Venus and Mars, on the other hand, are made almost entirely of CO_2 and so, pound for pound, they are much better at holding on to solar heat and warming their planets. In the case of Mars's wispy air, that doesn't add up to much warming. But it keeps Venus the hothouse of the solar system.

Molecules with more atoms and a more floppy structure make better greenhouse gases because they have an easier time vibrating to the infrared music. (Borin Van Loon)

As you know from the debate about global warming on Earth, when you start to look closely at planetary climate, it gets a lot more complicated and hard to predict. Gases don't just absorb radiation, they also make clouds. Clouds reflect sunlight into space, cooling a planet. Clouds can also absorb infrared radiation, warming a planet. The balance of these effects, whether clouds have a net heating or cooling effect, depends on such details as their altitude, composition, droplet size, and extent of coverage. These things depend on the mix of gases and atmospheric motions. Atmospheric motions depend on temperature differences, which are driven by radiation absorption in the air and the clouds. It's a tangled web.

Climate is complex. The reason we care so much about it is the close relationship between climate and habitability. Life, as we know it anyway, is based on temperature-sensitive chemical reactions, and so it can only exist within a narrow range of climate conditions.

Astrobiologists talk about a "habitable zone," a range of distance from a star within which temperatures are moderate enough to maintain liquid water (widely taken to be the elixir of life) at the surface of a planet. But, a planet's climate is not determined by distance from the sun alone. Location is clearly important, but any two planets at the same distance from their sun need not have the same climate. The boundaries of the habitable zone will shift inward and outward, toward or away from the sun, depending on just what kind of planets we are talking about.

We would like to have one theory that explains the climates of planets in some predictable way, telling us which ones will be likely to support life. Increasingly, however, it looks as though a lot depends on accidents of birth. Size and location are both important, but we still cannot predict a planet's fate without knowing more details about its birth and later life. In fact, when it comes to planets, the power of prediction may be too much to hope for. We are challenged simply to reconstruct their histories and explain how they got to be the way they are today.

ROLLING THE DICE

We generally assume that the inner, rocky planets all started life much more alike than they are now. We do have some evidence that supports this, but it is mostly indirect. Our belief in similar origins is largely based on our unconfirmed theories about how the planets formed.

Recall the story, described in chapter 5, of planets forming out of a disk-shaped nebula around the young Sun. The pattern of temperature variation through the original nebula led to a predictable distribution of planet-forming materials. Rocky planets formed in the hot regions near the Sun, and icy, gaseous planets formed farther out.

This is actually a gross simplification of a much more detailed theory that uses chemical calculations to make precise predictions of what each planet should be made of based on its distance from the Sun. This theory, called equilibrium condensation, can predict exactly what metals, minerals, ices, and gases should be available to make planets at any given temperature.* Then, using an educated guess about the temperature distribution of the preplanetary nebula, you can predict† exactly what raw materials went into making planets at any position in the solar system.

John Lewis invented this theory and worked out its implications for planetary formation at MIT in the 1970s. Lewis became well-known for this work while I was still in grade school. A decade later he was my doctoral thesis adviser at the University of Arizona.

When we started to understand more about the process by which planets actually assembled themselves, it caused trouble for this theory. It seems that planet formation is a much messier process than we once believed. Faster computers have allowed more realistic simulations of planet formation. We now believe that the final stage of planet making was a time of anarchy in the solar system. The growing gravitational power of all the young worlds caused them to throw each other around, displacing orbits every which way. In a few million years, the orderly pattern of chemical zones arranged with distance from the Sun, as neatly predicted by theory, was smeared out in a wild rumpus of giant collisions and orbital chaos. In the final stages of planetary growth much of the beautiful order created by the laws of chemistry got smooshed out of existence by the laws of orbital physics.

---

*If you make certain simplifying assumptions, you can basically calculate what molecules should be present, given a certain mix of elements, at any temperature. These are known as chemical equilibrium calculations.

†You may find it strange that we use the word *predict* to discuss events that happened a long time ago, before there was even an Earth. What we mean by *predict* in this context is "show the logical necessity of this outcome based on prior conditions." It may seem like a cheap psychic's trick to predict that which has already occurred, but we can test this type of prediction by seeing what else that has not yet been observed is also predicted by the same theory, and looking for these signs to confirm or reject our theory.

The dream of a universal theory to explain the nature of all the planets still eludes us. If such rules worked well enough, then, by understanding the formation of our own solar system, we would have the tools to predict the structure and properties of other planetary systems, from basic conditions at birth. This would allow us to say with some confidence where life like ours might evolve. Yet, it may be that the nature of planets is inherently resistant to such schemes, because the planets in their formation did not follow simple rules.

Even so, when the planets were made, some larger order underlay the chaos: the temperature-dependent segregation of materials does seem to explain the basic, large-scale structure of our solar system. Perhaps it can't predict the detailed differences among the rocky planets any more than an astrologer can tell me when I'll win the lottery, but it does make sense of the overall groupings of planets, the major architectural features of our planetary system. We can explain why we have worlds of rock and metal clustered near the Sun where these materials could stand the heat, and why we find ice moons and gas giants roaming the more distant, frosty regions of our system. This less ambitious application of equilibrium condensation remains the closest thing we have to a universal theory of planetary formation. We won't know how good our theories are until we get to examine a number of other planetary systems in detail. If our current theories are correct, then we would expect other planetary systems to conform at least to these more crude structural principles: little rock-worlds orbiting near a star, giant gas-worlds farther out.

We used to think that the solar system was sort of like a chemistry experiment. I always loved chemistry because, if you knew the original conditions and the rules, then you could predict the outcome. Instead, it seems that the solar system is more like a mix between a chem lab and a game of craps, a chemistry experiment where certain steps of the recipe were left up to chance, so that the amounts of certain ingredients and the steps included in their physical handling were specified by repeated rolls of the dice.

What we are doing with comparative planetology is examining the final results of such an experiment, and trying to figure out how much randomness was in the mix, and what structure we can discern beneath the muddy waters of chance.

When discussing the nature and fates of worlds, then, we have to add a third variable to the two, size and location, we have discussed. We have to add the influence of *luck*.

ODD BALL

Earth's peculiarly oxygenated atmosphere was made as pollution by photosynthesizing life. Here, plants used sunlight to turn CO_2 and water into oxygen and organic food. Life took over the cycles of carbon, nitrogen, sulfur, and water that dominate the planet's activities. The signs of life are so clearly evident in the makeup of our atmosphere and its differences from that of our CO_2-dominated neighboring planets that it is obvious that they do not possess our kind of life. This doesn't mean, in itself, that these planets are lifeless. But, if we ask, "Has life played the role on Venus and Mars that it has played on Earth?" the answer, my friend, is blowing in the carbon dioxide wind: an unambiguous no.

It is tempting to regard the comparative histories of Earth vs. its neighbors as a controlled experiment showing the effects of life on a planet, like a set of identical sterile petri dishes where living cells were injected in only one. Especially in comparing the lives of Earth and Venus, so similar in other respects, it sometimes seems like a case of identical worlds, where life was added to one, and 4 billion years later, voilà, you can observe the consequences.

If only it were that simple. The universe has not given us very many nearby planets to study, and our expanded appreciation of the role of accidental, large impacts and orbital chaos limits our ability to untangle the complex web of causality shaping the lives of planets.

Now we've seen how size, location and luck affected the birth and infancy of the worlds we know best, and sent them down their separate paths. Venus and Mars are Earth's only close siblings, but these three have all have gone their separate ways. Perhaps if we can understand the divergent life stories of these triplets then we can start to picture the environments of Earth-like planets elsewhere in the galaxy, and to ponder the prospects for life in such places.

11 | Venus and Mars

The Earth? Oh, the Earth will be gone in a few seconds . . . I'm going to blow it up. It's obstructing my view of Venus.

—MARVIN THE MARTIAN

The cities a flood
 And our love turns to rust
  We're beaten and blown by the wind
  Trampled into dust
  I'll show you a place
  High on a desert plain
 Where the streets have no name
—U2, "WHERE THE STREETS
HAVE NO NAME"

HOLDING WATER

What combination of fate and circumstance left Venus a dry, scorched, volcanic pressure cooker, Mars a frozen, windswept, barren desert, and only Earth a warm, wet, living oasis?

Remember, Earth was born in steam. At the same time, baby Venus and baby Mars arrived, also swaddled in thick, steamy air. They were littermates, fresh rocky worlds coalescing out of a single swarm of tussling planetesimals. Almost surely, they all began life somewhat waterlogged. How come only Earth has managed to remain so? Why did Venus and Mars lose their water while Earth retained what seems to us a healthy amount? The answer may lie in the different ways that each responded to the hot, steamy birth experience.

Recall how the young Earth had to fight to keep its water. When our planet's water was all puffed up in steam, it was helpless against the various forces ripping, stripping, and blasting our atmosphere off into space. Once the planet cooled enough, the water rained down into oceans that were much more secure against these atmospheric assaults. Young Venus and Mars suffered the same attacks, but each was ultimately much less successful—Venus because of location, Mars because of size—in holding off the falling rocks and solar breezes trying to steal its water.

Because of its lower gravity, little Mars was defenseless during the late stages of the primordial bombardment. This drubbing stripped Mars of its earliest atmosphere. Mars had surface water during its early history, and we know ice is locked beneath its surface today. But it probably never had more than a small fraction of Earth's original bounty, due to the ease with which a small planet can lose its early steam atmosphere.

And what of the fate of water on Venus? Our sunward sister does not have the Martian excuse of being too small to hold water. Yet Venus, although remarkably Earth-like in size, is seriously lacking in oceans or even puddles. Blame it on location. Where Venus sits, sunlight is twice as bright as on Earth. These twins may have started out nearly identical, but baby Venus got too much sun and became dehydrated. While Earth was cooling and enjoying the first rains after eons of choking steam, the water of Venus, heated by the nearby Sun, remained as steam. In the longer steam phase on Venus, much more water was lost to space. Eventually, rain came to Venus as well, and the remaining water condensed out on the surface, but we don't know how much water was left at that time. All we know is that now she's hardly got any. Four and a half billion years of evolution have left Venus a thoroughly desiccated place. Given all the similarities between Earth and Venus, and given the importance of water in determining so much of Earth's character and habitability, it is stunning that Earth today has about one hundred thousand times as much water as Venus.

Venus must have had a fair amount of water left when the rains finally came. There should have been oceans. There should have been seas, lakes, and warm little ponds. What happened to this water? It got hit by a runaway greenhouse. Again, location was key. The oceans of Venus, doomed by her proximity to the Sun, met a fate that awaits our own oceans in the distant future.

What causes a runaway greenhouse? Picture a planet like Venus with warm oceans, heated by twice the sunlight hitting Earth. Hot water evaporates, adding more vapor to the air. Water vapor is one of those floppy greenhouse molecules that absorbs infrared and helps hold in a planet's heat. The resulting greenhouse effect heats the surface, which evaporates more water, and so on. It's a positive feedback, a runaway train, and it will just keep getting hotter until much of the water is steam again. Then it is, once again, vulnerable to being lost to space. In the upper atmosphere, water is broken apart by solar ultraviolet, split into hydrogen and oxygen. Liberated from their clunky oxygen ballast, the fast and loose hydrogen atoms stream off into space. This, we believe, was the fate of most of Venus's remaining water.*

We don't yet know how long the oceans of Venus lasted. Our best models point back to hot oceans that persisted for hundreds of millions of years. It could have been billions. For a large portion of its history, our neighbor planet may have looked very much as portrayed in *Lucky Starr and the Oceans of Venus*.

One vital question that we can answer only through further planetary exploration is "Who kept its ocean longer, Venus or Mars?" We'll know the answers for Mars long before we do for Venus. Mars is easier to explore because the conditions there are less brutal on Earth-built machines, and Mars shows its entire history on its face. It's much harder to find evidence for ancient bodies of water (or ancient anything) on Venus, since big planets tend to eat their past. One consequence of the more vigorous geological activity on larger planets is that, above a certain size, they will always destroy their own rock trail as thoroughly as a politician with a paper shredder. The oceans of Venus may have been much larger and longer lasting than those of Mars, but the more active geology of Venus has erased all obvious signs of this watery time. One goal for future exploration of Venus is to search for more subtle traces of the lost water.

During the long oceanic phase, Venus and Earth may have been

---

*"What about the oxygen?" you might ask. Good question. The flight of the hydrogen probably left Venus with an oxygen-rich atmosphere. We should keep this in mind as we study the atmospheres of planets around other stars, since oxygen is usually cited as a possible "sign of life" on another planet (see chapter 14). Eventually, some oxygen might also have escaped into space, but the rest probably reacted with surface rocks, oxidizing various minerals churned up from the interior of Venus by the vigorous geological activity of that hot, young world.

nearly indistinguishable, only developing their individual quirks later in life. If we don't think that the origin of life on Earth was an unlikely fluke, then Venus should have had life. We may never know how far life evolved on Venus before she lost her oceans. When this catastrophic change came, the last drops of water disappearing into vapor, did life die out? Or did it adapt, as Earth life did repeatedly when faced with major global environmental catastrophes? If Venusian life wanted to survive the transition from warm oceans to hot CO_2, it would have had two choices: either find some new kind of metabolism, one that is not based on carbon and water and can thrive at nine hundred degrees, or migrate thirty miles up, into the cool clouds, and learn to love acid.*

VENUS INTERRUPTUS

Modern planetary exploration rudely interrupted our age-old dreams of an Earth-like Venus. The high surface temperature on Venus was the first significant discovery ever made at another planet with a visiting spacecraft. Venus, a metal-melting furnace with a corrosive atmosphere and clouds so acidic they could etch glass, was declared off-limits to life. The romance was over.

But (planets seem to be like this) every time we take a closer look, we see new sides to the place. Our understanding of the environment on Venus has again changed radically in recent years. In the 1990s we were finally able, with *Magellan*'s radar eyes, to peer through the clouds and map the entire surface in stunning detail. What we found was a much more varied and vigorous world than we had expected—a world where, as on Earth, the distant geological past has been consumed by the roiling present. Venus may be drier than Phoenix in June, but it's not dead yet.†

A big surprise is that almost the whole surface seems to have formed at roughly the same time. On most planets you can identify older areas and younger areas from global maps of impact craters. For example, on both Mars and the Moon you find ancient, heavily cratered highlands and younger (but still unbelievably ancient) volcanic plains with many fewer craters.

---

*I'll return to the subject of possible cloud life on Venus in chapter 17.
†I'm not sure I believe it, but some of my friends claim that, in my first invited conference talk, as a grad student, I nervously blurted out "Phoenix" when I meant "Venus."

On Venus craters are spread randomly around the whole planet. It looks as though almost the entire surface is the same age, and the total number of craters tells you roughly what that age is. Venus is accumulating a new crater about once every seven hundred thousand years. Since about nine hundred craters are on the planet, that means the surface is around 600 million years old.

We don't know of any other planet with a surface that formed all at once. What could have happened to suddenly wipe out all preexisting terrain? It sounds disturbingly biblical. Was God practicing on Venus? Is this catastrophic tale really the true story of Venus?

The evidence points to a period of massive volcanic flooding that covered most of the planet under thick flows of lava a mere 700 million years ago (give or take 100 million)—just last month in geologic time. In a striking case of planetary amnesia, almost all surface memories of older times on Venus have been lost.

Less than 1 billion years ago, the geology of Venus seems to have undergone a global pulse of rapid activity and then slowed to a crawl. This is disturbing. Venus on the inside should be just like Earth. Earth is stable and predictable. Isn't it? A billion years isn't all that long ago in planetary time. Why didn't Venus settle down a long time ago, as Earth did? Is there something we should know?

The answer may be that Venus oscillates every half billion years or so between spurts of furious global volcanic activity and long spells of relative inactivity. Instead of the smoothly running heat engine of plate tectonics that our planet enjoys, the internal engine of Venus might run in brief dramatic fits of intense activity. Like a motor, seriously in need of lubrication, that keeps getting stuck, the global style of Venus might be to occasionally lurch into action in massive geological tantrums that renew the entire surface all at once, letting a blast of heat out of the interior. If this fits-and-starts alternative to Earth's plate tectonics really does occur on Venus, we have to ask, "Why the difference?" Since thermal evolution, I've led you to believe, is controlled by planetary size, shouldn't Venus and Earth have the same overall behavior?

It may all come back, once again, to location, and the drying of Venus by the nearby Sun. The above analogy, of Earth being a well-oiled machine and Venus one in need of a lube job, may not be too far from the truth. In this case, though, it is water, not oil, providing the lubrication. Earth is so soggy that a large portion of the rocks in the crust and mantle are hydrated. A layer of water-softened rocks at the

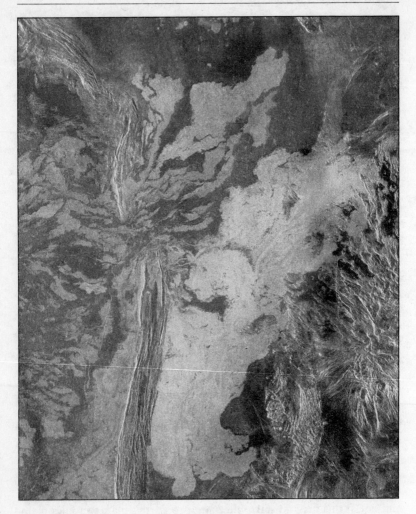

Massive lava flows running for hundreds of miles on Venus. (NASA)

base of Earth's crust allows the tectonic plates to slide around the surface of our planet nice and easy, smooth and slow. Models of Earth's plate tectonics, when altered to simulate dry rocks instead of the real waterlogged ones, begin to seize up and, instead of running smoothly, stop and start like a backfiring jalopy. Dehydrate Earth and it may begin to act like Venus.

Maybe then, the capacity for Earth-style plate tectonics on Venus was lost along with the water in a runaway greenhouse. If this picture is accurate, the lightly sleeping monster of Venusian geology may be getting ready to stir again. Pay attention, because the fireworks could start anytime in the next couple of hundred million years.

Actually, there is good evidence that the monster is not fully asleep. There are abundant atmospheric signs of ongoing volcanic activity. The mixture of gases in the Venusian air is "out of equilibrium" with the minerals at the surface. This means chemicals are primed to react, itching to get at each other. When we see such a condition in a planet's atmosphere, it's a clue that something is up. An atmosphere out of equilibrium is like a pile of hungry cats in a roomful of freshly opened cans of sardines. The situation is not static. A disequilibrium condition does not last long unless some energetic process is actively keeping it that way (a constant supply of fresh sardines, for instance, could explain why the cats haven't eaten them all). A disequilibrium mixture of gases, left to itself, would rapidly undergo chemical reactions and change into a different mix, more in equilibrium. So the atmosphere of Venus has not been left to itself. Something is regularly injecting fresh, reactive gases into the air. In particular, the amount of SO_2 is suspiciously high. This is strong circumstantial evidence of currently active volcanoes.

The most obvious, visible sign of something actively disturbing the atmosphere of Venus is the global clouds themselves. In an extreme form of acid rain, gases spewing from Venus's volcanoes are actively maintaining the sulfuric clouds and supporting the intense greenhouse climate. Without a continuous source of fresh sulfur gases from active volcanoes, the clouds of Venus would disappear in a mere 30 million years, as sulfur was consumed by reactions with surface rocks. The bright clouds of Venus are the smoking gun of active volcanoes on the surface in the geologically recent past.

From this, and the fresh appearance of the largest volcanoes in the *Magellan* radar pictures, many of us believe that Venus today has active volcanism. Yet most of the surface seems to have been formed during a bygone era when volcanic activity was much more intense. Since we've found that the current climate and clouds are strongly affected by the paltry amount of volcanism occurring now, this leads us to ask what the climate of Venus was like 600 million years ago when the global volcanic plains were forming, when lava was gushing onto the surface, and greenhouse gases were pouring out of the ground at hundreds of times their modern rates.

Inspired by our mentor Jim Pollack, my collaborator Mark Bullock and I have taken on this problem. It's not easy to tackle because it involves combining numerous techniques that have previously been handled by several subdisciplines that haven't always played well together. To follow the changing environment of Venus we need to be climate modelers, cloud physicists, atmospheric chemists, and volcanologists. No one can do all of this, but we've had lots of help.*

First we needed a good climate model, better than any previously constructed for Venus. We had to accurately simulate the numerous transitions of energy that occur when sunlight reaches Venus, reflecting off the clouds, filtering through the atmosphere, warming the surface, and reradiating as infrared, which heats the air. But, simulating the heat balance in the present atmosphere was only the starting point. We also need to be able to change the mix of gases in our model, simulating an episode of enhanced volcanic gases, and calculate the changes in surface temperature, cloud structure, chemistry, and so on. Fortunately, we had Jim Pollack to help us design the initial version of our climate model.

Pollack was Carl Sagan's first grad student at Harvard in the 1960s, and he cut his teeth on early climate models of Venus. For several decades after that, he oversaw an army of researchers at NASA's Ames Research Center in Silicon Valley and cranked out dozens of important papers on an astonishing range of planetary topics. One of his passions was climate evolution on Earth-like planets.

My first job after grad school was as a postdoctoral researcher at NASA Ames, with Jim Pollack as my adviser. Jim had a way of cutting through to the core of a scientific problem and helping you see clearly what needed to be done to solve it.

The major project Jim and I worked on during my apprenticeship with him was the construction of an improved model of the Venusian clouds. For this we used some wonderful infrared snapshots that the *Galileo* spacecraft had taken as it flew close by the night side of Venus

---

*As John Lewis says, twisting the "standing on the shoulders of Giants" line of Newton, we've been stepping on the ankles of midgets.

in February 1990, on the first leg of its wayward six-year journey to Jupiter. We used the pattern of the heat leaking unevenly through the clouds to nail down their structure and composition.

During this same period *Magellan* was in orbit around Venus and the first global surface maps were being assembled. The strange crater distribution suggested immediately that Venus must have had periods of intense volcanic activity. Jim was a global thinker and he encouraged the same in his colleagues—we had wonderful conversations about what this strange surface history might have meant for the atmosphere, clouds, and climate. He was a rigorous scientist who wasn't afraid to think bold thoughts—he encouraged me to follow my ideas, however fantastic, as long as I could back them up with physical models.

Around the time I was finishing my postdoc gig and packing up the old Corolla for the move to Colorado, Jim became ill with a rare form of cancer. When he became too sick to go to the office, he hooked up his first home computer and continued to hold court in cyberspace. Like a shark that needs to keep swimming, Jim needed to do science. He kept up his input on numerous ongoing projects, barely ever hinting at his worsening condition unless directly asked about it, right until his untimely death in June 1994 at the age of fifty-five. I've kept several of my final e-mails from him: strange electronic relics of a great mind and a kind soul.

Starting with the climate model Jim helped us to map out, Mark and I have added parts that simulate volcanic emissions of gas to the atmosphere, chemical reactions between gases in the atmosphere and minerals on the surface, diffusion of gases into the crust, the formation and destruction of clouds, and the escape of gases into space.

What we've found is that as the geology of Venus has gone through intense oscillations in activity, the climate has followed suit, episodically undergoing hundreds of degrees of global cooling, and then warming. These extreme temperature changes should have made surface rocks expand and contract, causing Venusquakes around the planet. In fact, in *Magellan* images we think we see the signs of climatically induced surface wrinkles that formed suddenly all around the planet after the epoch of massive volcanic outpourings. Unexpectedly, climate modeling may have helped solve some mysteries of Venusian surface geology.

SURVIVOR

When we look into the details of climate and geological evolution on Venus, we discover an interconnected maze of ongoing processes, all changing and mutually influencing one another. Volcanoes on the surface alter the atmosphere and clouds. This changes the climate, which affects the surface geology and chemistry and even influences the planet's interior. In turn, these internal changes eventually influence the rate of volcanism.

In this sense, we've been learning that Venus, despite having a surface so hot that it glows at night and an atmosphere that has quickly consumed every spacecraft we've dropped in there, is actually quite Earth-like in some profound ways. As extreme and hostile as the environment there seems to us, it represents a delicate and subtle balance of ongoing geological, meteorological, and climatic activity. Much planetary exploration involves studying dead worlds, surveying places that were once active but have long been still, and trying to reconstruct the events of billions of years ago. Venus and Earth are in a class of their own. They are both survivors.

In terms of its geology and climate, Venus, like Earth, is still alive and kicking. The rampant disequilibrium and complex climate cycles of the type we have found on Venus are often thought to be the hallmarks of planets with life. But the idea of life on Venus is not taken seriously because of our understandable obsession with water. Later, in chapter 17, I'll take a critical look at this consensus conclusion, as I consider a new way of thinking about life on planets. I'll ask whether life, rather than being something that happens *on* planets, might be more properly viewed as something that happens *to* planets.

Our interconnected "systems" approach to studying Venus has helped us develop an arsenal of modeling techniques we are now directing toward the general processes of terrestrial planet evolution anywhere in the galaxy or beyond. We wish to explore the balance of chance and determinism that goes into shaping worlds, in order to know how much we can safely generalize from the local examples. Understanding planetary evolution and the different paths it can take is a vital step in comprehending this universe's potential to make life.

IN THE ZONE

Earth's biosphere can be seen as an extension of our oceans, as a capacity that they have achieved, with ourselves as the semisentient ocean's primitive thought organs. So, though we run the risk of geocentric narrowness anytime we define specific criteria for finding life elsewhere, naturally we focus on searching for liquid water.

One key factor in maintaining a liquid water biosphere on a planet, over billions of years, is its distance from the Sun, or whatever star it may be orbiting. If it is in too close, then, like Venus, its oceans will boil away. At too great a distance a water world will at best have icy polar caps on the surface.

When I lived in Tucson, I loved the winding drive up to the observatory on Mount Lemmon, north of town, where grad students escape the summer heat and learn to use large telescopes. As you ascend from the hot town to the cool mountaintop, you pass through several distinct ecological zones, each existing only within a certain range of temperature and moisture and containing distinct species of life. You start off in low desert full of giant saguaro cacti, then ascend through mesquite grasslands, pine-oak woodlands, and ponderosa pines, and end up in a dense, lush forest of tall fir and spruce.

Does the solar system have a well-defined ecological zone, a range of distance from the Sun, where liquid water is stable? If so, the inner and outer edges of this zone are slowly drifting away from the Sun as our star ages and heats up. Earth is probably safe for another 1 or 2 billion years. After that, we will go the way of Venus, our oceans boiling off and fleeing back to space as the inner, hot edge of the "habitable zone" sweeps outward and leaves our spent planet behind.

My colleagues and I are now pursuing research to define the habitable zones around other stars. A zone of water-based life may be a common feature surrounding stars. Even if other, non-water-based chemical structures, unknown to us, can support life, it is reasonable to expect, given the strong dependence of chemistry on temperature, that each kind of life will have a limited temperature range within which it can thrive. Such alternate biospheres might have their own habitable zones at different distances from the same star. Traveling out through a solar system, from hot star to distant, cold edge, you may pass through several ecological zones, just as you do on the drive up Mount Lemmon to observe the planets and stars.

Assuming for now that water is what life needs, then the divergent lives of Venus and Earth can teach us where to expect the inner hot edge of the habitable zone for water-based life anywhere in the universe. What about the outer, cold edge? Is Mars inside or outside the habitable zone? I.e., is there liquid water on the surface? That seems to depend on when you look.

RUST IN PEACE

On the afternoon of August 8, 1996, I was a blissed-out, naked, buoy-ant extremophile, floating in the 103-degree waters of Orvis Hot Springs in Ridgway, Colorado. Among the relaxed, random chatter of other naked floaters I thought I heard a woman casually mention that she had just heard on the radio that "they" had discovered life on Mars. Her comment did not seem all that remarkable, you must under-stand, because this is the kind of thing you hear all the time if you hang around hot springs in southwestern Colorado, and after a while you learn to calibrate your responses.

Later that day, however, when we went into town in search of food, there it was, screaming from every newspaper box: the banner headline "Life Found on Mars!" "Holy shit," I mumbled, fumbling in my pocket for fifty cents. That'll teach me to hide from e-mail for a few days.

The discovery concerned a meteorite that had been found in Antarctica in 1984 and determined to be from Mars. This part wasn't new—we've known for over a decade that more than a dozen rocks in our meteorite collections come from the Red Planet.

Now, however, at a press conference held at NASA headquarters in Washington, a group of reputable scientists had announced with great fanfare that they had found tiny *fossils* in one of these Mars rocks. That same day, the president released an announcement praising the scien-tists, gushing about the significance of the discovery, and calling for a summit meeting to reassess our space goals.

What they had actually found were several separate oddities in the rock, none of which individually offered proof of life, but all of which, they felt, collectively pointed toward a signature of ancient Martian biology. These signs included organic molecules, strange carbonate deposits similar to those made by bacteria on Earth, and a few other chemical clues possibly suggestive of a biological origin. However, what really made a splash at the press conference were the pictures: photomi-crographs of tiny, segmented, wormlike structures that, you have to admit, do look like little creatures.

Everyone loves aliens—Martians especially. The fossils made news all

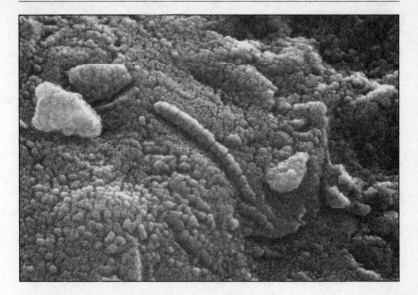

Possible microfossils in a meteorite from Mars. (NASA)

over the world. Plans for a series of new Mars missions, already in the works, were recast as an effort to find signs of life. Suddenly all NASA science was given a more biological spin. Astrobiology was born, amid surging scientific interest in the possibility that Mars might contain signs of past, or even present, life.

Mars today, on the surface, is a dead world: freeze-dried, geologically static, and irradiated daily with lethal ultraviolet.* The entire planet is covered with red dust that has been blowing in the winds, sandblasting every canyon, rock, and crater for eons. Mixed into this dry dust is, we think, a deadly brew of strong oxidizing chemicals such as hydrogen peroxide, which so eagerly attacks organic molecules, plundering carbon, that we use it on Earth to kill germs. Although Mars once flowed with water, the Red Planet ran dry billions of years ago.

However, visions of life on Mars are centuries old, and astrobiologists still dare speak of them. From orbit we observe valleys carved by ancient rivers and floods, and find hints of a bygone but familiar, com-

---

*Mars has nothing like Earth's ozone layer, which is all that stands between us and the sun's deadly UV rays.

forting, and life-sustaining flow. We imagine life that has receded into underground oases where the waters of Mars may yet run. What is it about Mars, or about us, that induces these recurring dreams?

Though, after Earth, Mars is the planet most intensely studied by humans, we are still having a hard time figuring it out. You'd think after sending forty-two spacecraft to Mars (and counting), eighteen of which have actually succeeded in sending back pictures or information, we'd have more definite answers. But while we have superb, detailed pictures of the entire planet, it remains cloaked in mystery. Despite its proximity and our relatively unobscured view of the surface, Mars has always been hard to see clearly.

Our powerful hopes for Mars, combined with great practical limitations on our knowledge, have seduced us down many blind alleys and dead ends. Every new, more detailed look gives us, along with greater clarity, new sources of confusion. Mars messes with our minds. It is a shape-shifter, revealing a different age and personality with each new look.

Through our naked eyes we see a willful bright red orb strutting through the passive starry backdrop. This confident march and its bright sanguineous glow have made Mars the macho warrior star of all the best and bloodiest splatter myths (and an equal number of lame movies).

Viewed with telescopes the warrior morphed into a shimmering fuzz ball with intriguing linear features that were either barely discernible or not there at all depending on who was looking. Shifting dark patches and seasonal polar caps suggested some dynamic presence.

From space, first photographed from fifty-nine hundred miles out by *Mariner 4* in 1965, it was a barren, cratered lunar landscape, disappointingly old and dead. Then, global views from Mars orbit showed a more varied and promising world of giant, extinct volcanoes, ancient flood patterns, sculpted canyons, and active wind streaks. Yet, captured with cameras landed on the surface, in five places so far, it is a desolate desert landscape with piles of rusty rocks and dusty dunes.

So, is Mars dead or alive? Right now I'm not talking biology. I'm asking whether, geologically, Mars is alive in the sense that its big sisters, Earth and Venus, are: hot on the inside and active on the outside.

In contrast, puny Mercury and our punk little Moon are cold and solid. If planets are made of butter, Earth and Venus are room-temperature soft, but Moon and Merc were left in the freezer. Their surfaces are rigid, immobile shells. They have been dead for billions of years.

Photographs of Mars from space show a world that is in many respects intermediate between the opposite extremes of Earth and Moon.

This Viking orbiter mosaic is one of my favorites, because it illustrates the dual character of Mars: static and dead on the inside, effervescent and shifting on the outside.

A southern hemisphere densely packed with giant impact craters betrays a frightfully ancient planetary surface. These are scars left from the early, intense bombardment that ravaged the solar system when all were young worlds. Like the Moon, Mars has large surface areas pockmarked by that primordial pounding. That we can still easily see the damage from this early epoch puts severe limits on the level of subsequent surface activity. Whatever action has been going down on the surface of Mars for most of its time, 4 billion years of it has been insuf-

Viking orbiter mosaic shows ancient craters on Mars as well as atmospheric haze. (NASA)

ficient to fill in those holes. Mars is a world that has been largely inactive throughout most of solar system history.

Geologically, Mars is only a shadow of its youthful self. At least in terms of internally generated geological activity, driven by the liberation of interior heat, like that which builds mountains and makes new crust on the Earth, Mars is long gone. It's been dead now for much longer than it was ever alive.

WINDS OF CHANGE

Look again at the above Viking orbiter mosaic. Mars is a rusted old hulk of a world. Yet, on the edge of the planet, at the border between the deep red of Mars and the deeper black of space, you can see a feature that clearly shows this is not the Moon or Mercury. There, an indistinct bright swath reveals an atmosphere thick enough to scatter light and hold aloft clouds and hazes. What's more, if you've got the attention span to watch Mars over days and months, you will see movement down there. Breezes blow. Dust swirls. Features are buried and unburied again. Storms gather and clouds roll by. Polar caps grow and shrink with the changing seasons. All is not quiet on the Martian front, yet these changes are all caused by the motions and transformations of the atmosphere, fueled by the heat of the sun.

Global movements of gas and ice do cause some geological activity. Glaciers freeze and thaw, carving channels and loosening rock. Underground deposits of permafrost expand and shrink in the polar regions. Mobile deposits of ice mixed with rocks cause "rock glaciers" to flow. Landslides tumble down slopes softened by the seasonal freezing and thawing. In all these ways the effects of the changeling atmosphere go more than skin deep. Geology on Mars today is driven by the sun, not by heat from the planet's interior.

Earth's surface gets it from both sides—from above and below. The internal heat induces most of the local geology (earthquakes, mountain building, volcanoes) that wipes out older surfaces. Earth's solar-driven atmosphere and hydrosphere constantly erode landforms from above.

On Mars, weather is very much a factor, but Mars today is not bothered by internally generated activity. Clearly, it wasn't always this way. Looking down at Mars from orbit, we can see that in the past it was more like Earth and Venus. We see widespread and varied volcanoes that once erupted, and giant faults that once caused mighty Marsquakes. Magnetic patterns in the ancient southern hemisphere hint at the past operation of something like plate tectonics. But these are all relics of ancient history. That we can easily see these structures, still exposed on the surface despite billions of years of dusty storms, is a further reminder of just how incredibly quiescent the planet has been for most of its existence.

A dead surface continually sculpted by a lively atmosphere is the dual personality of Mars today. The cryogenically preserved corpse of Martian geology, continually kicked around by its active atmosphere, can seem animated at first glance, but the body is cold and its soul long departed. The *lively* nature of the air currents and the shifting patterns of dust had many astronomers believing that we were seeing something *alive,* right up until the return of the *Mariner 4* photos provided a cold slap, awakening us to a new Mars. But the old Mars, some believe, may be there still, buried just beneath the dusty surface, and we are doing our best to find it. Recent pictures from the *Mars Global Surveyor* spacecraft have hinted that there may be life in the rusty old cadaver after all.

MEET THE NEW MARS

I first became aware that something new was up in early June 2000 when I heard from a colleague, a trusted senior scientist not given to hyperbole or idle speculation, that the cameras on *Mars Global Surveyor* (*MGS* for short) had "seen something" on Mars, that it was being kept secret and that "the president has been briefed." A few quick phone calls to some friends at NASA revealed that something really had been found, the White House had indeed been informed, and those who actually knew something—the scientists on the *MGS* camera team—were the only ones not talking.

What could possibly merit this kind of treatment? Keeping important scientific findings secret until publication is not unusual, but the White House briefing conjured up images of saucers landing in the Rose Garden. Naturally, speculation on the Net spun quickly into orbit. Active volcanic hot springs, tracks left by migrating Mars bunnies, or signs of crashed spacecraft were some of the putative discoveries being batted about by space and conspiracy enthusiasts.

Had they actually found life on Mars? That seemed quite unlikely for two reasons. First, whatever it was had apparently been seen from orbit. What we do know about Mars strongly suggests that any life to be found there will be hiding underground, protected from the lethal radiation showering the surface. Second, if life really had been found, then the prolonged secrecy would be irresponsible, fanning the ready flames of government conspiracy theories and general mistrust of science. Yet a cool new picture of some rocks or sand dunes would hardly merit an Oval Office audience.

Short of alien pyramids, a rock face with the likeness of Elvis, or an invading fleet heading our way, what could be behind the secrecy and the high-level briefings? With a little thought many of us quickly concluded that the big secret must have something to do with water. The current mantra of astrobiology is "follow the water"—i.e., water = life. A finding somehow indicating the presence of liquid water on Mars today, or in the recent past, would be big news. This seemed the only reasonable choice for all the fuss surrounding the secret new discovery.

It was interesting to watch the rumor mill spin into hyperdrive, spurred on by the hungry media monster. I began to get calls from local and national newspaper reporters about a week before the official announcement, asking if I'd heard anything. When I said, "Only rumors, but I don't really know anything," they would always ask me to speculate on what I had heard, and I would explain why I thought it must be some sign of water on the surface. What happened next was revealing, but not about Mars. News articles reported the discovery of water on Mars based on the speculation of scientists like me who didn't actually know anything but had reached this conclusion simply because we could not think of anything else that fit the profile of secrecy and hype. A reporter from *USA Today* called and implored me to speculate on the possible implications should the speculations of other scientists about published rumors prove to be true.

Such was the pressure to break the story in a timely fashion that several supposedly reputable news organizations published "scoops" from "inside sources" that proved to be completely false. Both MSNBC and BBC news ran a story reporting pools of standing water seen oozing from the bottom of the giant canyon Valles Marineris.

After the publication of these first false stories from "unnamed sources at NASA," the official release of the discovery was abruptly moved forward a week. Clearly, someone had decided that when it comes to Martian mysteries, too much anticipation is not a good thing. There was a real danger that the significance of the actual discovery would be buried in a dust storm of dramatic speculation.

At a hastily arranged press conference, *MGS* scientists showed stunning new pictures portraying their discovery: photos of numerous places with fresh-looking gullies running downhill, apparently seeping from the sides of steep cliffs. Many of these appear to flow out over sand dunes, demonstrating that the flow features are younger than the dunes (which are, themselves, presumably young, because dunes are ephemeral things). At several locations around the planet where the gullies were found, some liquid has apparently been recently seeping out of the ground onto the surface of Mars. "Recently," in this context, means less than a million years ago, and possibly last week. "Some liquid" probably means water.

These photographs are astounding, unsettling, and provocative. The second most amazing thing about them is simply that they look so hauntingly familiar. Here are close-ups of the face of Mars* looking

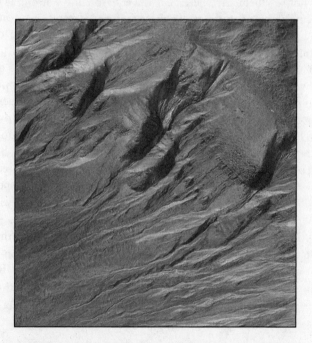

These gullies seen from the *Mars Global Surveyor* may be signs of recent water on Mars. (NASA/JPL/MSSS)

---

*Not to be confused with the face *on* Mars, which I'll discuss in chapter 21.

more like someplace you hiked one spring in Utah than an alien planet. The most amazing thing about them is the implication of recent water flow.

Most scientists see Mars as a good place to look for fossils from the good old days, long ago when the nights were warmer and the rivers flowed. Most do not expect to find life there today, but also admit that we cannot rule it out. Now, however, *MGS* has photographed several different places where it looks as if water might still be running. If there really is liquid water near the surface of Mars today, it truly does change the equation, shifting the odds in favor of present life.

But, here's where it gets strange: the sites where we see the seepage are among the places you would *least* expect to find flowing water, because they are, even by Martian standards, so damn cold. (This is not summer in San Francisco—we're talking one hundred degrees below freezing.) We do not see them anywhere near the equator, and not on the warm, Sun-facing slopes. Paradoxically, these features seem to shy away from the Sun, preferring frigid, high latitudes and shadowed slopes. This does make one wonder if perhaps there isn't some other substance, liquid at these temperatures, mimicking the action of flowing water.

All over Mars, we see ancient channels, rivers, and dried-up ponds and we think, "Water!" Since such features could not form under the present climate, we conclude that the climate has changed, that Mars used to be warmer and wetter, more like Earth.* This idea has a couple of major problems that we tend to sweep under the dust. One is that it is difficult to get the surface of Mars to stay above freezing no matter how we tweak our climate models. Remember, early on, the Sun was significantly dimmer than it is now, and this makes the problem harder to solve. We're working on it, but we haven't yet found a way to melt Mars. Maybe we're missing something. The other nagging problem is that we can't find the carbonate rocks that should be common on the planet everywhere if it once enjoyed a watery climate.

Even given these puzzling discrepancies with the wet-Mars theory, it seems nearly inescapable that Mars had liquid water flowing in the

---

*A cautionary aside: In the early nineties *Magellan* found many riverlike features on Venus. Since conditions on Venus are far too hot for liquid water, this possibility was never seriously considered, so we've dreamed up exotic liquids, such as carbon-rich lavas, that might flow under Venusian conditions as easily as water does here. No one anticipated the existence of such strange materials before the observations required that we think them up.

past. But running water today? In some of the coldest, most shadowed places? It seems wrong.

Several clever theories have been proposed to explain how the gullies could form without violating the laws of chemistry and physics. For instance, maybe the runny stuff is actually liquid CO_2. If you pressurize CO_2 to many times the surface pressure on Mars, it condenses into a liquid. Liquid CO_2 might exist in pressurized underground channels, sometimes breaking out and spraying onto the surface, causing loose material to run downhill and forming the gullies. Or maybe an underground heat source is pumping water around, or maybe the water is in some exotic mix with frozen carbon dioxide, giving it the right fluid properties.

Now that the cat has been let out of the bag and the president has gone back to bed, we are all free to gleefully theorize about these recent-looking flow features on Mars. Fresh surface water, if confirmed, would mean that, by the currently accepted rules, Mars should have life. It makes Mars, not Europa, Jupiter's watery moon, the most accessible place to test our "water = life" paradigm. It also emboldens those who urge manned—I'm sorry, "peopled"—missions to Mars in the near term, since an accessible water supply would make it that much easier for us to live off the land.*

SAME AS THE OLD MARS?

There is a cyclical quality to our ideas about Mars. Like pilgrims seeking tears on a stone Virgin, we are ready to announce, at the least provocation, that we see water running there still. For over a hundred years now, the watery channels of Mars have reappeared in different forms. So, when MGS scientist Ken Edgett excitedly exclaimed at the press conference, "This is not your mother's Mars!" I thought, "No, but it might be my grandmother's."

I am not convinced that the gullies are evidence for surface water on Mars today, but they've sure shown us that things are happening there that we don't understand. Mars, viewed with unprecedented detail in the MGS camera, has once again showed us new sides of its personality.

---

*One of Asimov's best SF stories, "The Martian Way," involves the struggles of Martian colonists to achieve water independence from imperialist Earth. They achieve this in the end by mining water from the rings of Saturn.

While the hints of current water flow remain provocative and puzzling, less controversial are numerous features providing the best evidence yet that flowing water really did affect much of the surface in the distant past. Numerous old impact craters and other topographic depressions contain subtle deposits that look like dried-up lakes. Perhaps most striking is the wide occurrence of terrain made of layered, sedimentary rocks. With the improved clarity of the *MGS* cameras, the layered structures that have been there all along jump out at us.

These deposits look familiar to terrestrial geologists, and for that matter to any terrestrials who have visited the Grand Canyon or any number of places where the layered structure of sedimentary rock dominates the landscape. Such thick tableaux of layered rocks have been laid down over extended periods of geological time, usually by liquid

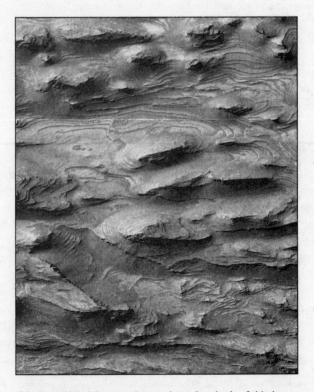

This *Mars Global Surveyor* image shows hundreds of thin layers of rock. (NASA)

water. Though some have proposed that on Mars these layers could have been built up out of wind-borne sediments, they look exactly like sedimentary rocks on Earth that we know were laid down beneath shallow seas of water. More recently still, the *Mars Odyssey* spacecraft, which arrived at Mars in October 2001, has revealed vast fields of permafrost, frozen water mixed with dirt, in large areas surrounding both poles. The Red Planet may once have been blue.

But was it ever green?

Jonathan Eberhart, the superb writer who covered planetary exploration for *Science News* and was a fixture at our conferences for years until he retired in the early nineties, was also known to many as an excellent singer/songwriter. We had fun rocking out together at some planetary meetings in the eighties: Jonathan sang and played piano, I played guitar, planetary geologist Dave Pieri played bass, and Kelly Beatty, the editor of *Sky and Telescope*, played drums. It was the ultimate geek jam, a regular planetary Nerdstock. Jonathan's song "Lament for a Red Planet" is an ode to all the lost visions of a vibrant, comfortable, and living Mars from the past of science and science fiction. I first heard him sing it in the pressroom at the Jet Propulsion Laboratory the week of the *Viking* 2 landing on Mars. In a deep, sea-shanty voice, he sang:

> *Ten thousand times a hundred thousand dusty years ago,*
> *Where now extends the Plain of Gold did once my river flow;*
> *It stroked the stones and spoke in tongues and splashed against*
>   *my face,*
> *Till ages rolled, the sun shone cold on this unholy place.*
> *Your ochered cliffs and rusted sands stand regal and serene,*
> *But oh my wan and wasted world, I miss your blues and greens.* *

When it comes to life on Mars, we've got a feeling we can't hide. Our intuition that Mars holds life is strong, recurrent, and seemingly impervious to data. Every time we learn a new way that Mars is deadly, it represents only a temporary setback for this stubborn belief, until we can invent new ways that Mars might, after all, live on. This pattern suggests that our persistent belief fulfills needs that go beyond scientific evidence. Decade after decade, both before and during the space age,

---

*Jonathan's album *Life's Trolley Ride,* which has this song, is available from Folkways Records. Sadly, Jonathan passed away in 2003.

our science has been swayed by this hope. The more we learn, the more likely that we won't get fooled again, but if we ever do discover life on Mars, one of the lessons might be that we should trust our intuition about these things. The picture of Mars that is emerging from our latest missions of exploration is oddly reminiscent of Percival Lowell's dying, dried-out world. Gone are the canal builders, but visions of rebel bacteria colonies holding out in some underground cave will be much harder for our science to rout out. The deeper we probe, the deeper they seem to retreat into the bowels of the planet.

As my planetologist friend (and former office mate at NASA/Ames) Kevin Zahnle, a keen observer of the scientific condition, wrote in 2001 in *Nature*, "The most interesting information remains right at the limit of resolution. . . . Always life on Mars seems just beyond the fields that we know."

12 | Growing Up with Europa

All these worlds are yours—except Europa.
Attempt no landings there.
  —ARTHUR C. CLARKE'S *2010: Odyssey Two*

Lime and limpid green the sound
resounds the icy waters underground.[©]
  —SYD BARRETT,
  Astronomy Domine

THE TRIALS OF GALILEO

So far our search for alien life isn't going too well. Although we cannot count Venus or Mars out yet, neither looks as promising—at least for our kind of life—as many believed when we began exploring. But from this we do not conclude that the universe is lifeless.

We're not capable of reaching such a conclusion—logically or emotionally. In this sense, aliens are like gods. No scientific result would convince us that alien life does not exist. Yet, the opposite—proof positive of life on another planet—would be accepted by everyone except those yahoos who think the lunar landing was faked.* The hopes and faith we place in our dreams of alien life are not in any danger of being dashed by science. At worst, they will be displaced at the pace of our exploration. At best, they will be confirmed in a dramatic discovery of something unmistakably alive.

---

*If you are one of those yahoos, I apologize for my insensitivity.

Our reluctant letting go of warm oceans on Venus and Mars simply forced us to broaden our definitions of habitability. We widened our search, traveling ten times farther afield to other worlds, which seem, on the surface, even more exotic.

Just the idea that there *are* other worlds has been a difficult and painful concept to get into our little carbon-based skulls. Galileo was the first to actually find any new worlds, and look what happened to him. Recently our search for life has taken us back to the site of that first decisive Copernican battle: Jupiter's Galilean moons.

We know them as the four lovers of Jupiter: Io, Europa, Ganymede, and Callisto.* It only took us four hundred years to progress from discovery to exploration. Now, we've launched ships to cross wider seas and explore Galileo's new worlds. Out there we've found unforeseen wonders that feed our hopes for a new revolution of Copernican proportions. Could we possibly find confirmation, among the moons of Jupiter, of our suspicion that we are not alone in the universe?

Our first good look came in 1979 when the two Voyager spacecraft whipped through the Jupiter system, frantically snapping pictures as they sped by. It all happened fast; each Voyager, with no way of slowing down, spent just two days in the vicinity of Jupiter and its moons before being gravitationally flung back out into the void toward a rendezvous with Saturn.

In the summer of 1979 I was a space-freak college freshman with a way-cool summer job: assistant to the Voyager Imaging Team at the Jet Propulsion Laboratory (JPL) in Pasadena during the *Voyager 2* Jupiter encounters. When I wasn't running off to Grateful Dead shows, I was closeted in a big, air-conditioned cubical building with the men and women who had planned and built these early ships of deep space, delighting in every new picture as the expanding, approaching dots became worlds. It was then that Galileo's four strange moons first took on their distinct personalities: the manic volcanic face of Io; the dark lines winding like turnpikes across Europa's bright, icy landscapes; the densely grooved terrain of Ganymede; and the ancient, icy, cratered surface of Callisto.

I will never forget the heart-stopping moment when the first close-up pictures of Europa arrived on Earth. On the morning of July 9, we—the imaging team and assorted students, assistants, and hangers-on—were

---

*Jupiter is bisexual. There is a lot of him to love.

stuffed into a small, dimly lit room, staring up at a dark screen in communal, nervous anticipation. It was as if we were all riding along on the spaceship and gathered together to stare out of a porthole near the bow, awaiting our first glimpse of an unknown shore. A film crew from *Cosmos* was crammed in there as well, adding to the suspenseful air.*

As the face of Europa began to appear on the screen, slowly, in thin, vertical strips, gasps of disbelief erupted from this crowd of generally reserved scientists. Never before and never since has our knowledge of another world taken such a great leap in a few brief revelatory moments.

Our theories predicted an ancient, dead, cratered landscape on the small ice-planet. Instead, we saw a smooth, bright surface crisscrossed by strange dark lanes. The immediate impression was of a fresh, active world. At first nobody spoke. The appearance of Europa was so truly unexpected that no one, in a room full of verbose rocket scientists, could offer anything intelligent. The first to speak was Carl Sagan: "Percival Lowell was right!" he exclaimed. "Only, the canals are on Europa!"

While they're not canals, those enigmatic markings might be signs pointing to life. This took a while to sink in. The Voyagers' trajectories were not optimized for Europa viewing. Each ship had a date with Saturn that dictated its hurried path through the environs of Jupiter.†　On a flyby you only get one chance, so coverage of Europa was spotty and incomplete. We saw only half the planet in photos taken from thousands of miles away, leaving us with only imagination and theories to fill in what we could not glimpse.

We had to wait nearly two decades to get back to Jupiter. During that time I finished college, moved out West, toured with a reggae band, cut off my Afro, and became a comparative planetologist funded by NASA. In 1989, the year I got my Ph.D., a spacecraft bearing Galileo's name was launched toward the moons that he had first seen on a cold January night in 1610.

Even before its launch, *Galileo* was a hard-luck spacecraft. Always at the edge of disaster, it has at times seemed jinxed. The spacecraft has

---

*I actually appear briefly here in the television show: I'm the skinny geek sporting an improbably huge Afro, wide-eyed in the back of the room.
†*Voyager 1*'s path through the system four months earlier had not permitted any close-up pictures of Europa.

followed a tortuous path and faced trials that rival those of the man for whom it's named. Budget-cutters in Congress threatened repeatedly to cancel the mission. Protesters tried to derail the launch over concerns about the fifty pounds of plutonium that it carried for power at Jupiter. In December 1986, the *Challenger* blew up, killing seven brave explorers and shutting down the space shuttle program for three years. *Galileo* had been slated for the very next launch.

The irrepressible craft finally made it off the ground in October 1989. Upon reaching deep space, its twelve-foot-diameter main antenna, folded tight for launch, was supposed to spread open like a metallic beach parasol. But when the moment came, it just sat there, stuck stubbornly in the folded position. Ground controllers tried to jar it loose by running its motor in various stop-and-start sequences. They tried spinning the spacecraft in hopes that it would fling itself free. Months went by, and increasingly desperate efforts failed to open the umbrella, which we needed to establish a communications link between Jupiter and Earth. Gradually the reality set in that it would not open. The main antenna, vital portal for all the riches expected at Jupiter, remained in a crumpled-up, completely useless configuration for the entire mission. Imagine the frustration. We had the most sophisticated spacecraft ever built finally sailing toward Jupiter, able to photograph the mysterious moons in unprecedented detail, but we had no way to send the pictures homeward.

Houston, we had a problem.

NASA engineers marshaled all of their legendary, save-the-day ingenuity and found solutions. A puny "low-gain" antenna, not designed for this kind of work, was pressed into service—a little like finding that your new high-speed modem is dead and being forced to settle for two tin cans and a piece of string. The expected rushing stream of data was reduced to a feeble drip. The mission never fully recovered from the main-antenna failure, but clever work-arounds, combined with *Galileo*'s stubborn longevity, allowed us to get 10 percent of the originally planned pictures of Jupiter and its moons.

After suffering enough technical glitches to give the most even-tempered engineer an ulcer, *Galileo* made it to Jupiter in late 1995 and, seemingly against all odds, it worked. Unlike the Voyagers, which were flyby missions, *Galileo* is an orbiter. When it got to Jupiter, it fired its powerful rocket perfectly, slowed down, and established permanent residency, becoming (as far as we know) the first artificial moon of

Jupiter. For eight years it circled Jupiter, caroming among the Jovian moons, scanning and photographing each of them on repeated close passes.

Now *Galileo*, its mission completed, is crippled but triumphant, near senile from radiation damage and hobbling through its final orbits with that pathetically crumpled main antenna still in tow, its failing gyros barely able to keep the ship oriented. As of this writing, *Galileo* is scheduled to be intentionally crashed into Jupiter in September 2003. Why don't we just leave it in orbit? Because it might someday smash into Europa, contaminating that world with flecks of plutonium and, conceivably, some stowaway bacteria. It seems unlikely, but who can say? Diving *Galileo* into Jupiter while we are still able to control it is the responsible thing to do.*

A SURPRISE INSIDE

Galileo carried the first digital camera ever flown in space and, while bouncing among Jupiter's moons, made photographs with a level of sharpness and detail new to space exploration. The strange beauty of these distant worlds raises some questions: Why should these places, where no terrestrial eye can ever before have wandered, be beautiful to us? What structures, deep within our brains or deep within the physical universe that created both Gaia and Europa, are resonating when we gaze upon an alien landscape for the first time and find exhilarating beauty? Would alien souls feel it, too?

What we have found here is something much more profound than simply a random collection of odd worlds. These large, complex moons are really planets in their own right, and *Galileo* is humanity's first exploration of a new planetary system. As you might expect, many comfortable preconceptions have been completely overturned. Much of what we thought we knew about comparative planetology turns out to be wrong. This includes our previous notions of where interesting geological—and biological—activity may be found in the universe. The moons of Jupiter, as glimpsed by the Voyagers and explored by *Galileo*, turn out to be a much more active and diverse gang of worlds than our previous theories of planetary evolution would have led us to believe.

---

*Wouldn't it be ironic if in our effort to protect the Europans we ended up nailing some poor Jupiterian gas creature?

We thought that only large, rocky worlds like Earth could have active geology, and that we would find only dead ice worlds out this far from the Sun. But who needs a star when you've got a giant planet like Jupiter? Several of those moons are seething with internal heat, restlessly churning inside and out. It is Jupiter's influence that creates this heat and activity. Just as our Moon tows the oceans around the Earth, the huge gravitational pull of Jupiter yanks the insides of these moons around, creating the internal heat that drives the furious volcanic activity of Io and continually warms and cracks the ice on Europa.

This surprising activity is facilitated by the tight, polyrhythmic orbital dance continually executed by the three innermost giant moons. Io orbits Jupiter twice for one Europa orbit, and similarly Europa laps Ganymede twice each orbit. They are locked together tighter than the tightest rhythm section in Jamaica. Every time they pass, they grab at each other with gravitational arms, but Jupiter pulls them back into line. This rhythmic back-and-forth keeps them flexing, pumping energy into their interiors. The heat of the dance keeps their insides hot and their faces young and fresh. The greatest heating goes to those caught most deeply in Jupiter's gravitational spell. Thus we see an evolution of planetary style, each moon getting progressively colder, and its surface older, as we travel outward from innermost Io past Europa, Ganymede, and Callisto. The only moon that receives no internal heating from this intricate multipartner dance is outermost Callisto. Like an older chaperon watching the young ones dance and bloom, Callisto comes closest to our original expectations of all the Jovian moons—an ancient, dead, heavily cratered ice world.

The active nature of these moons was actually predicted right before the Voyager discoveries. In a triumph of theory, and an example of good timing, three planetary scientists[*] modeled the effect of tidal heating on Jupiter's moons and predicted that Io would be volcanically active. Their paper was published in *Science* one week before *Voyager 1* arrived at Jupiter. When *Voyager 1* flew past Io, dang if they weren't right there: giant volcanic plumes shooting off into space from the edges of the little red moon. The plumes astonished everyone (including the scientists who had made the prediction).

Europa, too, gets a workout on the inside from those Jovian tides, but we don't know exactly how much heating it gets. Some models pre-

---

[*]Stan Peale, Pat Cassen, and Ray Reynolds.

dict that the icy surface is just a thin shell, a few kilometers thick, over a deep liquid ocean. Others suggest that the icy shell is tens of kilometers thick, but increasingly, nearly everyone accepts the likelihood of a deep ocean beneath the ice.

Ever since Voyager we've wondered just how active Europa might be and what really goes on there, and science fiction writers have riffed on the theme of Europan life. Now, seen through *Galileo*'s digital eyes, it turns out to be arguably the weirdest world we've yet explored. On first viewing, its planetwide system of dark fissures and raised bands looks like a global tangle of roots. The face of Europa seems eerily unnatural, with animated landscapes suggesting arrested flow. At every scale, from hemispheric to close-up, the globe is covered with what could almost be veins, like a giant, frozen mutant leaf.

Europa looks alive. Its bright, lineated surface is composed of freshwater ice crisscrossed by a global system of fissures and cracks, formed as the surface pulsates with the tide. *Galileo* has now shown us the surface at a magnification a thousand times Voyager's best. In the most detailed photos we see icebergs, which have apparently broken loose, floated, and jostled before freezing in place again. This adds to a mounting pile of evidence for an ocean of liquid water beneath the ice. In fact, we now think that Europa may have more water than all of Earth's oceans combined. And, on Earth at least, where there is water, there is life. Does anything swim through Europa's icy seas?

As far as I know, the possibility of life in underground oceans on icy moons was first mentioned in 1975 by Guy Consolmagno, another John Lewis graduate student, in his master's dissertation at MIT. Guy (who is now a Jesuit priest as well as an active planetary scientist) was studying the thermal evolution of ice moons. His calculations showed that they should sometimes develop layers of liquid water within. The appendix of his thesis ended, "But I stop short of postulating life in these oceans, leaving that to others more experienced in such speculations."* This passing comment in a student dissertation went mostly unnoticed, but the possibility of life on Europa became a hot topic after Voyager's discoveries. Arthur C. Clarke made it one of the main themes of his novel *2010*, published in 1982 when the new images from Voyager were fresh on our minds.

---

*Reflecting on this now, Guy wrote to me, "So I guess that makes me the first person NOT to predict life in these oceans!"

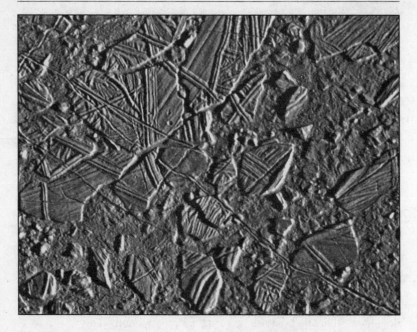

This "chaos terrain" on Europa has many "icebergs" that have floated around and refrozen in place, suggesting warm ice or liquid beneath. (NASA)

With *Galileo*'s further hints of a deep, dark ocean, Europa has emerged as the most promising place to look for water-based extraterrestrial life in our solar system. Indeed, this bizarre little moon can serve as a test bed for our current assumptions about life.

We are currently planning a mission that we hope to launch sometime in the next decade to orbit Europa. It will determine definitively whether an ocean is flowing within. We'll also look more closely at the surface for life-revealing chemicals mixed into the ice. If we do confirm the presence of liquid water, the next step will be to go ice fishing. We'll go back and land a self-disinfecting probe that can slowly melt through the icy crust. Then, when we break through to open water, we'll go exploring in our solar system's deepest ocean.

GRAVITY'S REIGN

We thought the Jovian system would be a fossil, a freeze-dried remnant of the early solar system. Instead close-up spacecraft investigations revealed a living, evolving planetary system. The unanticipated activity and youthfulness of these worlds implies that gravitational energy may be of equal or even greater importance than solar energy in fueling life in our universe.

The likelihood of oceans on ice-moons like Europa suggests a whole new concept of habitable zones. If there can be water, and perhaps life, on Europa, then the same could be true on moons orbiting giant planets anywhere in the universe. This would mean that distance from a star matters less in defining habitability. Such "gravitational habitable zones" could be common and large, implying a universe with a lot more real estate having favorable conditions for life as we know it.

The idea of life on Europa has got us all thinking about life that could be independent of starlight. Meanwhile we've discovered extremophile organisms on Earth that live deep below the surface and have little interest in the Sun. This may brighten the corners for life and planetary habitability.

We animals are completely enmeshed in the part of the Earth's biosphere that lives off the Sun. We're just a minor outgrowth of the green world and the ubiquitous, hidden microbial world—we're those weird brainy things that crawled out of the compost heap after the oxygen waste piled up. Its hard for us to imagine life that is truly isolated from the solar influences that drive and so thoroughly shape our world.

The idea of life underground is not new, of course. Subterranean creatures have filled our mythologies and occasionally our scientific fantasies as persistently as have creatures living beyond the sky. Hell, in the seventeenth century Fontenelle speculated about microbes living underground on the Moon or elsewhere, slowly eating rocks. Yet now we have new reason to wonder if perhaps his conjecture was prescient, and surface life is not the only game in town.

Life may be something that frequently, or even ubiquitously, happens *inside* planets. Perhaps biology can be a purely internal planetary phenomenon. If life begins underground, it makes the origin of life anywhere else in the universe less dependent on surface conditions. If all you need is some internal energy and some liquid water, then most planets must have the right combo at some point in their history, since planets are born hot and wet. The classical concept of the habitable zone starts to seem like a bourgeois notion invented by self-centered, Sun-worshiping surface dwellers. Can a planetary biosphere that loses its surface water retreat inward and persist, subterranean and homesick, but still alive? In some cases an "inhospitable" surface may be only the thick, protective skin covering a thriving underground world.

Astrobiologists agree that there are three essential ingredients for habitability: (1) organic molecules (the building blocks), (2) liquid water (the medium), and (3) a source of energy (the spark of life). Based on our studies of comparative planetology, how rare might such a combination be? Organic matter seems to be ubiquitous, falling from the sky everywhere in the universe, so this requirement serves only to rule out places with environmental factors that directly destroy organics. No, on second thought, it doesn't even do that. Otherwise, Earth, with its organic-burning oxygen atmosphere, would be lifeless. Opportunistic life might find a way to derive energy from whatever it is that destroys organics, as we do on Earth with high-octane poison oxygen.

The belief that water is necessary *and* sufficient rests on the assumption that suitable energy sources will be common. How reasonable is this? Well, the presence of liquid water actually implies some kind of energy source, so the two requirements aren't completely separate. Our two examples of oceanic planets are Earth and Europa. Earth is wet because it is in the right place to soak up plenty of solar energy. Europa is wet because of the release of tidal energy. So, both ways that we know of for a planet to stay watery also come with built-in, long-term energy supplies.

But is tidal energy sufficient to drive a biosphere, or does life need a sun? Picture the recently discovered dense colonies of bottom-dwelling life clustered around the "black smoker" volcanic vents on Earth's ocean floors. The seafloor on Europa may also have volcanic vents, driven by internal tidal heat, which could provide the chemical fuel for a native biosphere.*

Another idea was proposed by Chris Chyba, a former student of Carl Sagan's, now at the SETI Institute in Mountain View, California. Chyba's idea makes use of the punishing radiation at Europa's surface. The powerful magnetosphere of Jupiter whips charged particles into a frenzy; all of the inner moons exist within a raging storm of radiation. At Europa, an astronaut in a space suit would receive a lethal radiation dose every twelve minutes.[†] Could a Europan biosphere somehow har-

---

*Indeed, Europa on the inside may be like a less hyperactive version of fiery, volcanic Io, doused with an outer shell of ice and water.
[†]Which is one reason why crashing a small amount of plutonium onto the surface may not really be as bad as it sounds.

vest this potent energy source? Radiation breaks chemical bonds, which is why it is often lethal. It must be inducing chemical changes in the Europan crust. Chyba proposed that the radiation rips apart water ice molecules, liberating hydrogen to drift off into space, and leaving behind oxygen and various oxidized compounds. Perhaps this oxygen is eventually mixed down into the ocean where indigenous creatures, their fragile bodies protected from the radiation by miles of ice, breathe, harvesting the chemical fruit of the irradiated surface.

The intense radiation at the surface of Europa can be regarded as a threat to life or as a source of energy that could drive a biosphere. There is an important concept here that we can generalize to help us think about life in the universe beyond the biases of our terrestrial perspective. Paradoxically, a deadly environmental factor may create opportunity, if you can control the slide toward destruction. We live by burning ourselves in oxygen, but slowly, slowly. It's a fine line between a deadly chemical or radiation and a bountiful source of energy. Life may adapt by staying at a safe distance, reaping the flow without being destroyed by it. Just as our biosphere runs on nuclear energy, keeping its distance from the solar reactor, underwater life on Europa might use the intense radiation at the icy surface while avoiding its direct effects. A whole category of potential adaptations might make use of such "dangerous," "lethal" energy sources. Life would need to take advantage of the inevitable flows induced by any source of energy, while avoiding the destruction that comes from getting too close to the fire. One thing we know about life is that it is inventive. What seems like a death ray to us may be a meal ticket for suitably adapted creatures.

BIOSPHERE TWO

That first voyage by robotic submarine through Europa's briny sea, sometime in the next couple of decades, with all the folks back home eagerly watching for something to wriggle through the floodlights, will forever change the way we think about life in our universe. If we find not so much as a chemical trace of life, then we'll know that life needs more than just any watering hole to thrive. We'll have the first concrete hint that life on Earth is a more rare, more lucky phenomenon than we had surmised. At the very least, we'll know that our assumptions about the requirements for a good home, extrapolated from our one example of a living world, were wrong.

Of course it will be impossible to prove definitively that Europa (or any other planet for that matter) does not have life. What if the life there is sufficiently alien that we just don't recognize it? A negative result on Europa would cause us to rethink how we go about defining and looking for life, as did the slightly ambiguous negative results from the Viking mission to Mars in 1976, the first mission ever sent to look for life on another planet.

What if we do find something swimming under the Europan ice, something that grows, breathes, reproduces, or dances the boogaloo with an unmistakable signature of life? Just as Galileo's original discovery of Europa and company provided undeniable proof of a new worldview, the discovery of life there would cause us to take a new look at ourselves and how we fit in. The existence of another living world so close to home would carry implications of a widely fertile, densely inhabited universe. It would strongly suggest that life is no accident but an inevitable step in Cosmic Evolution following from atoms, galaxies, stars, and planets. Our pursuit would then be transformed from a narrow study of our own single biosphere to a comparative study of life in the universe. When we have a second example we can begin to study life as a general phenomenon.

All Earth creatures use the same chemical machinery: DNA to pass life's recipe into the future, proteins to build tissues and regulate chemistry, and so on. Does it have to be this way? Is this commonality profound or accidental? Are cats, begonias, squid, slime and people all virtually the same, chemically speaking, because proteins and DNA are unique solutions to problems that nature cannot solve in any other way? Or is this all the result of arbitrary early choices that have been passed down to all life and become so enmeshed in the fiber of Earth life that they seem essential?

The best way to answer such questions is to examine life that we're *not* related to. If there is a biosphere subsisting in the subsurface ocean of Europa, then it is almost surely isolated and independent of Earth. (Such a claim cannot be made for Mars, since we know that Earth and Mars frequently chuck rocks at each other.) The Europans, so far away and buried under irradiated ice, have effectively been quarantined. If we find that the Europans use DNA, then DNA is instantly elevated in status. It is revealed as a cosmic standard, not just a terrestrial accident or convenience.

If not DNA and proteins, then what do we share with our Europan counterparts and what does Earth share with Europa? Does life transform all of its home worlds in ways that transcend the local details of evolution? What features do biospheres have in common? The answers would bring some much needed sophistication to our efforts to define and search for life elsewhere.

It is, of course, possible that we will discover that we are, after all, related to Europan life, that we share common ancestors even with these distant, iced-in creatures. Then we will have some thinking to do. Such a finding would force us to completely reassess our agreed-upon story for the origin of life on Earth. We might have to give another look to some ideas that make scientists squirm, such as panspermia and even intelligent design.*

Although our discussions of Europan life tend to focus on possible bacteria, we might even find larger, more complex, advanced life there. We know of no physical or biological law that rules out the Europan equivalent of giant squid or sperm whales with songs of their own. Even if we find only simple life there, such a discovery so close to home would greatly brighten prospects for finding other intelligent life. If Europa, too, is alive, then our universe must be bursting with life. And in a universe like that, surely we will soon find someone else to talk to.

Our smashed preconceptions about the Jovian moons came from applying local rules of planetary evolution, which we mistook for universal laws, beyond their jurisdiction. We missed something fundamental when we neglected to consider the tidal energy created within small moons of giant planets. There may be an important lesson here for astrobiology. Complex phenomena, such as planets or living organisms, do follow physical laws, yet they cannot easily be predicted. The outcome of natural "experiments" in planetary or biological evolution depend on too many factors. Speculation and theorizing are fun, but we learn the truth through exploration.

In some ways our current knowledge and beliefs about the oceans of Europa are reminiscent of our pre-space-age ideas about Venus. The bright Europan ice is playing a role that the bright clouds of Venus played before 1962, deflecting deeper inquiry and providing cover for

---

*If we reject these theories out of hand just because we find them ideologically repulsive, then we are practicing pseudoscience.

rational fantasies of alien life. As far as we know, anything may lurk beneath the ice.*

We are the only generation who will see the planets transformed from objects into places, and who will ever experience the first rush of knowing, for sure, that other worlds are out there among the stars. We might even get to be the ones who learn whether Earth has living company in the solar system. We are tasting the fruit promised four hundred years ago by Galileo's revolution. If we keep exploring and stay healthy, with a little luck you and I may live to go ice fishing on Europa and, just possibly, by finding the first real aliens, plunge into a bracing new sea of cosmic awareness.

---

*Another interesting parallel is that sulfuric acid, the stuff of the Venusian clouds, has been identified by *Galileo* in the surface ice of Europa, causing us to wonder just how acidic the ocean there may be. The recent discovery of acid-resistant extremophiles on Earth opens the question of whether highly acidic conditions would preclude life on Europa, or on Venus.

13 | Enter the Exoplanets

Are you ready now? Then close your eyes, and tap
your heels together three times. And think to yourself—There's no place like home; there's no place
like home; there's no place like home.

    —GLINDA, THE GOOD WITCH OF THE NORTH

What do you do when you know that you know
    that you know that you're wrong?
You've got to face the music.
You've got to listen to the cosmo song.

                —SUN RA

MEDIOCRITY

Hey, have you heard the news? We've finally found other planets out
there, orbiting distant stars. Lots of them. This discovery is recent and
surely profound. But what does it mean for our place in the grand
scheme?

    Is Earth the best of all possible worlds, just an average world, or the
only world there is? In the last four hundred years science has narrowed down the options, dispatching the third one to the dustbin of
history. We know that our world is one among many. But is it *only* one
of many, or is it something more as well? Does Earth belong in the special and gifted class of planets? As befits a species stumbling through
adolescence, we are wondering where we fit in, seeking our peers, looking for answers.

    Expectations about more distant planets have been, of necessity,
based on the local crew. If our planetary system were somehow typical,

it would be a gift to science.* It would make our job so much easier, because if we can safely generalize from our own solar system, then we already know what goes on out there in the rest of the galaxy. If our planetary family is ordinary, then it may not be all that unusual for one or more terrestrial planets to end up in the habitable zone where liquid water sloshes onto ripe shores. Some of these watery worlds will become biospheres, perhaps following the same biochemical pathways that life has found here, creating oxygen-rich atmospheres, ozone layers, forested continents, animal crackers, and Internets. If it happened once, it can happen again, right?

Today, we all know that we are nothing. We learned in grade school that we inhabit an ordinary planet circling an average star floating inconspicuously and, until recently, innocently in a backwater arm of a galaxy containing hundreds of billions of other stars, a humdrum galaxy that is itself undistinguished among more galaxies than you could shake a stick at in a zillion and a half lifetimes.

Jeez. Maybe there is no reason to think that our planet is special in any way at all. This is an extreme application of a general principle in cosmology, known as the principle of mediocrity, which says that the universe is basically the same everywhere. Wherever you go, there you are. There is nothing unusual about our position, and the view from anywhere else should look more or less the same as ours. This thought is comforting to scientists because we like to believe that our conclusions do not depend on where in the cosmos we happen to be born. We are not looking for regional truth. What we seek transcends location, and that's a much easier trick if location isn't important.

The principle of mediocrity is a close relative of the arguments by analogy that have made people believe in a plurality of inhabited worlds for hundreds of years, at least. Can we use it to deduce the existence of habitable planets and life out there? From the remarkable uniformity in all the qualities we can observe in distant realms, should we deduce a sameness in the features we cannot yet observe? It has worked before. We assumed, by analogy, that planets were around other stars. Now we know we were right. Do you suppose that life might be another of those features found more or less everywhere? As Fontenelle, in 1686, said about the view of nearby Saint-Denis from the towers of Notre Dame,

---

*Not to look a gift universe in the mouth, but if we were *too* typical, it would be highly suspicious.

everything one can see there resembles Paris: steeples, houses, and so forth. So we conclude that Saint-Denis, like Paris, is inhabited. If the universe tends to repeat itself, in gross architecture and building materials, does that mean that it is fitted with the same fine trim and landscaping elsewhere, achieving an economy of detail that we just can't quite make out yet?

During our first decades of exploring our solar system, we learned that planets evolve in ways that largely depend on accidents of birth, and chance occurrences. As a result, planetary fates cannot be predicted from their initial states. If planets were like electrons or atoms, it would be different, because if you've seen one electron, you've seen them all. Atoms are somewhat more complex than electrons. They are not all the same, but they can be grouped into a hundred or so types (the elements) that are easily classified (in the periodic table) and understood. If planets were like atoms, by studying a small sample we could know them all.

Will we ever be able to assemble all the planets we know into something like a periodic table of planetary types? No. Because planets, and apparently entire solar systems (by which I mean stars with retinues of orbiting planets), are more like people. Each is a complex individual, shaped by the unique experiences of a long life in a chaotic, changing environment. In the details, no two will ever be alike.

Imagine if you were trying to understand human beings as a general phenomenon but you only had intimate knowledge of one person.* If from studying your subject you deduced that everyone needs to sleep, everyone experiences puberty, and that people respond to music, you'd be right. But, depending on the peculiarities of your randomly chosen subject, you might also conclude that everyone is way into heavy metal, that Falun Gong exercises at lunchtime are a daily ritual for all humanity, or that everybody must get stoned. In generalizing from your random sample of one, you would draw some correct inferences and be led astray with others.

Let's face it. We need more planets.

MORE PLANETS, PLEASE

In truth, it would have been shocking and disturbing if they weren't out there. Nevertheless, their number and nature were entirely unknown

---

*I suppose you could say that this is the situation we are all stuck in.

until October 1995, when the efforts of patient and diligent observers paid off, producing one of the most marvelous discoveries of the twentieth century.

The discovery was made by Swiss astronomer Michael Mayor and graduate student Didier Queloz. Using Haute-Provence Observatory in southern France, they were attempting, in late 1994, to find planets by taking detailed spectra of Sun-like stars. They knew they could not see the planets themselves with any telescopes yet built, but they knew also that large planets, as they orbit, should cause their stars to wobble slightly, first approaching us and then receding. By watching for the subtle spectral shifts induced by a star's rhythmic motion toward and away from us, they hoped to turn up clear signs of a planetary companion. Others had tried this and failed, but Mayor and Queloz had developed new spectroscopic techniques that could detect velocity changes, in a distant star, of just thirteen meters per second. This would be barely sufficient to detect Jupiter if observing our Sun from a nearby star, so they thought they had a shot at it.

In late September they began observing a star called 51 Pegasi.* This star is nearly a dead ringer for the Sun. It is located about fifty lightyears from here: star number 51 in the constellation Pegasus, the Winged Horse. By December of that year, they had noticed irregularities in the star's spectrum. It seemed to shift slightly every time they observed it. Yet the shifts were large, amounting to velocity changes of up to sixty meters per second. They first suspected a problem with their equipment. By January 1995, they saw a pattern to these shifts, yet something was strange.

The star's velocity was varying periodically, as would be caused by an unseen planetary companion. Yet the variation was far too fast to be any kind of planet we knew or expected to find. The period of oscillation was just four days. By contrast, in our solar system, the fastest planetary orbit is that of Mercury (the winged messenger), which takes a spin around the Sun every fifty-nine days. Following universal laws worked out in the sixteenth century by our old friend, the freaky genius Kepler, close-in planets orbit more rapidly than distant ones. Thus in our solar system Mercury's year is the shortest and Pluto's the longest. Kepler's third law of planetary motion tells us exactly how far

---

*Or just 51 Peg to its friends.

away a planet is from its star if we know the orbital period. If the four-day period they saw was really the signature of a planet, then it was orbiting a mere 5 million miles from its star. By comparison, Earth orbits at a safe distance of 93 million miles, and Mercury, the inner-most planet in our system, is 36 million miles from the Sun. The data seemed to indicate something very strange indeed: a large, Jupiter-class planet following a tiny orbit, only one-seventh the radius of Mercury's orbit.

They kept quiet and kept gathering data. By March, Mayor and Queloz were confident that they had something, but they were reluc-tant to announce their discovery. They wanted to be careful because for many decades astronomers have announced discoveries of extrasolar planets (those orbiting other stars), only to be proven wrong. Such false alarms have harmed many careers. The apparent strangeness of their new planet only increased their caution. In March they had to suspend operations as 51 Peg disappeared behind the Sun. They planned a set of observations for early July, when the star would reemerge. They calcu-lated exactly the velocity 51 Peg should exhibit when it reappeared. This would be the final test. If it had the right velocity in July, then there was no other explanation and they would announce to the world that they had found an extrasolar planet.

In the first week of July, Mayor and Queloz returned to the observa-tory. This time they brought their families, and some champagne and cake, along to the telescope dome. Late on the first evening Pegasus rose over the horizon and they pointed the telescope at 51 Peg. A few minutes later they had their answer: the velocity matched their predic-tion perfectly. They uncorked the bottles: 51 Peg is being swung to and fro by a planet-size orbiting companion.*

Mayor and Queloz did not go public with this finding until they found out that the report they had submitted to *Nature* was accepted for pub-lication. In October, Mayor made the stunning announcement at an astronomy conference in Florence. After that, word traveled quickly. It soon reached the Swiss team's greatest competitors, Geoff Marcy and Paul Butler at San Francisco State University. Marcy's reaction, upon hearing of the discovery, was mixed. He had been searching for extraso-

---

*Mayor has proposed the name Epicurus for this new planet. As of this writing its official name is still "the planet orbiting 51 Pegasi."

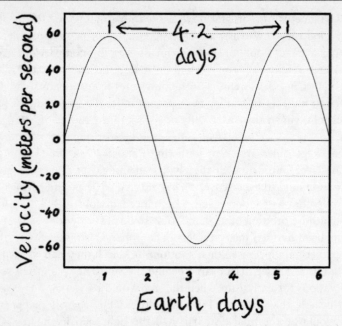

Evidence for an unseen planet is revealed in the rhythmically changing velocity of the star 51 Pegasi. (Anthony Cooper)

lar planets, using a spectroscopic technique similar to that of the Swiss team, for eight years. He had been scooped in a major way. He could have made this discovery had he looked at the right star and analyzed the data.* On the other hand, he was thrilled at the idea of a solid planet detection. He was also skeptical, given the history of false alarms about extrasolar planets. Just five days after Mayor's announcement, Marcy and Butler went to the Lick Observatory, in the mountains east of San Jose, to observe 51 Peg. Within a few nights they had gathered enough data to confirm the Swiss discovery. The planet was real.

What happened next was strange. Marcy and Butler issued a press release that began: "Astronomers at the University of California at Berkeley and San Francisco State University have confirmed a recent

---

*Marcy and Butler had actually excluded 51 Peg from their search because of an error in a star catalog, which misclassified this star as an unstable red giant, rather than the stable Sun-like star that it is.

report of a planet orbiting a nearby star—the first time a reported discovery of a planet around a normal star has withstood scrutiny."

The next morning the astronomers began receiving a barrage of phone calls from every major media outlet in the world. The day after that, their discovery was front-page news on papers all around the globe. Later that week ABC canceled their planned *Nightline* and Marcy appeared as a guest, talking about the new planet.

Back in Switzerland, Mayor and Queloz were fuming. In the press reports announcing that American astronomers had discovered a planet, some mentioned that this object had also been observed by a Swiss team, and some left out this detail. This was not the fault of Marcy and Butler, who, whenever they had the opportunity, were careful to mention that the discovery had actually been made by the Swiss. But Mayor and Queloz were forbidden from talking to the press by the editors of *Nature*. Those are the rules: if you want to be published in *Science* or *Nature*, you have to refrain from publishing your results elsewhere so that the journal can have the scoop. This created an absurd situation where the actual discoverers suffered under a gag rule while those who had confirmed their detection got the glory and became media stars.

After the first discovery, several new extrasolar planets (or exoplanets) were soon found by using the same spectroscopic technique. Marcy and Butler made many of these detections, and despite the media confusion surrounding the first discovery, their reputation as leading planet discoverers is completely deserved. In fact, they did not even have to go back to a telescope to discover more planets. They found them by "data mining," simply by analyzing data that was sitting in their hard disks from the past several years of observations. They already had evidence of more planets but had not found them because they had not bothered to look for the signature of planets with very short orbital periods, since everybody knows that giant planets cannot exist that close to their star. Often in science the really exciting discoveries are made when what everybody knows turns out to be wrong. In this case, once Marcy and Butler realized that giant planets can be on tiny orbits, they looked through old data and found the signature of several giants with orbital periods of just a few days. By 1996, they had found four more extrasolar planets.

Since then, in eight years we've found more than one hundred planets orbiting other stars. The number of known extrasolar planets now dwarfs the nine planets in our own system. Reports of new exoplanets have been pouring in so rapidly that individual detections have ceased to be big news, even as the expanding collection of new worlds continues to delight astronomers. For the first half dozen or so planet detections, each was given big headlines. After that, as planet detections have increased, newspaper headlines about extrasolar planets have decreased, but persisted.

The 103rd new planet is just not as exciting as the third, and not only the press has become jaded. Scientists, too, have gotten used to hearing about new planets. By 2002, it had reached the point where, at least on a busy day, when you see the headline "15 New Planets Announced at Conference," you might not even click on it to read more. Only headlines proclaiming new kinds of planets are guaranteed a click.

What the scientists know is that nondetections are important, too. Every time we observe a star and find no telltale wobble, it puts new limits on the number and type of planets that may be out there. The nondetections do not make headlines ("Scientists Discover Nothing"), but cumulatively they are as interesting as the detections. When we're wondering how unusual our own system is, and how this changes our estimates of the prevalence of life in the universe, it is the *pattern* that's important—the overall distribution of planets of different types orbiting stars of different types. This pattern will gradually be revealed over many years, and not in one headline-making discovery.

Planet finders are still holding press conferences, but only when they find new planets or planetary systems that stand out from the growing herd. In most cases, this involves systems that are, in some way, more like our own solar system than any other previously found.

A glance through the headlines of articles about extrasolar planets written by *New York Times* science writer John Noble Wilford reveals this progression. By 1998, individual planet detections no longer merited their own articles until June 26, when "New Planet Detected around a Star 15 Light-Years Away" announced one closer to the Sun than any others. On April 16, 1999, "At Long Last, Another Sun with a Family of Planets" described the first discovery of a system where three planets orbited the same star. On March 30, 2000, "Two Relatively Small Planets Are Found" heralded the discovery of two planets with masses slightly smaller than Saturn's. Most previous discoveries had revealed planets much larger than Jupiter and Saturn. By June 14, 2002, Wilford was proclaiming, "Astronomers Detect Signs

That Jupiter Has Distant Cousin." This article described the first detection of a planet whose mass and orbit are similar to that of Jupiter. We can expect that this parade of "firsts" will continue indefinitely, as improving observing technology, and more time to look, turns up systems that are closer and closer to our own in various ways.

THE FIRST HUNDRED

In part, this great emphasis on seeking systems with characteristics closely resembling ours arises because the first one hundred planets discovered are almost all completely unlike anything we had expected. This collection certainly does not resemble the planets of our solar system. Like the planet that Mayor and Queloz found orbiting 51 Peg, these are mostly "hot super-Jupiters": giant planets, mostly more massive than Jupiter, locked in tight little orbits, many of them closer to their home stars than Mercury is to our Sun. Not only don't these planets resemble the ones around here, but the architecture of all of these newly found planetary systems seems to be fundamentally different from that of our own system. They are built on a different plan.

Maybe we shouldn't be surprised. There are obvious "observational selections" at work here—aspects of the search techniques that bias our findings. The first ones you discover are the easiest ones to see, and they are not going to be representative of all that is out there. When you look at a distant redwood forest, you first see sequoias and miss the undergrowth, but that doesn't mean that the only inhabitants of the forest are the giant trees that dominate your initial view.*

Keep in mind that we don't "see" these worlds at all. Instead, we infer their presence by the gravitational sway they induce in their central star. It's like watching a hula-hoop contest from across a wide valley. We're too far away to see the thin little hoops, but why else would those big hips be shaking like that? What we actually notice is a big boogying star, doin' da gravitational bump with a slight, unseen companion.

We don't know much about these new planets, but we can tell a few things. We can guess their weight by the size of the stellar wobbles they cause. Giant planets are easiest to detect because, with their massive

---

*It's like pot smokers—the ones who call a lot of attention to themselves are a minority who tend to fit a certain negative stereotype. This creates a skewed demographic sampling for those who don't look closely.

gravitational pull, they really get their stars moving. So, of course the first ones found are preferentially "super-Jupiters." Most of them are more massive than any planets in our own system.

It's also easier to find planets on orbits that hug their stars closely. The undulations they induce are both larger and faster. Kepler's third law ensures that close-in planets will orbit faster than their more distant siblings anyplace where gravity follows the laws of gravity.* Stars with close-in planets wobble faster, so you don't have to observe for nearly as long to bag the planet. Who's got the time to wait around to observe the slower and smaller effects of a more distant Jupiter that takes twelve years to orbit?

Once we do find a planet, this same law allows us to deduce, from its orbital period, its distance from its star, which allows us to estimate its climate. The planet orbiting just 5 million miles from 51 Peg must have a surface temperature of nearly two thousand degrees Fahrenheit, which makes even Venus and Mercury seem temperate by comparison.

Given these artificial selection factors, the first hundred extrasolar planets we've found are naturally dominated by hot super-Jupiters, that is, by giant planets orbiting close to their star. We're beginning to get used to this population of oddball planets, but initially they came as a complete surprise because there weren't supposed to be worlds like that.

Six years before the first extrasolar planet was discovered, I finished grad school, where I learned the rules for building planets and solar systems. Back then, the expectation was that planetary systems elsewhere would, in their overall basic architecture, resemble ours. They would have a few gas giants out in the cold outer regions and, in the inner hot zone, a few rock balls, possibly one or two in the zone with watery swimming holes and maybe even something in there doing the backstroke.

Given the characteristics of the first hundred new planets, these expectations were clearly too narrow. Most of the giant planets discovered are orbiting close in, where theory told us it would be too hot for gas giants to form. More recently, several planets have been discovered with orbital distances and masses more closely resembling those of Jupiter and Saturn. However, unlike Jupiter and Saturn, many of these are not on well-behaved, near-circular orbits. Many follow wild elliptical orbits that would probably not allow for Earth-like planets to share their planetary system.

---

*And, dude, that's, like, everywhere.

Okay, so the first hundred new planets do not conform to our old expectations. What can we learn from this, other than (surprise, surprise) that we were naive? How can this ensemble of worlds help our effort to contextualize Earth and Earth life?

At this point, "Rare Earth" thoughts may start popping into your head: What if we live in a completely deviant solar system, and our presence here indicates that such an unusual location is required for something like us to come along?

The extent to which the earliest exoplanets found defied our expectations and differed from our own system has been seized upon by some to draw conclusions about our planet's uniqueness. In describing the extrasolar planets discovered up until 2002, the authors of *Rare Earth,* Peter Ward and Don Brownlee, wrote in their next book, *The Life and Death of Planet Earth,* that "astronomers were surprised to discover that most solar systems do not have the larger planets far from the Sun in well-behaved, nearly circular orbits as ours does." They used this to support their contention that systems, and planets, like our own are rare and unusual. Yet this statement is flawed. We simply don't have the data to draw such a conclusion. We cannot yet say anything about what "most solar systems" are like.

We are seeing an astounding variety in exoplanets—and much more variation in the basic architecture of planetary systems than we ever thought likely. From this, we are tempted to conclude that ours is not a garden-variety solar system, but we don't know this. What we can say is that the galactic garden has gone to seed and is overgrown with a surprisingly diverse bouquet of wild suns and their unruly planets. This certainly doesn't tell us that there aren't many other systems out there that closely resemble ours. Far from it. In fact, more recently we've found a few that do seem to resemble our own. It appears that they are not the majority, but even this appearance may turn out to be due to the observational biases that make certain kinds of planets much easier to find.

We have now looked around at a few hundred stars and found that around 5 percent have planets that we can detect with our current technology. At this point, systems like our own are still difficult to detect, and 90 percent of solar systems could look exactly like ours. We won't know definitively how typical our own planetary system is until we take a more thorough census of the planets in our stellar neighborhood.

All we now know for sure is that nature is, as always, more imaginative than we are, and planetary systems are much less uniform in construction than we expected. We're learning that the galaxy is not constructed like a Soviet-era Russian (or modern-American) housing development where each unit is identically built as far as the eye can see. Like houses in some urban neighborhood with a storied past, planetary systems are a hodgepodge of style and size.

Is there a silent majority, or even a noisy minority, of well-behaved, "normal" planetary systems obeying the rules and living quiet lives? Is there a vast population of systems like our own that haven't yet called sufficient attention to themselves to be picked up with our crude techniques? Statistically, even if our kind of solar system *is* quite rare, say one in one hundred thousand, then there must still be many millions of them, given the 10,000,000,000,000,000,000,000 stars the universe has to work with. Even if "good" solar systems are a tiny fraction of the whole, they might still be quite numerous.

And then, of course, we have to ask, what makes us so sure that the good ones are like ours?

ON THE EDGE OF KNOWING

For four hundred years we've believed in extrasolar planets. Now, the universe has shown us a sign and our faith has been rewarded. This is the beginning of the end of our long isolation in the realm of the Sun. The galactic distribution of planetary systems is being revealed to us, but not all at once. It's more of a striptease than a grand unveiling. With every new detection (or nondetection) our view improves slightly.

For astrobiologists, the most interesting extrasolar planet discoveries won't be made until we know about the harder-to-find planets, until we can overcome our strong observational limitations and uncover the true range and prevalence of planetary environments in the universe. The picture will come gradually, and the most mind-blowing revelations may never make the news.

What does it all mean? The most secure answer right now is "Come back and ask us in a few years." But we're not like that. Because we are nerds. Even when the data is piling in so fast that we know that it will all become more clear in months, we can't resist speculating, interpreting, pontificating even.*

For the next few years, the new discoveries will continue to be giant

---

*It's a common but little discussed problem: Do you suffer from premature speculation?

planets and the excitement will be over finding "other Jupiters" and Saturns. But what about finding "other Earths," small, rocky worlds that might harbor biospheres like our own? They are out there, too, and we'll start finding them before long.

A satellite called *Kepler* is going to find them. It will not have to travel far, only into Earth orbit, for a clear view of the stars. *Kepler* will find planets by watching for the faint flickering of far-flung suns. The tiny gravitational wobbles induced in other stars by alter-Earths are beyond the limits of our current searches. But we don't need to see the wobble if we can catch the shadow of the planet itself, as it passes our line of sight to its star.

Which reminds me of an effect I've sometimes noticed looking out of airplane windows. When I fly, I always try to get a window seat. I just love to be that high and watch the world go by, and I especially enjoy pondering and analyzing and explaining anything that appears mysterious.* On some of my first few trips back East after moving out to Arizona, I noticed something odd. As I watched the ground on a nighttime approach to an East Coast airport, I saw the lights madly flicker, in a way that they don't on approaches to cities in Arizona.

After noticing this on several flights, I decided that I, being Joe Scientist and all, should be able to explain it. Then it hit me like a branch in the face. It's the trees. The lush Northeastern forests versus the sparse Western deserts. Massachusetts, New York, and greater D.C. are much more heavily treed than Arizona—at least in the flat parts where they build major airports. Approaching Dulles or Logan at night, the flickering lights you see betray the branches of ubiquitous trees hiding in the darkness. Each little twig is briefly revealed in a twinkle as it occults the lights behind it. And that, my friends, is just how *Kepler* will find the little planets that our other searches cannot detect. If the galactic forest grows thick with small planets, *Kepler* will catch the occasional twinkling when one passes in front of a star.

It amounts to eclipse watching, from light-years away. Orbiting just beyond the murky, turbulent distractions of Earth's atmosphere, *Kepler* will continuously monitor the brightness of one hundred thousand stars for four years, watching for the brief flickering of distant partial eclipses. If such a dimming is due to an intervening planet, it will occur

---

*For this reason, if you ever fly with me, I recommend you bring headphones, so you can pretend you're listening to music.

at regular intervals as the planet orbits its star. The amount of darkening will tell us how large the planet is, and the orbital period will tell us its distance from the star. The biggest drawback is that a planet must be orbiting in just the right orientation so that it crosses the line of sight between us and its star. Inevitably, we'll miss many more planets than we'll catch with this technique. Still, we're quite sure to catch a small but reliable fraction of them. About half a percent of planets should, at random, be oriented in the right way to be caught by *Kepler*. This is not the way to take a complete census, but it's a great way to get solid statistical data on how common planets of various sizes and orbits are. Currently, *Kepler* is scheduled for launch in 2007. So if all goes well, later in this decade we should be getting some good information on the demographics of Earth-size planets in our home galaxy.

When we do find these "other Earths," they'll be too far away to see directly. We'll know their sizes and orbits and what kind of star they haunt. Can we learn anything else about them? Perhaps. If some of them are found around nearby stars, then we should be able to identify gases in their atmospheres with spectroscopy. This is no easy trick—analyzing reflected light from dark little objects so far away and so near their bright stars. You need an extremely precise way to separate two sources that are so close together and vastly different in brightness.* We don't have the ability to do this yet, but we've got plans like you wouldn't believe. An ambitious mission called Terrestrial Planet Finder is scheduled for launch in 2014 (don't be shocked if this schedule slips). This amazing machine will be able to see details one hundred times smaller than the Hubble Space Telescope. It should be able to measure the atmospheres on any Earth-size planets up to fifty light-years away.

We don't even know exactly what gases to look for in these distant atmospheres. Remember, we still don't have a *Galactic Life Detection Manual*. Until someone beams one down to us, we'll have to wing it. We certainly want to look for oxygen, water, methane, ozone, and CO_2. We'll look for things that seem anomalous, such as gas mixtures that are out of equilibrium, begging for reaction and explanation, or any other characteristics that seem surprising or unlikely to be caused by geophysical processes acting alone. For example, if we find a planet with both oxygen and methane, then something must be actively sup-

---

*For example, the Sun is 1 billion times brighter than the Earth.

plying one of these, or they would rapidly consume each other. That something could be life, or even alien technology.

With a range of techniques, we are zooming in on the demographics and lives of solar systems in our galaxy. Soon we will have some solid knowledge of the numbers and types of planets inhabiting local space. We'll even know a little bit about what some of these worlds are made of. Yet it will be a long time before we have a comprehensive understanding of the lives of entire planetary systems. This fully developed theory of comparative planetology is way off in the future—we'll only be able to do it once we've had an opportunity to explore dozens of systems close up. That's right, I'm talking about interstellar exploration—not remote observation, but actually going there. This kind of exploration seems quite far off now—probably as distant as voyages to the planets seemed at the beginning of the twentieth century. Yet by century's end planetary exploration was being taught in history class. Interstellar exploration could begin by mid-twenty-first century if things go well.

Given this timescale, studying planetary evolution for a living involves a certain intellectual self-flagellation. Why devote your career to a problem you know will only be solved after you're gone? I don't know. Maybe it's that cathedral thing. Whenever I see a giant cathedral that took generations to make, I think of the people building it who knew that they would not live to see it finished. You spend your life working on the flying buttresses, knowing that the roof won't go up for a hundred years. Why bother? Some worked under duress. But some did it because they believed that they were serving a higher purpose and contributing to something larger than themselves. You feel you are a part of something built to last. Perhaps that's why we do what we do— wrestling with the unsteady foundation stones of a theoretical edifice that will be completed by some later generation.

THE GOOD NEWS FOR MODERN MAN

The recent discoveries of extrasolar planets are quite encouraging for our ultimate chances of finding extraterrestrial life. They confirm a mode of thinking that all our theories for life elsewhere depend upon. As seers, natural philosophers, and astronomers have long surmised, for reasons ranging from mystical to mathematical, there are indeed many worlds out there. Score one for arguments by analogy. You do get the impression that we really can figure out some of the details of the rest of the universe even while confined, as we are, to our tiny little cosmic corner. Even when we had no way to find other planets, most of us believed, because it seemed so unlikely that our star alone should have orbiting companions. Like our belief in extraterrestrial life today, this was based more on informed intuition than direct evidence. Can our strong faith in the existence of other living worlds make the same extraordinary passage, from deeply felt belief to confirmed knowledge? Whether or not there are a lot of places like Earth in the details, we've already discovered something important for life out there. We know there are a huge number and variety of environments on various kinds of planets that we have barely begun to detect, catalog, or explore. Each of these, in our current ignorance, should be regarded as a potential niche for some kind of life.

Extrasolar planet detections have contributed to a new air of excitement and optimism about alien life, and a new respectability in scientific circles for the serious study of exobiological questions. Following our nonlinear childlike curiosity, we approach the truth of the universe through successive approximations. Every answer leaves us unsatisfied and provokes in us new questions. Planets elsewhere? The answer is yes. Yet they are not what we thought they'd be. Are we allowed to change the question? Can we get one more wish? What we really want to know is if other worlds are as friendly, fecund, and funky as our gurgling, green Earth. In searching, we'll repeatedly be surprised by avantgarde solar systems that we would never have thought up, and nontraditional orbital arrangements that we didn't even know were legal.

Life is a capacity the universe has to make small pockets of intense order and beauty. It thrives on certain physical conditions found on Earth. How specific and rare the required conditions are, we don't know. All of our ideas about life elsewhere are still educated guesses based on an extrapolation from this single example. Yet, any way you look at it, no matter what the planetary census ultimately shows, these exoplanets popping up around numerous stars in our galaxy are comforting on many levels. The question "Are we alone?" is really many questions. In answer to one of them we can now definitively say, "No. Ours is not the only family of planets." Phew. Didn't think so, but it's sure nice to know. It's reassuring when our widely accepted conjectures become empirically supported truths. Sometimes we need a sign to show us that we're not just lost in the wilderness following our own footprints, chasing our shadows. The universe is telling us that we are on the right track.

14 | Exobiology: Life on the Fringe

Surely one of the most marvelous feats of the twentieth century would be the firm proof that life exists on another planet. All the projected space flights and the high costs of such developments would be fully justified if they were able to establish the existence of life on either Mars or Venus.

—STANLEY MILLER AND HAROLD UREY,
WRITING IN *Science,* 1959

The most important scientific discovery that could be made in this century is the discovery of life elsewhere in the universe.

—ED WEILER, NASA ASSOCIATE
ADMINISTRATOR FOR SPACE SCIENCE,
QUOTED IN *Science,* 2002

DETECTING AND PROTECTING LIFE

From the beginning, scientists advocating space exploration have been motivated by the search for extraterrestrial life and have used this evocative topic as a means of rousing public and official support. Yet they have had to be cautious. Few scientists have been willing to predict short-term success, especially after the "disappointing" results of our first few planetary probes. Do we want to stake the future viability of planetary exploration on a long shot of uncertain odds?

Born at the fringes of the space program in the late 1950s, exobiology focused on two questions: detection and protection. How do we detect life on another planet? We can only search using a framework of preconceived ideas about what we are looking for, yet the ultimate object of our search is to test these same ideas. This paradox has forced us to accept fluid definitions and provisional search strategies.

The "planetary protection" question makes visible some ethical and practical conundrums raised by our new ability to travel across space: How do we protect other planets from contamination by organisms hitchhiking on our spacecraft? How do we bring back samples of other planets while protecting Earth from even a remote possibility of dangerous contamination? Is new knowledge worth a small risk of interplanetary ecological disaster?

The first respectable modern scientist to publicly voice questions about such cross-contamination was Nobel Prize–winning biologist Joshua Lederberg. In 1958, he published an article in *Science* expressing his fear that our impending Moon probes might inadvertently contaminate our natural satellite. That year, spurred by Lederberg's early activism, the National Academy of Sciences adopted a resolution calling on scientists to "plan lunar and planetary studies with great care" and encouraging "the evaluation of such contamination and the development of means for its prevention."

At that time the issue was couched in language that was more practical than ethical. The fear was not so much that we might commit an unforgivable and irreversible moral blunder by sacrificing one planet's biosphere to another's clumsy curiosity. Rather, the danger was of ruining future opportunities for scientific research presented by the study of uncontaminated worlds. The worry was about a crime not against nature, but against science.

Lederberg convinced the powers that be that we should take appropriate precautions as we began sending our contraptions beyond Earth's gravitational embrace. The problem of protection was given equal time with detection in a paper Lederberg published in *Science* in 1960 called "Exobiology: Approaches to Life beyond the Earth." If there is a founder of exobiology, it is he.

Winning a Nobel Prize is often said to mark the end of one's productive scientific career. Perhaps this is most often said by scientists who have not won Nobel Prizes. Certainly Nobel laureates have less to prove and are more free to indulge what may seem, to their more insecure colleagues, like crazy whims. In a sense, exobiology owes its birth to such Nobelian self-indulgence. Lederberg, commenting on the effects of his Nobel Prize, said that it had enabled him "to stay in a nonreputable game. Not disreputable, mind you, but nonreputable. It might have been very, very difficult otherwise and it would have been very hard for a capable young scientist who's had a lot of risks to take in his career to hitch it to something as uncertain as exobiology."

Since the early 1960s a handful of scientists have dared call themselves exobiologists. In the spring of 1960, recognizing that biology would be an important component of space exploration, NASA established an Office of Life Sciences. In 1962 an Exobiology Division was established at NASA's Ames Research Center in Silicon Valley. Ames has remained an important center of exobiological research ever since then, and in 1998 it became the headquarters of NASA's new Astrobiology Institute.

COSMISM

I've been describing all this from an American perspective, but on the other side of the Iron Curtain, plans were also being drawn for robotic visits to the planets. In fact, the Soviet scientific establishment was more open than American scientists to the study of extraterrestrial life. The ground was prepared by the philosophical tradition of Russian Cosmism.

Cosmism, or cosmic philosophy, was a space-related philosophy that thrived in the late nineteenth and early twentieth centuries. It combined ideas from Eastern and Western spiritual traditions (including a dash of Russian Orthodox Christianity) with a scientifically constrained understanding of the origin, history, and future of the universe. The cosmic role of humans and other sentient beings was emphasized. Many Cosmists believed that the entire universe, from tiny individual particles to stars and galaxies, was alive and conscious. Humans, it was believed, and other more highly evolved intelligent creatures, had the highest concentration of this consciousness. With this realization came a great moral responsibility to encourage the further peaceful development of advanced consciousness in the universe, uniting the spiritual and technological aspirations of humankind.

Several Russians who are better known to us for their scientific accomplishments were also deeply involved in the formulation of the spiritual, historical, and ethical ideas of Cosmism. One important contributor was Vladimir Vernadsky, who invented the concept of the biosphere (and conceived of most of the essentials of the Gaia hypothesis) in 1917. Vernadsky then went on, in loose concert with Teilhard de Chardin (who, though not Russian, was kind of a Catholic Cosmist), to elaborate the concept of the noosphere—a sphere of intellectual activity rooted in the Earth, and expanding into the wider cosmos.

The first person who explicitly connected cosmic philosophy with space travel, alien life, and the search for extraterrestrial intelligence was Konstantin Tsiolkovsky (1857–1935), a near-deaf, self-educated, rural-village schoolteacher of limited means and unlimited imagination.

Though isolated from the professional scientific community, Tsiolkovsky created designs for rockets and orbiting stations that greatly influenced modern space technology. He worked out the basics of multistage, liquid-fueled interplanetary rocket flight years before the Wright brothers took their first short, winged hop across a North Carolina field. Nearly a century later, many of his technical ideas still seem futuristic and feasible, such as his plans for domed habitats on the surfaces of asteroids.

Tsiolkovsky's numerous volumes of brilliant designs, drawings, and calculations were all in the service of a utopian vision of the human future. Space travel was a necessary step in our evolution toward cosmic perfection and infinite happiness. Ethical principles were built into the physical laws of the universe—a universe full of advanced extraterrestrials who had perfected interstellar travel. They left us alone, he felt, out of a sense of moral obligation. For Tsiolkovsky, the concept of God was indistinguishable from that of advanced extraterrestrial intelligence of the kind toward which humankind and its space-faring descendants would inevitably evolve. In his quasi-Buddhist notion of time, death, and rebirth, the essential nature of life was immortality, and death was an illusion.

He believed that human beings would move into giant, self-contained orbiting space colonies and later colonize the asteroid belt. Finally, he declared, we will evolve into "Homo Cosmicus," beings able to utilize solar energy directly, live forever, and freely roam the universe. The emergence of human intelligence on Earth, he believed, is an example of a wider cosmic movement in which the universe evolves toward sentience. Tsiolkovsky advocated an end to nationalistic thinking and urged people to think of themselves as citizens of the cosmos. He saw this change in perspective as necessary preparation for humanity's joining a cosmic community of intelligent beings.

American space-heads like me grew up reading about the technical achievements of Tsiolkovsky without learning of his extensive spiritual

Konstantin Tsiolkovsky, 1857–1935 (State Tsiolkovsky Museum of the History of Cosmonautics)

and philosophical writings. This was largely because these aspects of his work, and of Cosmism in general, were suppressed by the Soviet Union out of distaste for all things even quasi-religious. I knew of Tsiolkovsky's pioneering work on rocketry and his ideas about space colonization (probably first from reading and listening to Sagan), but I only recently discovered that he also wrote books on *The Natural Foundations of Religion* and *The Unknown Intellectual Forces of the Universe*.

When I first learned of Tsiolkovsky's philosophical ideas, as an adult, they seemed strangely familiar. This may well be because many of the space age thinkers I admired as a teenager[*] were influenced by his ideas. His mixture of Cosmic Evolution, belief in extraterrestrial intelligence, and the essence of Eastern and Western spirituality stripped of stale doctrine seems so obviously right on to me that my response on reading his ideas is to smile and say, "Yes, of course." Cosmism is the closest thing to a scientific religion I've yet seen.[†]

---

[*]Asimov, Sagan, and Clarke, for starters.
[†]Next time I am asked for my religion on a census form, I'll write in Cosmism.

The eminent Soviet astrophysicist and SETI pioneer Iosif Shklovskii (1916–85) was also working in the tradition of Cosmism, with its spiritual side deemphasized to a degree acceptable to Soviet authorities, when he wrote the classic work *Universe, Life, Mind,* in 1962. This far-reaching treatise extrapolated from modern astrophysics and biology to explore the role of life and intelligence in Cosmic Evolution. Stripped of its politically incorrect religious overtones, Cosmism blended smoothly with Communist doctrine.* Soviet ideologues didn't much care for the spiritual part, but they liked the idea that mankind was perfectible. In their hands, Cosmism mutated into a socialist technocratic utopianism in which, throughout the universe, historical progress must lead to life, intelligence, and ultimately the withering of the state and the creation of ideal workers' utopias ringing every stable star.

Shklovskii's book was updated and expanded by Carl Sagan, who became a coauthor of the English edition, published in 1966 as *Intelligent Life in the Universe.*[†] This book became a bible of exobiology. In a sense, it was a 1960s version of Fontenelle, meant to persuade "philosophers" (now called scientists) to adopt a pluralist vision, but written mostly in language that anyone literate could enjoy.

In *Intelligent Life in the Universe,* Sagan's comments are set off from Shklovskii's by triangles ∇ like this ∆. Occasional tensions are revealed between these two visionary thinkers, who saw eye to eye about cosmic history, but inhabited opposing Cold War societies. Shklovskii would begin a statement with "In accordance with the philosophy of dialectical materialism . . . ," and Sagan would come back with ∇ However, Kant's positivistic philosophy teaches . . . ∆. Sagan always got the last word, as in this passage:

> ∇ At this point in the Russian edition of the present work, Shklovskii expresses his belief that civilizations are not inevitably doomed to self-destruction, despite his description of contemporary Western literature as filled with details of atomic holocaust. He expresses his belief that as long as capitalism exists on Earth, a violent end to intelligent life on the planet is probable. There is reason to assume, he asserts, that future peaceful societies will be

---

*At the end of his life, Tsiolkovsky considered himself a Communist.
[†]Shklovskii revealed in his autobiography that he only found out about this arrangement when he saw a published copy of the book with the names of two authors on the cover.

constructed on the basis of Communism. I am able to imagine alternative scenarios for the future . . . Δ

The influence of both Tsiolkovsky and Shklovskii can clearly be seen in Sagan's writings. Many of the hip, cosmicomic, superballistic flights of scientifically constrained fantasy in *The Cosmic Connection,* Carl's first widely successful book, were distilled and spiced-up riffs on themes seen in *Intelligent Life in the Universe.* Sagan's early work was a major conduit through which the Western world was given a dose of Russian Cosmism.

Given this national philosophical tradition, exobiology (or astrobotany, as it was then called in Russia) did not have a huge credibility problem. In the 1950s, Russian astrobotanists used spectral studies of Mars and Venus to make detailed inferences about the nature of the vegetation that they believed existed on both worlds. After *Sputnik* and before the first Moon probes, concerns about biology, both detection and protection, entered into Soviet exploration plans.

The responsibility of planetary protection was taken seriously in both Moscow and Washington, even when it meant scaling back exploration ambitions. These inherently global concerns may have helped remind the Cold War adversaries that, as far as the health and safety of our planet is concerned, we all live in the same country.

WOLF TRAPS AND SETBACKS

In 1961, NASA awarded its first financial grant in the area of exobiology: $4,485 to the Yale microbiologist Wolf Vishniac to develop a prototype device for detecting microorganisms on Mars. The practical task of designing and building instruments to find extraterrestrial life forced scientists to sharpen their ideas. Vishniac assumed that life "always will be based on carbon," although he acknowledged, "It may turn out that we are deluding ourselves—that we are simply limited in our imagination because of our limited experience."

Vishniac's experiment, nicknamed the Wolf Trap, was designed to search for the presence of alien bacteria by looking for the changes in acidity and turbidity (light absorption) that, on Earth at least, always accompany bacterial growth. A later version of the Wolf Trap won one of four slots for microbiology experiments to be flown to Mars on the Viking lander. This was an exobiologist's dream. Wolf was going to

Mars. But, when the Viking project budget started to soar, cuts had to be made to save the program. To the dismay of Vishniac and the rest of the team, the Wolf Trap was removed from the spacecraft design.

Vishniac froze to death in Antarctica in 1973, while trying to use the Wolf Trap to prove that life could exist in the extreme near-Martian conditions found in the Antarctic dry valleys. Now we know that he was right. This harsh environment does indeed support microbial life, including clever things* that bask in the faint sunlight penetrating just below the surface of rocks. But it is not a place where complex animal life can easily survive, as Wolf Vishniac—the first martyr to exobiology—tragically found out.

Another scientist who took an early leap into exobiology was Gil Levin. His background was in public health, and he had already built instruments for automatic life detection that he used to test for microbial contamination in drinking water. In 1961 he received a NASA contract to develop an instrument called Gulliver, which consisted of a sample chamber in which Martian soil would be fed a broth of (hopefully) tempting organic compounds labeled with radioactive carbon. Any organisms that took the bait would exhale labeled carbon dioxide, which could be detected with a Geiger counter. If the Geiger counter started clicking, it meant something in the soil was eating the broth and breathing out the carbon as CO_2. Gulliver evolved into the Labeled Release instrument, which was part of the Viking biology package landed on Mars in 1976. Of all the original Viking biology investigators, Levin alone remains convinced to this day that Viking did find microbial life on Mars.

Other early ideas for detecting life on Mars involved using simple cameras and microscopes. It was recognized early on that one of the most promising ways to find life without making too many assumptions about its nature was just to use our eyes.

From 1960 to 1964, optimism and excitement reigned within the small community of scientists involved in exobiology. The field may then have been, as Lederberg said, "nonreputable," yet there was small but seemingly steady support from NASA.

Then came *Mariner 4*. Most press coverage of the mission had played up the drama of the search for life, which only served to heighten anticipation of spectacular imagery with revolutionary implications. The

---

*Or at least well-evolved things.

lunarlike cratered surface, and the absence of green fields, forests, streams, or any recognizable Earth-like surface features, was a letdown for both scientists and the public.

Mariner 4 caused the first major backlash (since 1920 anyway) in what has been a cyclical pattern of raised hopes followed by disillusionment. NASA's critics sensed blood in the water and went in for the kill. Congress held hearings and several scientists, most notably the microbiologist Barry Commoner and Philip Abelson, the editor of *Science*, ridiculed NASA's goal of searching for life on Mars as a wasteful, irresponsible, pseudoscientific pipe dream. In an editorial published in *Science* in 1965, Abelson warned, "In looking for life on Mars we could establish for ourselves the reputation of being the greatest Simple Simons of all time."

In the 1960s planetology was itself struggling to achieve respectability as a new scientific discipline. There has always been an insecurity about the future of our field since we are dependent on an expensive ongoing program of exploration. Nobody wanted to see planetology suffer further ridicule on the floor of Congress, and after *Mariner 4,* exobiology became further marginalized within the planetary science community. Percival Lowell's excesses had discredited planetary studies for much of the century, and the fragile support for planetary exploration could not afford to be held hostage to new Martian fairy tales.

Exobiology climbed back up a notch in credibility in 1971 when *Mariner 9* revealed Martian surface features that were clearly carved by flowing liquid. Mars was again teasing us with an almost Lowellian glimpse of a warmer, wetter past, and daring us to fantasize about Martians who had somehow survived the great drying. The Viking program of two orbiters and two landers, which arrived at Mars in the bicentennial summer, 1976, was sold, mostly, as search for life.

VIKINGS ON MARS

With Viking, exobiologists finally had a chance to try out some of the life-detection experiments they had been developing since the early 1960s. The most precious cargo—the essential core of Viking—was the biology instrument: a thirty-pound cube about the size of a microwave oven. It was one of the most complex machines ever built: a miniature, self-contained, automated laboratory complete with sample chambers, nutrient containers, canisters of radioactive gases, mini grow lights, valves, and Geiger counters. That little cube contained about forty thousand parts.

A mechanical sample arm scooped Martian soil into the biology instrument, which contained three separate experiments. One was Gil Levin's Labeled Release experiment, which tested to see if Martian microbes would eat our offerings of radioactively labeled nutrients. These "nutrients" were a broad mixture of organics including amino acids, nucleic acids, and vitamins—the kind of glop you could buy today at Vitamin Cottage. This Martian health food, jokingly called chicken soup, seemed so good that any organic Martians would not be able to resist a taste.

The second experiment was Gas Exchange, in which wet nutrients were mixed with soil and any gases given off were analyzed so that the tiniest breath or belch of a growing microbe would be detected. The third, Pyrolytic Release, exposed soil to gases with labeled carbon, then heated the sample to a high temperature, essentially cooking any life within,* and sampling the gases given off to see if any of the labeled carbon had been eaten by some form of life.

One aspect of these experiments seems strange in hindsight. Except for Pyrolytic Release, they were designed to find life that would thrive in Earth-like conditions, rather than Martian conditions. They were "wet experiments" using liquid water to encourage growth. Because liquid water cannot exist under Martian surface conditions, the samples had to be heated and pressurized to get the water to stick around.

Why go to Mars and look for something that can only survive in a warmer, wetter climate? The designers of the experiments were influenced by the idea that Mars goes through periodic climate changes. We know that the Martian spin axis wobbles over long timescales, but we really don't know how dramatic the climate swings are. The theory was that if Mars oscillates between its current severe climate and much more moderate conditions, it might have creatures that can persist in dormant form, for tens of thousands of years if necessary, holding out for the return of warmer, wetter days. The Viking experiments simulated the Martian conditions expected during such a thawing. As Sagan put it in *The Cosmic Connection*, Mars could harbor "sleeping beauties, awaiting a somewhat wet kiss from Viking."

---

*Pyrolysis is a technical, scientific term for "burning the crap out of something."

In addition to the three experiments in the biology instrument, the search for life was aided by the cameras and an instrument called the GCMS, for Gas Chromatograph Mass Spectrometer, which searched for traces of organic molecules in the Martian dirt. Cameras carried one important advantage: the only assumption they required about the nature of the life there was that it should not look exactly like rocks or dirt. The disadvantage, of course, is that most life we know of is microbial, so it's possible to land on a planet that is thick with life and not see any of it.

Sagan successfully agitated for photographic sequences optimized to search for moving creatures. Carl actually hauled snakes and turtles out to the Great Sand Dunes National Monument in Colorado and photographed them crawling across the sand with the spare Viking camera, to practice for the possibility of finding Martian "macrofauna"—large, visible organisms. Shortly after this, while my parents and I were visiting Carl's Ithaca home, he pulled out his new pictures of desert animals bizarrely smeared by the slow-motion camera and showed us how the distortion revealed the speed and direction of motion. If Viking did see turtles, snakes, or, as he liked to say, "silicon giraffes," Carl was not going to be caught unprepared.

The Viking cameras worked wonderfully. Seeing those pictures materialize in real time, strip by strip across a screen, was unforgettable. Just thinking about it brings me right back to being sixteen and in love with space. But the Vikings never did catch sight of anything crawling through the dust.

The instrument that, in the end, seemed to most surely rule out the presence of life was not part of the biological package or the cameras. It was the GCMS, which gave us the ability to detect even trace amounts of organic molecules mixed into the dirt. At each of the two Viking lander sites, separated by forty-two hundred miles, it failed to find any at all, down to the parts per *billion** level. That incredible, disconcerting absence of organics at the two Viking lander sites was damning for life on the entire planet. The fitful winds of Mars are constantly blowing dust around the globe, so Viking sampled much more than just those two sites. A planet with life anywhere, at least life that resembles ours at all, should, it seems, have some organic material spread around its surface. Mars has, to a close approximation, none.

---

*Pronounced "*bill* yun."

The Viking results at first seemed puzzling and ambiguous. For each of the three biological experiments, the criteria for a positive response—meaning that there was life—was established in advance. Such an approach is necessary to keep us honest. If we didn't have agreed-upon requirements for success, then it would be easier to twist the results to reach a desired conclusion. Ironically, all three experiments in the biology package produced positive results at first. Viking detected life according to these previously determined criteria!

When fed our "chicken soup," something in the Martian dirt immediately began exhaling labeled carbon dioxide. The Geiger counters went crazy, but then quickly settled down. This created some brief excitement, and a media feeding frenzy. Yet, ultimately, when the red dust settled, the results as a whole didn't add up to the signature of life so much as an unexpected kind of chemistry in the soil. At least we *think* we can distinguish a biological growth pattern from the pattern of a chemical reaction that just burns up the available fuel and then stops. Further, if you first heated the Martian soil to killing temperatures (320°F) and then repeated the experiments, you still got positive results. That doesn't seem like the response of living organisms. So, unless the life there is so different that we just don't know how to approach it, Viking found no life on Mars. A highly oxidizing surface was considered the most probable explanation for all of the Viking results, though this consensus conclusion is not by any means proven. Gil Levin still shows up at astrobiology meetings, insisting to anyone who will listen that his Labeled Release experiment *did* indeed detect microbial life in the Martian soil.

Norman Horowitz, a veteran exobiologist and lead investigator for the Viking Pyrolytic Release biology experiment, wrote in *To Utopia and Back,* his poignant book about the Viking search for life, "The failure to find life on Mars was a disappointment, but it was also a revelation. Since Mars offered by far the most promising habitat for extraterrestrial life in the solar system, it is now virtually certain that the Earth is the only life-bearing planet in our region of the galaxy. We have awakened from a dream."

The Viking biology experiments illustrated just how hard it actually is to search for life on another world. In retrospect the experiments might seem naive. But let's face it, any exobiology investigation will always be a bit of a stab in the dark. Most of them will fail, until someday, one succeeds spectacularly. It takes courage to so publicly, and expensively, put your ideas to the test. Viking now seems like a rite of passage we had to go through to bring our thoughts about life in the universe to a new level of sophistication. The next time we send experiments specifically meant to search for life somewhere, be it Mars, Europa, or elsewhere, we'll try to cast a wider net.

In the aftermath of Viking, there was another backlash, and throughout the 1980s and early 1990s exobiology persisted as a fringe element within planetary science, tolerated but not tightly embraced. It took a while for it to again become permissible to talk about life on Mars. But, clearly, we can't give it up for long. Soon, the question became, how can there be life that would not have shown up in Viking's instruments? Of course there are answers. In fact, after *any* negative test for life, we can always ask, "What kind of life might there be that we would have missed?" and we will always be able to come up with answers. What if Viking had only scratched the sterile surface of a world brimming with bugs who stay a few feet underground, hidden from the deadly radiation, the oxidizing surface chemicals, and Viking's sample arm? There might even be some underground hot springs where life is beautiful all day long. The idea was born that life can persist, hiding out in isolated oases. This idea still dominates contemporary discussions of life on Mars.

UNCLE CARL AND THE LITTLE GREEN MEN

One who was never afraid to keep pushing exobiology as a reason for planetary exploration was Carl Sagan. By the late 1960s he was gaining prominence as both a scientist and a popularizer, spreading his vision of modern astronomy infused with cosmic philosophy (stripped of its explicit spirituality). Carl promoted planetary exploration and the search for life with the passion of a true believer. In 1971, *Time* magazine called him "exobiology's most energetic and articulate spokesman."

Throughout exobiology's fickle fortunes in the 1960s and 1970s, Sagan stood firm in his belief that biological concerns were central to both public and scientific interest in space exploration. His aggressiveness and skill as a communicator kept exobiology in the face of the planetary community during the years when many hoped it would just go away. He reminded generations of planetologists that life is really why we care.

Carl helped planetary science and exobiology erase the Lowellian stain, but he somehow managed to generate a backlash of his own within science. College professors would take cheap shots at him in class. At Brown in 1981 my cosmology professor loved to point out in his lectures that he thought Carl's science was all wrong. This Sagan-bashing among scientists began after *The Cosmic Connection* became a best-seller and peaked after the *Cosmos* television series aired in 1980. There were several reasons for this. Popular works, no matter how innovative, creative, or substantial they might be, are suspect in the academy. There is a regrettable but pervasive attitude that any work that can be understood by a nonscientist is a waste of time for profes-sionals. There was also the unavoidable jealousy engendered by celebrity status in a world where teachers are criminally underrewarded and thinkers think themselves underappreciated. There was resentment toward someone who had become a multimillionaire popularizing dis-coveries that were often the result of someone else's hard work. But undoubtedly much of this scorn was simply the price for embracing exobiology.

At times, Sagan took a fair amount of slack from his colleagues, even as his visibility and popularity among the wider public soared. Many who have now enthusiastically embraced the new astrobiology move-ment were among those who snickered at Carl and dismissed exobiol-ogy as too flaky to be considered real science. Once, at a conference in Silicon Valley in the eighties, we were sitting at a table during a coffee break: me, Carl, his graduate student Chris Chyba, and scientist X—a prominent and accomplished planetary theorist. Carl expressed enthu-siasm for the exobiological implications of Dr. X's new results. X said, rolling his eyes, "Well, Carl, I like little green men as much as anyone, but . . . " and went on to give his more sober assessment of the true importance of the work, which had nothing to do with exobiology. Now X has become a recognized leader in astrobiology.

Sagan's experience was emblematic of the scientific ambivalence toward exobiology. Scientists enjoyed the public interest but feared the potential for ridicule. Nobody wanted his research to be seen as a search for "little green men." At planetary science conferences the occasional presentation about exobiology was sometimes met with gig-gles in the back of the room. Most scientists who dabbled in exobiol-ogy were careful to cover their asses (and their research assets) by also doing some more "serious" science. In NASA's official statements of goals and objectives, looking for life was always mentioned as one goal, but never as *the* goal.

Students excited about the search for life quickly learned that this was not the way to gain the respect of established scientists. When I was in grad school, which wasn't all that long ago, exobiology was semitaboo. I felt pushed and pulled in different directions. Sagan urged me to take my work in the direction of exobiology, but some of my professors at Arizona took a dim view of this. When I told Carl that I was doing some modeling of the effects of "impact-generated dust clouds" (dark dirt splashed up into the atmosphere when big rocks land from space) on the early Earth climate, he encouraged me to consider the implications for the origin of life. When Carl was excited about something you were doing, it was impossible not to respond, and we ended up collaborating on a study of large impacts and the conditions for early life. I would write up my results and send them to Carl in draft form. A month or two later I'd get my draft back, marked up in Carl's precise, minuscule handwriting, contributing everything from persnickety punctuation points to paragraphs of precise prose. We concluded that the early Earth, around the time of life's apparent origin, was frequently shrouded in a cloud of dark dust, which would have kept the surface frozen much of the time.

In this kind of science you have to make some fairly crude assumptions, because there is so much we don't know about the environment of the early Earth. The payoff is a new and interesting conclusion about the environment where life first emerged. Much of exobiology (and now astrobiology) depends on assumptions that are hard to test, and which either lead you astray or lead you to some new answers on the really big questions.* Depending on your attitude, this kind of science is either flaky or cutting edge.

One time I got up my courage to ask the Great and Powerful Professor Don Hunten—a leader in the theory of planetary atmospheres—for his opinion of some calculations that I had done for an upcoming talk. It was to be one of my first invited talks, at an international conference in Helsinki, and I wanted to make sure that I had done things right. The topic was "The Effect of Asteroid and Comet Impacts on the Early Terrestrial Environment," and Sagan was my coauthor.

---

*You know, like how do you pronounce Io and Uranus, and is the hokeypokey *really* what it's all about?

Hunten had a scary reputation but he is one of the smartest guys there is, and I always found it rewarding to get his scientific advice when my ego could stand it. After a few moments of silent prayer in the hallway, I knocked on his door and said, "I have something I'd like to get your comments on, but I'm worried you might think it's a little flaky." He glanced at the Grinspoon and Sagan abstract, cleared his throat menacingly, and grunted, "Well, considering who the two authors are, I wouldn't doubt it," without cracking a smile. I suppose I set myself up for that. Though I temporarily felt two inches shorter, as usual Hunten followed the sarcasm with a thoughtful critique that helped me improve the work. It ended up expanding into a chapter of my Ph.D. thesis.

My thesis, completed in 1989, touched on exobiology in several places. Half of it was about how terrestrial planets get their oceans, and the other half was on the climate of the early Earth. In my flowery introduction, I related my research to the habitability of planets and the search for extraterrestrial intelligence. In the rest of the thesis I stuck to the "serious" scientific problems at hand. I did, after all, wish to graduate. As I reread my dissertation now, it looks like a work of astrobiology, only we didn't call it that back then. Times have changed.

Ultimately, the public attention gained for planetary exploration helped offset the perceived drawbacks and risks of exobiology. Scientists took note of Sagan's clear success in helping maintain official backing for our planetary missions by tapping into the public's sense of wonder. Although it is hard to avoid some sarcasm when one of your own starts showing up as a regular guest on Johnny Carson's show, alongside Tony Bennett and the dancing ferrets, most people in the field came to appreciate the role that Carl had played in relating planetary exploration to the masses.

Exobiology is over forty years old now. It was born at the same time as planetology, and though we have sometimes tried to ignore it, it has been here all along. The ideas promoted by Sagan throughout his career are just now, seven years after his death, being embraced by many at NASA headquarters and in the planetary science community as the central themes motivating our solar system exploration strategy.

15 | Astrobiology

It will be especially interesting to see whether it is astronomy that absorbs biology, or the other way around.

—FRED HOYLE

Nothing takes the past away like the future.

—MADONNA,
"NOTHING REALLY MATTERS"

SAY YOU WANT A REVOLUTION

True story: During the First Astrobiology Science Conference at NASA's Ames Research Center in April 2000, President Clinton was, by coincidence, landing at the adjacent Moffett Field Air Base, where Air Force One parks when the president comes to the San Francisco area. In a scene right out of *The X-Files*, one of his Secret Service men, who had stopped a suspicious-looking scientist, was heard to shout anxiously into his walkie-talkie, *"What the HELL is astrobiology!?"*

It's a good question. Popular books and magazines announce it as a new scientific revolution. In the introductory chapter of *Rare Earth*, entitled "The Astrobiology Revolution and the Rare Earth Hypothesis," we read, "A whole new science is emerging: astrobiology, whose central focus is the condition of life in the Universe. . . . We are witnesses to a scientific revolution. . . . It is very much like the early 1950s, when DNA was discovered, or the 1960s, when the concept of plate tectonics and continental drift was defined."

But where is the revolution, and will it be televised? What is the new world-melting idea? We think of Copernicus banishing Earth to the margins of space, Darwin linking all past and present life together,

Einstein warping light and time, Bohr dissolving solid matter at its smallest into shadowy waves of quantum probability. These ideas destroyed our notions of space, time, matter, and life. They left us permanently changed. What about the "astrobiology revolution"? What's the big idea? That other planets may be inhabited? That is certainly a pretty rad thought, but we've been kicking it around for at least twenty generations.

Well, then, is all this "revolution" talk based on the premise that we are perched on the edge of a devastating new discovery that will rock our universe? Thirty years ago, Carl Sagan wrote in *The Cosmic Connection* that the study of extraterrestrial life had "finally come of age." The preface to David Darling's 2000 book, *Life Everywhere: The Maverick Science of Astrobiology,* stated, "Poised on the brink of a momentous breakthrough that will change forever how humankind thinks about itself and the universe around it, astrobiology is quickly coming of age." Which causes me to wonder: Just how long can we stay poised on a brink? We don't know if we're any closer to finding life now than when Viking was on its way to Mars in the 1970s. When we do make the big discovery, bag an ET, which could be tomorrow afternoon in time for the evening news, or two centuries from now*— *that* will be the start of the real revolution.

If astrobiology is not a revolution, then what is it? Perhaps a new discipline, a novel field of research. Try telling that to anyone who was already pursuing exobiology before the new hubbub began in the late 1990s. Some recent descriptions of astrobiology tend to ignore exobiology's checkered history and present it as a virgin birth, sprung immaculately into the world. In truth it is more of a resurrection. Is that all it is? Have we just dusted off exobiology, given it a new name, and sent out press releases?

No, it's more than that. Astrobiology may not be, as advertised, a scientific revolution, but it is an important new movement—marking a shift in attitude about ET life. Exobiology had always survived on the

---

*In general I say the sooner the better, but I'm not sure how I'd feel about a major discovery between now and the publication date of this book. . . . When Kubrick was working on *2001: A Space Odyssey,* he actually tried to buy an insurance policy against the possibility that extraterrestrial intelligence would be discovered before the film opened, rendering his work obsolete.

fringes of planetary society. It was fed scraps but made to sleep outside. Reborn as astrobiology, it has rather suddenly been invited inside the main house and embraced as the mascot of our space science enterprise, receiving official encouragement and generous funding in the bargain.

In 1998 the NASA Astrobiology Institute (NAI) was started with an initial annual budget of $5 million. By 2002 this had increased to $15 million (this does not include the much higher cost of spacecraft missions). As NASA has funneled money into astrobiology, planetary scientists are discovering a latent interest in the astrobiological implications of their research. Biologists, chemists, and earth scientists are joining in the feast. Nothing like a new watering hole to get all the jungle animals to pay attention, come together, and talk about life. For some, it was a chance to finally receive funding—and community approval—for research they had always wanted to do. For others it was a chance to branch out into an area they had never considered working in. Since the late 1990s, planetologists, biologists, and others have converged several times a year at large astrobiology conferences, and two new journals have started up, *Astrobiology* and the *International Journal of Astrobiology*. Now astrobiology is going worldwide. Centers affiliated with NASA's institute have sprung up in Spain, France, Australia, the UK, and Japan. In May of 2001 the First European Workshop on Exo/Astrobiology (a title that hedges its bets) was held in Frascati, Italy.

Increasingly as well, astrobiology has become the public face of NASA, in press releases, schools, TV documentaries, and museum exhibits—astrobiology is the new hook. Across a wide spectrum of activities, aliens are *in* at NASA, like never before.

But, why now? The mid-1990s saw a convergence of several surprise breakthroughs. Each, by itself, would have generated sparks, but combined, they ignited a conflagration. These were, in order of importance, (1) the discovery of possible fossils in the Martian meteorite ALH84001, (2) the discovery of the first planets outside our solar system, and (3) *Galileo*'s images confirming the likely presence of an ocean on Europa. When all these occurred within a two-year period, the excitement was cumulative. Something happened.

What happened was that Dan Goldin, NASA's administrator at the time, took heed of the lavish press coverage and the enthusiastic public reaction to these discoveries. Halfway through his nine-year tenure at

NASA (1992 to 2001), Goldin seemed to become convinced that exobiology, rechristened as astrobiology,* should largely define NASA's mission and its public image. We were given a green light to write press releases and funding proposals highlighting the question of alien life.

This fanned a positive feedback loop between media, science, and government, all egging each other on. An astrobiology spin helped get their science in the papers. The new wave of public visibility pleased NASA administrators and made it easier for them to sell their programs to Congress. This translated into increased funding. Suddenly, astrobiology was not only hip but profitable. A renewed research focus, more headlines, and more funding all followed.

It helped that, by the nineties, the second generation of planetologists was becoming well established in the field.[†] The insecurities about being taken seriously by other sciences, and the reluctance to be associated with the search for alien life, were fading away. Also, since the time of Viking, planetary science (and science in general) had become much more media savvy. Scientists realized, especially after the Cold War fizzled, that they could not take public support for granted. No American planetary missions were launched for a decade between *Pioneer Venus* in 1978 and *Magellan* in 1989. Through the crucible of near extinction, planetary scientists became better adapted to the media age. In the nineties it became de rigueur to issue a press release anytime a paper was published that might possibly be seen as newsworthy.

When "the Mars rock" was greeted with global headlines in 1996, it suddenly seemed that the only angle that mattered was the life angle. Stories about all areas of planetary science were being reported with an alien-life spin. Any result that could be cast in terms of the search for life had an excellent chance of making the news.

Meanwhile, out at Jupiter, Galileo had recently entered orbit and had its first close encounters with Europa. The spacecraft, sharp-eyed but autistic, was slowly sending down a stream of new images that increased the circumstantial evidence for an underground ocean slosh-

---

*Wes Huntress, the former NASA associate administrator for Space Sciences, apparently suggested this title for NASA's renewed commitment to what had formerly been called exobiology.

[†]It cracks me up to see my grad school contemporaries leading planetary missions, chairing important committees, and pontificating at meetings. They act like real scientists, but I am not fooled.

ing beneath Europa's cracked icy face. *Galileo* scientists were encouraged to talk about what this could mean for life there, in a way that would have been frowned upon seventeen years earlier during the Voyager encounters. Times had changed and attitudes had shifted.

Planetary encounters in the late 1990s were much more strategically packaged for the public than they had been in the 1970s. The rocks at the *Mars Pathfinder* landing site, where the little *Sojourner* rover romped in July 1997, were given cute, media-friendly cartoon names like Scooby-Doo and Yogi. The rocks at the Viking lander sites had mostly just been assigned numbers.* The new rocks weren't any cuter than the old ones, but a later generation of scientists, raised on cartoons, video games, MTV, and computers, was more tuned in to the rhythm and the value of catchy sound bites.

At times we have gone overboard, conspiring with the media to exaggerate or distort the significance of our results. It gives them easy headlines and us an ego boost, visibility, and easier access to funding. For a while, it seemed that nearly every discovery of the *Galileo* mission, from magnetic fields to intriguing surface patterns on the moons, was somehow spun for the press with an extraterrestrial-life angle. I realized that this had gone too far a couple of years ago when I received an e-mail from my eleventh-grade English teacher, Martie Fiske.† An avid follower of science news, she was annoyed by the twentieth story she had read that year with the headline "*Galileo* Discovers New Clues to Possible Life in Europa!" Martie asked me, "Why do you people keep feeding us the same story over and over again?"

Though it can be taken to questionable extremes, all in all it's a welcome change that scientists have become more aware of how their work relates to the concerns of John Q. Public. Newsworthy angles are now sought for results that might otherwise seem obscure. Sometimes the right spin can reveal a human-interest angle lurking in the most arcane research.

The movement was encouraged by the (accurate) perception that new funding was available for old projects successfully recast as astrobiology. To some extent, success in securing funding has gone to those

---

*Actually, four rocks at the *Viking* 2 site were named Mr. Badger, Mr. Mole, Mr. Rat, and Mr. Toad after characters in *The Wind in the Willows,* but these never received the prime-time star billing that the *Pathfinder* cartoon-character rocks got twenty years later.
†A teacher who changed my life by getting me to read the right books at the right time.

who could most eloquently describe, in astrobiological terms, whatever research they were already doing. This is not a bad exercise to have to go through—relating your work to the big picture. At any rate, it's what we call good academic survival skills, and if you don't have them, you're probably doing something else.

We've all done some of this repositioning. For over a decade, with funding from various NASA research programs, I've been making computer models simulating the evolution of the environments of terrestrial planets. These models can also be used to explore questions about the early habitability of local planets, as well as the habitability of Earth-like planets (still hypothetical, but not for long) around other stars. Recently I've received money from both NASA and the National Science Foundation to use my models in the service of astrobiology—something I used to mention only as an aside in my proposals.

VIVA!

Astrobiology may at times have been falsely hyped as a scientific revolution or a brand-new discipline, but it is a refreshing and encouraging development. A revolution really is going on—not a scientific revolution, but a revolution in the *culture* of science, one that is healthy for science in a number of ways.

First, the biocentric tilt of NASA allows us to come clean about our true reasons for wanting to explore and understand the cosmos. Questions about life in the universe have always been behind our exploration of space. We just haven't always been free or willing to admit it. The rise of astrobiology as a highly visible backbone of NASA's space exploration efforts represents a new alignment of our public priorities with our private dreams. Even when we explore desiccated, sun-broiled Mercury or frigid and fragile little comets—places we don't really expect to find life—we are searching for ourselves out there, trying to understand just how we've arrived on the cosmic scene. By throwing its official weight behind the search for life in the universe, and sending the signal loud and clear that it is respectable and safe for serious scientists to think about such things, NASA has moved these questions from the fringes to the center, where they belong.

Of equal significance, astrobiology is bringing our space research more into line with the public's desires for NASA. You could look at this as merely improved marketing, but NASA administrators are encouraging us to pay more attention to what people respond to. As well we should. It is your tax dollars that pay for our science and exploration. We need to avoid the "Europa Effect" and not pander or issue near reruns of press releases to boost our ratings, but by focusing on the question of life we are giving the people what they want.

A revolutionary discovery of life elsewhere could arise from these efforts, but astrobiology's certain radical potential is in the way it bucks two deep trends in modern science. One is the tendency, in recent decades, for science (like everything else) to become much more market-driven. Profit is hot. Pure knowledge is not. An increasing portion of research is corporate-funded, which often blurs the lines of scientific ethics. Particularly in the biosciences, corporate support has led to troubling conflicts of interest between scientists' pursuit of knowledge for the sake of humanity and the pursuit of private gain.* At the same time, government-funded research has increasingly had to justify its existence in practical, economic terms. Fortunately, Darwin and Einstein did not face this pressure when they developed their useless theories.

Swimming against this stream is astrobiology. It is not for profit and can't pretend otherwise. We explore space for reasons that are romantic and idealistic. The universe beckons. We want to go because we want to know. With astrobiology there is no fronting that the rationale is practical or the benefits material—we do it out of curiosity and longing, to satisfy the human need to know the cosmos that spawned us. Fancy that: a scientific movement that is justified on fundamentally spiritual grounds.

Astrobiology is also potentially revolutionary in its attempt to reverse the slide toward increasing scientific specialization and isolation. We want to blur the borders and tear down the walls that modern academia has erected. Astrobiology at its best is a step toward the reunification of science and, perhaps, the rebirth of natural philosophy (see chapter 16).

Right now, this cultural shift is more important than any new scientific result. Scientists from different fields are working harder at communicating with one another. We have to. When I was invited to give a

---

*Some mad scientists out there think they own parts of your genome!

talk on planetary atmospheres last year at the University of Washington Astrobiology Program in Seattle, my audience consisted of biologists as well as geologists and astronomers. I had to think more carefully about the language that I used, avoiding jargon that might slip in if I was speaking to a roomful of fellow planetologists.

One cool thing about planetology has always been the chance to learn a lot of different kinds of science. Now this includes biology, too. For this reason I love going to astrobiology conferences. You never know what you're going to hear. The official support for astrobiology is making scientists braver in attempting to bridge disciplines. I say "attempting" because we're out of practice being interdisciplinary, and so there is an aggravating side to it, too. The enticingly eclectic mixture of disciplines can also be a recipe for frustration because we don't all speak the same language. All scientific conferences provide some mixture of fun and exasperation. Astrobiology meetings have more of both.

The interdisciplinary intentions of astrobiology are a challenge not just for scientists but for many others who round out the cultural enterprise of scientific research and communication. A life sciences editor at *New Scientist* magazine recently told me that astrobiology stories are always falling through the cracks because the life sciences editors want to give the stories to the space sciences editors, and vice versa. We are all used to driving along in well-worn ruts.

Astrobiology is embraced by many in the planetology community, but others feel that it is being forced down our throats. Some think that we were doing just fine before astrobiology became all the rage, and that it is strategically dangerous for the planetary community to identify too closely with the search for life. Believe me, no small amount of grumbling and cynical joking about astrobiology goes on in the hallways of astronomy departments and research institutions.

Yet, there may be no turning back. NASA has thrown itself into astrobiology, and our administrators have let the planetary science community know that we are to be astrobiologists. We need the biologists now. By making ourselves dependent on astrobiology we're placing a lot of trust in that relationship. This is no longer just a flirtation—we're committed to an ongoing dance with biology. A divorce at this point would be messy, embarrassing, and costly. After all, add biology to astronomy and you get astrobiology, but try to remove the *bio* from astrobiology and what are you left with? *Astrology:* an astronomer's worst nightmare!

HIVE-MINDED

In its interdisciplinary nature, astrobiology may be breaking essential ground for the future of all science. Our progress in understanding the universe is hindered by our inability, and/or reluctance, to cross the artificial boundaries imposed by our institutionalization of science. But if interdisciplinary work remains a huge challenge, it is not only because of bad attitudes. There are good reasons why we specialize.

We need to be generalists, but not dilettantes. Different scientists need to pursue separate specialties for us to collectively maintain, and increase, our physical understanding of the myriad "small picture" problems that help us test our ideas about the universe in concrete ways. This is why we are reductionist—we have to break the universe into little pieces to have a chance at understanding any of it. The picture only gains solidity from the detail work, which must often be done at close range, with a narrow focus. Then we need to step back and take in the larger view.

We know too much. Our squishy little brains can't handle it all. Our knowledge has increased exponentially and nobody knows everything.* We can only transcend the limits of individual knowledge, and narrow disciplines, to the extent that we can talk to each other and trust one another. Like a distributed computer network that is much more powerful than any single node, we can create a larger whole, and the combined effort is something that no individual, or individual field, could pull off.

The problem, then, is not specialization. It's isolation. We can see more by dividing the universe into pieces, but only if we are able to put it back together again. Forging a common scientific language is hard, but necessary. When we do this, the scientific community functions as a sort of "hive mind," in which the capacity of the collective is greater than that of the individuals involved. By using our ability, however limited it may be at present, to achieve a kind of group mentation, we get smarter and extend our powers, as a species, to comprehend the cosmos.

We also extend our capabilities by becoming machine/human hybrids. Integral to the Astrobiology Institute is an effort to use new telecommunications technologies to facilitate virtual collaboration

---

*And the ones that do are too annoying to talk to, unless they're also very funny, like Isaac Asimov, but his kind do not come along often.

between distant research groups. Novel forms of scientific visualization allow us to extend our limited senses, rendering visible previously hidden patterns of nature. With all of our telescopes, laboratories, and interconnected computers, we are as cells in an evolving superorganism in which the planetary/human hive-mind reinvents itself, grows new organelles of metal and glass, and sprouts veins flowing with electrons and light. Like a baby reaching for mysterious objects floating over the crib, we extend our hands away from Earth.

This global mental activity encompasses not only the science and technology that allows us to gain new insights about the cosmos, but also the collective awareness of this expanding view by the entire species. We are gathering knowledge for all mankind, for the noosphere, but only if we get the word out. So outreach beyond our own professional communities needs to be an integral part of our science.

If astrobiology is going to be a new metadiscipline, communication must play an unprecedented, central role. The skills we need to communicate with other scientists—to make our work comprehensible to all—are the same skills we need to communicate with the public. A spin-off from the move to mix disciplines is that, as we get better at talking to those outside our little science cliques, it spills beyond the boundaries of the professional world and out into the streets, where compelling, comprehensible science is wanted and needed.

JUST GOTTA POKE AROUND

In the past few years I've been willingly sucked into the maelstrom of committees, administrators, congresspeople, scientists, engineers, lawyers, journalists, activists, and bureaucrats that is space policy. This started in 1998 when I was asked to join NASA's Solar System Exploration Subcommittee (SSES), a group of twelve scientists that reports to the associate administrator for space sciences, providing scientific input for NASA's space exploration plans, and making policy recommendations.

Right now there is a hot new word in space policy: *astrobiology*. A debate is going on over whether our exploration strategy should be "biocentric." Should "the life question" be the stated rationale behind our entire exploration program? Funding for exploration always feels precarious because of the constant danger that it could be declared an unnecessary frill by the powers that be. Along with art and education, exploration is one of those activities that a society can briefly convince itself it can do without.

Some believe that astrobiology can save planetary exploration by giving our program an attractive and exciting new focus. Others feel that this strategy would be unwise. Mindful of the cyclical history of public support, and scorn, for our search for "little green men," they wonder at the wisdom of putting all of our eggs in this alluring but potentially fragile basket.

While the scientists debate this, NASA and the last two presidential administrations have already decided that our exploration program is to be focused on astrobiology. A biocentric approach to exploration is even specifically mandated in President Bush's 2003 budget request for NASA, which states that from now on our missions will have "clear science priorities that support key goals in understanding the potential existence of life beyond Earth and the origins of life."

So, there you have it. The president wants us to find life. That's cool. I don't have a problem with that. But how do we actually go about exploring in a biocentric way? This means different things to different people. For some, it seems to translate into "Explore Mars and Europa and everywhere else can wait."

When we propose new space missions, we have to make the case as compelling as possible, because our proposals are just chirps in a crowded nest of hungry little birds, beaks open wide and desperately hoping for that big worm from Mama NASA. These days everyone knows that if we want to get fed, we had better squawk loudly about astrobiology. We all want our planetary missions to be as sexy as possible, and "to seek new life" has an enticing ring, whereas simply "boldly going where no one has gone before" sounds like a rerun. Making this connection is sometimes seen as a great challenge for planets that are generally regarded as big biospheric losers.

It's hard to think of two more different planets than Venus and Pluto, yet politically they have in a way ended up in the same boat (I hope they're not sharing a cabin because they'd be fighting the whole time over the thermostat.) Pluto's biggest problem is that it is not Europa. Venus's biggest problem is that it is not Mars. Both Venus and Pluto should factor into a broad biocentric exploration plan. Venus is, in many ways, our best hope for learning about the ongoing functioning of complex Earth-like worlds. As a representative of a completely unexplored realm, Pluto is an unopened time capsule dating from the earliest days of the solar system. Pluto will teach us about the history of planetary ice and how some of it became water and then life on Earth. Pluto also seems to have an active interchange between its surface and atmosphere, so we are almost guaranteed to find some kind of complex phenomena there that will surprise us when we finally see them. It is also the case, if you are attached to "life as we know it," that even a liquid water ocean in the interior of Pluto cannot be ruled out. Only a misguided notion of what it means to be biocentric would deter us from going to Venus or Pluto. Right now things are looking good for both the little cold one and the big hot one. The first ever mission to Pluto—a flyby that will hopefully launch in 2006—is being funded, and for the first time in many years plans are being drawn up for a major new American Venus mission. Yet, these missions have both faced major uphill battles. Each has been accused at various times of not going to Mars or Europa.

In my capacity as an adviser to NASA, I have argued that astrobiology should not change our strategic exploration plans to a large extent. Never mind "still haven't found what I'm looking for": we still don't know *what* we're looking for, and we won't know until we find it and hear ourselves ask, "How come no one thought of that!?" We should simply continue to explore the solar system widely, seeking a more complete understanding of the planets, always keeping one astrobiological eye open for the strange, the anomalous, the complex, the improbable, the "unnatural" signs of life. Certainly we should have an openly biocentric *attitude* as we explore and keep thinking about the kinds of features and patterns that *might* indicate life. If we do find something that really seems like a sign of life, we will immediately alter our strategic plans. If it looks like somebody might be home somewhere, we'll toss our plans out the window and go back for another visit.

Approached from this perspective, *biocentric planetary exploration* is just a slightly longer and more compelling phrase for *planetary exploration*. Having said this, I don't have a problem with calling our exploration program biocentric. If it speaks to why we do what we do, why not say it?

The problem is, we want to tap into the public fascination with the question of life to build support for our missions of exploration. It's much easier to do this by building a specific expectation of finding life. Even when we don't do this deliberately, it's natural for journalists to want to play up this angle. The "search for life" makes better sound bites than does "a wide effort to illuminate the mysteries of the solar system," but the latter might end up teaching us more about life.

Of course it would be worth almost anything to find out what life really represents in our universe. But we have limited resources. Investing money in a targeted search for life elsewhere is more risky than any bet in Atlantic City, since we don't really know what kind of game we're playing. We have to base our strategy on educated guesses. We want to go and look anyway because, well, how could we not? This argument is not sufficient to sway congressional committees, or the White House Office of Management and Budget. It doesn't do to say, "Okay, Madam Senator, we admit we are stabbing in the dark, but if you give us a lot of money, we'll certainly learn much of value, and who knows, we *might* make the discovery of the millennium." Our missions to other planets are in part scientific experiments, but in large part they are just poking around the neighborhood to see what we dig up. So we ask for money for exploration and seek to justify it as science.

Now that *exo* has morphed into *astro,* is biology here to stay as a linchpin of our space exploration plans? If life continues, like quicksilver, to elude our grasping hands, will we stay on task? By casting our lot in with astrobiology we are expressing faith in its longevity. Astrobiology is hot now, but historically both scientific interest and government support are cyclical. It remains to be seen whether we can commit for the long haul. But for our society this is good practice at thinking on long timescales, which as a survival skill is as essential as learning to think globally.

WHAT ABOUT THE ROCKS?

And what of the little worms who started all this astrobiology fuss, the "microfossils" in the Mars rock? In the eyes of most of the community, they have been demoted to strange mineral deposits that were probably not made by living organisms.

In the seventeenth century Johannes Kepler described oceans and an atmosphere on the Moon. The excitement generated by Kepler's reports helped spur scientific interest in a plurality of worlds, and that interest retained momentum long after better telescopes and more objective observers revealed the Moon to be a dry, battered ball of rock.

Similarly, in the late twentieth century, astrobiology was given its biggest impetus by a discovery most practitioners now regard as highly dubious. As with the Viking biology results two decades earlier, the confusion underscores the difficulty of identifying, or agreeing on, signs of life.

Personally I'm not sure. I don't think we can entirely rule out the possibility of some biological products in the Martian rocks, but I am highly skeptical. One problem is that we know nothing of where on Mars these rocks came from, since we received them as random shrapnel sprayed by some long-forgotten impact. Assuming we can find more rocks with similar features on Mars, it will be much easier to interpret them when we can examine the environments where they formed.

There are, in fact, living and fossilized terrestrial bacteria that look very much like these things, with the same segmented, elongated shapes. However, the Martian "worms" are incredibly tiny compared to any organisms we know of on Earth. They are about one hundred nanometers long and ten to twenty nanometers wide. (A nanometer is a *billionth* of a meter.) The smallest bacteria on Earth are hundreds of times larger than this. Life-forms using our type of chemistry probably cannot fit into a package so small. You need to have a container large enough to hold the genetic material that describes how to make the container.

Today, the general attitude in our field is surprisingly scornful of the Martian fossils, even among many who are enjoying the funding and interest in astrobiology that they sparked. Scientists agree that we need more data from Mars. It is commonly said that we won't be able to answer the question definitively until we return carefully chosen samples from Mars and analyze them in our labs on Earth. Even this may be optimistic. It is quite likely that once we do return the first Martian samples, the answer about the "fossils" will still be "maybe."*

The scientists who claimed that they had found fossils in ALH84001 did us all a big favor, and I don't just mean the increased funding for research in astrobiology. They got us all thinking. When we asked ourselves, "Could this be real?" we realized that there is no reason why we shouldn't find ancient fossils on Mars, even if there are none in these

---

*Returning a small sample from Mars is turning out to be much more difficult, technologically, than we once thought. When I joined SSES in 1998, the central goal of our Mars program was to return Martian samples to Earth by the year 2005. Then it got pushed back to 2014, and now 2020 is looking more realistic . . .

particular rocks. And, the reasoning continues, since the extremophiles on Earth show us that life can adapt to a surprising range of conditions, couldn't this ancient life have somehow adapted to modern Martian conditions?

Biological or not, those nanoscale wormies in that little four-pound Mars rock made NASA and the planetary community turn and face our true purpose. By exploring space we are embracing life, not running away from it.

16 | Is It Science Yet?

LONDON—An elaborate, 155-year-old hoax was revealed Monday, when the Royal Astronomical Society confessed that the planet Neptune does not exist. "It appears to have begun in 1846, when Johann Galle needed a big discovery to give his career a jump start, so he fabricated this new planet," said Royal Astronomical Society president N. O. Weiss.  "Ever since, every astronomer who's wanted some attention has come up with some new report on 'Neptune' and made up some rubbish to support it. I swear, we meant to come clean eventually, but the whole thing just kind of snowballed."

—*The Onion*

WEIRD SCIENCE

We scientists recoil against alien stories and beliefs that ignore standards of evidence and common sense.* Yet, extraterrestrial life is among the most difficult of subjects to approach scientifically. Astrobiology is not just another science. It has sometimes been derided as a "science without a subject," or even more harshly as a pseudoscience. I disagree. But, to recap, here are four ways in which it is weird science:

1. We ourselves are part of the phenomenon we seek to study. Can a camera photograph its own lens?

2. We can't say what it is exactly that we are looking for on other planets, but we think we'll know it when we see it.

---

*Admittedly, sometimes science rules against common sense.

3. We rely entirely on a single example—life on Earth. Could you imagine trying to develop a science of botany if you were locked in a room with only one seed? This does not make for "good science" the way we usually think of it. In medicine, for example, we are suspicious of studies with small numbers of test subjects.

4. We are blatantly biased by an overwhelming desire to find certain answers. Life good. Dead universe bad. Science, we have been taught, is supposed to be neutral, like Switzerland. We're not supposed to take sides or rig the game. Yet, when it comes to life in the universe, we've unabashedly forsaken our neutrality. We can't hide our love away. We want life.

Science demands "well-posed" research questions that we can frame as specific hypotheses for us to test. We can't go to Mars and do an experiment designed to "search for life, whatever that might be."* But we also can't just sit here on Earth and fret over our philosophical conundrums while Mars stares down at us licking its chops, guarding its precious secrets. So, we make educated guesses about which aspects of Earth life will be universal. We go to Mars and search for water, for organic molecules, for biogenic gases, for microbial mats. We look for signs that on Earth would point to life. We employ widely accepted, but unproven, criteria for inhabited planets. To "do good science" we have to pretend we know the answers to the big questions "What is life?" and "What kinds of planets are living?" This is fine, as long as we don't forget that we are still following hunches.

Astrobiology is intellectually unrestrained—some would even say flaky—compared to other fields. It is not too hard to come up with a conjecture about the origin or early evolution of life, or the possibility of life in some alternative planetary environment, and get a publication (and a newspaper story). Peer review still applies, but our filters are somewhat loosened, as there is a wide acknowledgment that outside-the-box thinking must be encouraged, up to a point, since we don't really know the shape or size of the box. Inevitably, some junk slips through. When reading astrobiology papers, you have to set your bull-

---

*Although simple cameras come the closest to such "open to anything" experiments, which was why Sagan was so keen on using the Viking lander cameras to look for macrofauna.

shit detector on a slightly higher setting than you would when reading in an older, more established field.

Lately, I've been thinking of astrobiology as a resurgent branch of natural philosophy. Calling it this may help us remain mindful of the limitations and assumptions necessary to do our science and the larger questions that we sometimes sweep under the rug.

NATURAL PHILOSOPHY

The fence we've built between science and philosophy is a recent construction. Up until about 150 years ago, the study of the natural world was not called science, but natural philosophy. There were no "scientists," only natural philosophers. Many famous philosophers, including René Descartes and Immanuel Kant, had important physical insights that contributed to the foundations of science. Likewise, none of our classical Western scientific heroes—(e.g. Galileo, Darwin, Newton, and Kepler)—called themselves scientists.* They were philosophers with a particular interest in understanding nature. Starting with Galileo, some philosophers developed an experimental method of establishing truths. We who inherited this approach call ourselves scientists, and we generally forget that what we assume to be the ground rules—the obvious, unquestionable truths about nature—were part of a new and radical philosophy just a few hundred years ago.

Bertrand Russell said that "science is what you know and philosophy is what you don't know." Indeed, science grows at philosophy's expense, continually siphoning off the known to sprout new disciplines. For example, the study of consciousness and the mind used to be part of philosophy, but now it's psychology and neuroscience. Studying the heavens and pondering other worlds was once something that philosophers did. When we started to learn something definite, these musings became the science of astronomy. Only those areas in which we have no solid answers at all are left as a sticky residue for philosophers to roll around in.

Along with the switch to "science" came a narrowing of scope. Science

---

*None of them would today be considered "skeptics" either: Newton was obsessed with alchemy. Kepler practiced numerology and based his conclusions about extraterrestrial life on mystical principles. Herschel was convinced, for metaphysical reasons, that there was life on the Sun, Moon, and planets and devoted his astronomical career to finding evidence for it. Darwin was a creationist.

became professionalized. It became a full-time day job and splintered into a few, and then many, separate fields and specialties. Natural philosophers did not carve up the universe like that, and they didn't separate science from questions about the limits of science. In becoming scientists we've accepted an invisible framework. We don't worry about the rules about how to do science, because these seem obvious, beyond question. Paradoxically, none of us can say exactly what these rules are. The history and philosophy of science are not part of a scientific education. Rather, scientists learn the stories and rules through an intuitive cultural assimilation, the way a child picks up the rules of social interaction.

Philosophers of science have competing ideas about what "the scientific method" is, or even whether there is one.* They argue about the way science really works, but they all agree that no theory about science is remotely scientific. Meanwhile, working scientists are far too busy *doing* science to worry about what it is.

Scientists ask, what can we know of nature? To this, natural philosophers added, how can we know this and what *can't* we know? Science, as it has grown in confidence, has lost the ability, or at least the desire, to question its own authority. Before science became Science, thinkers kept these questions closer in mind. A new natural philosophy approach would rejoin science with philosophy of science as a common area of inquiry.

In some ways our present science cannot completely handle the question of extraterrestrial life. In this field, more than any other, we must keep our grand ignorance of the ways of the universe constantly in mind and be open to anything. Yet, this kind of approach flies in the face of what is usually considered to be good science, because science requires that we make specific predictions to test hypotheses. The odd status of astrobiology in the suite of sciences can, I think, be understood by realizing that it is not yet science, exactly, but still natural philosophy.

An awareness of the limits of science is especially important when we skirt close to its edges. For an honest consideration of the questions raised in this field, we must bravely sail beyond the edge of the scientif-

---

*Historically, these ideas have included the empiricism (letting nature speak for itself) of Galileo, the experimental, inductive approach of Francis Bacon, the pure reason of Kant, the provisional truths and falsification of Karl Popper, and the "normal science" and revolutionary paradigm shifts of Thomas Kuhn.

ically mapped world back into the realm of natural philosophy where, in addition to chipping away at some relevant, well-posed scientific questions, we should constantly ask ourselves, "Why do we believe what we believe?" What is the interplay between common sense, evidence, intuition, and faith that forms our beliefs about life in the universe?

None of this is meant as a put-down of astrobiology. I think natural philosophy is a fine thing indeed and may be what science needs now.

By helping rejoin the splintered communities of science, and rubbing our noses in the limits of science, astrobiology can help us to rediscover the study of nature as a whole.

Another problem plaguing science these days is a sometimes deserved reputation for arrogance. But modesty is called for when facing the huge unknowns in studying alien life. A natural philosophy approach could also infuse our quest with a much needed dose of humility, which can only help science.

During the Enlightenment, science grew out of natural philosophy and took on a life of its own. In an unconstrained field like astrobiology, where our ignorance so outweighs our knowledge that we are not even sure how to ask the right questions, we can benefit from hearkening back to the earlier approach. Our innocence in the ways of the universe demands that we be natural philosophers again.

OUT OF THE NEUTRAL ZONE

We have been taught that science is largely value-free except for a sense of integrity to the truth. We are supposed to dispassionately interrogate nature and accept the answers, whether we like them or not. Yet, in the past, natural philosophers often mixed their spirituality, ethics, and values in with their science.

Our neutral stance is not just a cop-out. It arises from an attempt to maintain a high standard of objectivity—a commitment to letting nature be the judge. Undeniably, it is methodologically dangerous to want a certain answer too badly. When hunting wabbits on Mars we have to be vewwy, vewwy careful not to scare them up out of nowhere. It's so easy to get carried away by wishful thinking. History shows that this has happened many times before in science. In particular, the data-starved, emotionally charged science of extraterrestrial life has been, and remains, vulnerable.

Despite the good intentions behind the "value neutral" stance, scientists sometimes take this position too far. It can lead to an "it's not our problem" approach toward the serious societal issues created by our rapid advances in knowledge. With power comes responsibility. Science has allowed the technical advances that have created many of the most crucial issues facing our species, such as global warming and the promise and peril of genetic engineering.

Maybe the fact that astrobiology is so blatantly value-laden will be good for us, help us get over the "value free" myth that allows science to weasel out of its share of responsibility for many challenging aspects of modern existence. With astrobiology, there is no hiding from societal issues that go far beyond the strict search for physical truth. In the first paper in the new journal *Astrobiology*, published in spring of 2001, NASA scientist and administrator David Morrison addressed the societal responsibilities inherent in NASA's astrobiology program:

Astrobiology research has implications that are felt beyond the confines of the laboratory. As our understanding of living systems and the physical universe increases, we will confront the implications of this knowledge in more than just the scientific and technical realms. To understand the consequences will require multidisciplinary consideration of areas such as economics, environment, health, theology, ethics, quality of life, the sociopolitical realm, and education. Together we will explore the ethical and philosophical questions related to the existence of life elsewhere, the potential for cross-contamination between ecosystems on different worlds, and the implications of future long-term planetary habitation and engineering.

Clearly, ethical questions are an inseparable part of this challenging new scientific movement. This can only help, because we must somehow learn to do science that is guided by values but not distorted by wishes. It's a tough but necessary distinction. If we want to survive long enough to become a player on the galactic stage, we are going to need to incorporate humanistic and biophilic values into our scientific culture. Only through acknowledging this can we help humanity to use our growing technical prowess wisely. And we can help science, an institution plagued with critics and doubters, to win friends and influence people.

The possibility of contact with ET life presents many new ethical quandaries. For example, what if we discover aliens who are so alien that we cannot tell if they are "intelligent" or not? How should we treat them? What rights would they have? In my house and yard, I've used chemical poisons to kill wasps and rodents, and traps to remove other pests. Yet chemical warfare and kidnapping are both appropriately seen as criminal acts. Personally, I believe that mistreatment of dolphins, including involuntary confinement, is criminal. Where do we draw the lines? Are "primitive" life-forms on other planets to be exterminated if they interfere with our plans?

A current intersection of technical and ethical questions is in planetary protection. This is the name we give to our efforts to behave like responsible, grown-up members of our solar system, and avoid having unprotected curiosity. While we explore the planets, we must vigilantly guard against "forward contamination" (the accidental spreading of Earth germs to other planets), and "back contamination" (the accidental infection of Earth with alien germs).

We know that some forward contamination has already occurred. Earth's Moon is littered in a few places with unsterilized spacecraft and human waste. The documented survival of Earth organisms on our Moon junk demonstrates that we could inadvertently infect other worlds—like birds propagating seeds in their scat—if we carelessly leave our shit around the solar system.

The only way to practice "zero tolerance" of accidental contamination would be to never go exploring. And one way to make sure that you never hit anything or anybody with your car would be to never, ever, take it out of the garage.

In the case of back contamination, the worst-case scenario is pretty bad: by introducing alien organisms we could cause an unstoppable global ecological disaster.* This first became a consideration in the sixties when we brought lunar samples back to Earth. The Moon rocks were all dead as, well, rocks, and quite a bit deader than any Earth rocks. Now, as we make new plans to retrieve samples of comets and freshly picked Mars rocks, back contamination is once again on the front burner.

NASA actually has an official "planetary protection officer." Is that a cool job title or what? I even thought about applying when the job opening was announced a few years ago. Could a guy like me really

---

*Oh, well. You win some, you lose some.

become planetary protection officer? Why not? I've got a Ph.D. in planetary science, a bunch of NASA grants, a respectable publication record, no criminal record to speak of, and besides, I really care about my planet and all the others. Unfortunately the job carried one major drawback—it was located in Washington, D.C., an understandable place to put the NASA Office of Planetary Protection, but outside of my habitable zone.

NASA's current planetary protection officer, the man now responsible for protecting Earth as well as the rest of the solar system, is John Rummel, an exobiologist with a background in ecology. Rummel is perfect for the job. He is actively involved in astrobiology and takes his responsibility seriously, but he is also possessed of an impish sense of humor and is not above using the absurdity of his title for some good laughs. I've heard him introduce himself at a meeting by saying, "John Rummel, Planetary Protection Officer. I protect the planets. All the planets, all the time."

Some mock our concerns about planetary protection. It's an easy subject for ridicule because you can convincingly argue that the chances of an interplanetary infection are so low as to seem virtually impossible. And besides, some even say, who cares if we infect other planets? They'll thank us later. We'll be doing them a favor by sharing the gift of life.

In particular, some advocates of immediate human exploration and colonization of Mars are quite dismissive of the need for caution about forward and/or back contamination. They see concerns about planetary protection as irrational, annoying impediments to the speedy realization of our "manifest destiny" on Mars. Mars Society president Bob Zubrin published an editorial in the July/August 2000 issue of *The Planetary Report* called "Contamination From Mars: No Threat," in

NASA PLANETARY PROTECTION
All of the Planets,
           All of the time....

which he referred to the threat of back contamination as "not only illusory but hallucinatory" and said that the kindest thing that could be said about this concern is that it "is just plain nuts." (What is the unkindest?) Many of his arguments, taken individually, are perfectly sound. He notes that it is highly doubtful that Martian organisms, if they do exist, could infect terrestrial organisms.

Further, Zubrin points out that Earth and Mars are not presently isolated systems. About a ton of Mars falls upon the Earth every couple of years, so if Mars bugs pose a threat to Earth life, it is too late. They are already here.

Zubrin is absolutely right about these points. If our current concepts of biological and planetary evolution are correct, then no danger of back contamination exists. Yet he does not consider the magnitude of the irreversible disaster that could be caused if we are wrong. We might not get another chance. His arguments are reasonable, but they are not iron-clad. He assures us that "it is the opinion of experts" that Mars bugs, if they exist, have already naturally been transferred to Earth. But should we risk it all on the "opinion of experts"? I am an expert of sorts who basically agrees with his arguments, if not his conclusion. If I had to guess, I'd say that further research will come closer to proving that there is no danger. But that's not good enough. What if we're wrong? Zubrin says that we should stop worrying and just go and get the Mars rocks, set people loose on Mars without even washing their hands, just get on with it. He concluded his editorial by saying, "Back contamination mavens need to back off. Their warnings have no rational basis and are being used to urge crimes against science."

This is not the way to win folks over to the cause of Mars exploration. Though I share the dream of future human expansion into space, on to other planets and then outward to the galaxy, when I read these arrogant, reckless, irresponsible arguments, I find myself thinking, "Maybe we're not ready."

One argument against the "back contamination mavens" is that organisms are always uniquely evolved to take advantage of their own environments. Thus, invaders from a different world will always be at a competitive disadvantage and will never thrive. Yet look at all of the places on Earth where introduced species have outcompeted native species and run amok. In some rain forests in Hawaii, almost no "native" species remain. And what about those vicious cane toads in Australia, or the damn killer bees? The analogy to planetary protection is not exact: all terrestrial species are related, making it more plausible that, even if they've never met, they could enjoy each other's habitats. But it is possible that *not* being related could conceivably be an advantage to an invader, since we would have no defenses against it. Imagine some unknown, tough, primitive Mars organisms that multiply like crazy in any warm, wet environments and voraciously consume any organic molecules they find. To them the human body, or the cells of any terrestrial organism, might seem like an all-you-can-eat buffet.

I would place money on Mars dirt being safe. If you handed me a spoonful of Mars and dared me, I'd take a nibble to see what it tasted like.* I don't think a little Mars would get me sick. In my heart, I am not too worried about contamination from, or of, Martian organisms because I do not think there is life on Mars. But I could be wrong. Extraordinary risks require extraordinary caution.

Looking at the history of science, how many times have we thought we had a complete understanding of some aspect of nature, only to find out later that we were looking at a tiny fraction of some larger truth? When making decisions that might affect our entire biosphere, we owe it to ourselves, our ancestors, and our descendants to keep this history in mind. Do we want to bet the entire farm on our belief that our current concepts of biology are correct? We should *assume* that our understanding of biology is wrong and proceed from there. After all, we've never been able to study alien biochemistry. We could be dead wrong, so let's proceed with caution.

To ignore planetary protection is the scientific equivalent of refusing to use a condom. It might make it is easier or faster to get what we want. But we've got to show that we're a lot smarter, and more respectful, than that if we want the public, who pay for our exploration of the solar system, to entrust us with its protection. If we don't want the people mad at science, then we shouldn't act like mad scientists.

I'm proud to say that NASA has always taken planetary protection issues seriously. We make and follow rules designed to protect against even remote chances of contamination. Of course implementing these rules will involve educated guesses. Nothing is 100 percent safe. But we can't be paralyzed by fear. Do not fear the universe. Don't be afraid to leave the house in the morning, but remember your rubbers.

If any modern human enterprise could be called "applied natural

---

*I know researchers who have clandestinely tasted Moon dust. It tastes like dust.

philosophy," it would be planetary protection. In designing planetary missions and formulating policy we cannot avoid questions like "What is life?" "How much is new self-knowledge worth?" "What is the intrinsic worth of preserving nature?" and "How does this change if it is living or inanimate?" We are stepping up to our ethical responsibilities as scientists, and as sentient residents of the solar system. This could be a model for other sciences. Doctors, when they complete their education, take a Hippocratic oath and pledge to "first, do no harm."* Perhaps we need a similar pledge for scientists.

SONG OF OURSELVES

Our efforts to anticipate and find life out there require us to start with some image of what that life should be like. Seventeenth-century writers imagined men and women in European dress occupying a universe of alternative Earths. Now the likenesses we paint out there take the form of squirming proteins swimming in Earth-like seas.

On every evolutionary level, from the taproots to the tips of the tree of life, our thinking about alien biology is extrapolated from the local example. We assume that other life will be built of the same basic biochemical parts, that it will need liquid water, will be carbon-based, and may even use proteins and DNA. Up in life's high canopy, we imagine that other complex, intelligent creatures will develop mathematics, science, and technology similar to ours and will sing songs of their own in prime-numbered pulses of radio waves.

Everyone assumes that carbon chemistry in water environments will be *the* way that our universe makes life. The justifications for this view rest on the unique qualities of carbon chemistry and the strange properties of water.

Carbon is unusual in the way it forms molecules that are endlessly variable, so flexible, and yet so stable. Its modular self-construction makes it the ultimate molecular shape-shifter, forming itself into the right tool for the job, whether that happens to be building a cell wall or facilitating a chemical reaction between two other, essential carbon molecules, themselves optimized for the jobs they do. Water, too, has strange properties. On Earth, all biology takes advantage of some sub-

---

*Rumor has it, this is being replaced with the HMO, the Hypocritic Money Oath: "First, do nothing unprofitable."

tle features of water that are not at all obvious from reading the periodic table and predicting chemical behavior. Water's triangular shape,
with the two H's both pulling to one side of the O, gives it a slightly
negative charge on one end and a slightly positive charge on the other.
This makes water a "polar molecule." The polar nature of liquid water
creates the possibility for a whole new kind of bond: "hydrogen
bonds," which operate independently of the elaborate rules for electron
trades and collectives that most chemistry relies on. Water's polar properties also allow it to form attractions and repulsions to small parts of
large carbon molecules, helping to orient them and shepherding large-
scale organic structures. In chapter 8, I described how the first cells
grew out of the self-organization of lipid molecules, which line up in
orderly fashion in response to water's opposite pull on their two ends.
The polar nature of the water molecule allows these organized structures to form. The list of unusual, biofriendly properties of water goes
on. Life has evolved to make use of many of these unique features.

It's hard to think of another chemical system having carbon-in-
water's essential, life giving properties. I can't think of a good substitute
and, I'll bet, neither can you. But does that mean that such alternate
biochemical systems do not exist, or merely that we are currently too
ignorant or too blinkered by our assumptions to imagine them? We
should not confuse the limitations of our own tiny intellects with limitations on the creativity of Cosmic Evolution. Our universe is full of
varied environments and we have no idea what kinds of chemistry are
occurring, no inkling of what's crawling in alien seas of unknown composition.

Our evolution has expertly exploited the idiosyncrasies of carbon
and water. After 4 billion years of adapting to and building on these
peculiarities, of course these materials seem pretty special to us. We are
built of carbon molecules, floating in an ocean of water. We still live in
that ocean, only now we carry it around within our cells. You can taste
it in our blood, sweat, and salty tears. How might this immersion warp
our perspective?

We have many ways of justifying our "carbaqueous" assumption,
but sometimes I wonder if the real motivation is not simply that *we'd
be lost without it*. With it we have narrowed the scope of possibilities
enough so that we may apply science to the problem. Once we've narrowed life's needs down to one condition—water—we can look around
the universe for water worlds and apply our theories toward predicting

where they should be found. Our "life needs water" paradigm is born of pragmatic necessity, not solid scientific deduction. We need it so that we can do science.

That's fine. We do have to start somewhere, especially when taking on a subject as wide-open as the question of extraterrestrial life. But, when we repeat something frequently enough, it can work its way into our psyches. We forget the shallow basis on which we've reached our tentative conclusion that life needs carbon and water. What started as an educated guess becomes a consensus reality.

Unfortunately, the justification boils down to "that's the way it works here and we can't imagine any other basis for life." It is like someone who does not know how to design watches finding an unbelievably exquisite watch, dissecting it, and declaring, "This is so perfect, it has to be the only way that a watch can be put together."

There's no way that we could have designed something that works as well as our own biochemistry, so how can we state confidently that there's no other substance in the universe suitable for this kind of construction? If our metabolism and structure were based on some chemical system other than carbon chemistry, on a planet without liquid water, would we know anything about this carbon potential? In many ways, some obvious and some subtle, our world has been remade by life to look like one in which carbon is the only "natural" source of real complexity. Whether or not this is really true, the world would seem this way. So we might be fooling ourselves about carbon.

By the time any kind of life—made by any chemical system, carbon or not—finally evolves consciousness, it will be stunningly well adapted to its world, and its world thoroughly changed (as our world has been) by its biochemistry. This life, upon first examining the universe, will conclude that life can only evolve using its own peculiar kind of chemistry. Curious scientists of any chemical construction would observe many features of their universe that seemed to confirm this view.

I know it won't sit well with some of my carbon-based, and carbon-biased, friends, but I choose to remain an agnostic with respect to the carbon religion. No, agnostic isn't right. I believe in carbon. I worship it. Sign me up for the carbon church. I'll show up every Sunday morning and do the DNA dance. But to believe, must we swear that carbon is the true and only way, forever renouncing all other elements?

Some scientists, coming from a biochemical background, even talk about what we can learn by sequencing alien DNA and comparing it to our own. To me this seems as ridiculous as dissecting crashed saucers at Roswell. Expecting to find DNA elsewhere is like expecting a *Star Trek* universe with humanoid aliens who speak English and insist that we join them for dinner at eight. A chemical hereditary system, like a language, is the result of a complex evolutionary process filled with randomness, contingency, and frozen-in accidents.

Might life use completely alien chemical systems elsewhere? The question is not exactly science, because it is difficult to think of a scientific way to address it. It is really a question of natural philosophy.

We have to be careful when stating what is impossible. Different chemical environments may breed unforeseeable sources of chemical order. And what about life that doesn't need chemistry at all? Why not life on different spatial or temporal scales? Who are we to say that the universe couldn't make some kind of complex, self-organizing, evolving structures using its gravitational or nuclear forces, forming living structures that are too large or small for us to notice? Life at the scale of molecular clouds or even galactic superclusters? Why not life, and even civilization, at the level of elementary, subatomic particles, where empires that dwarf any in human history rise and fall in a nanosecond, at a level completely invisible to us? Life so fast or slow that we don't notice it? Why not? No reason. No answer.

Time and again we think we know more than we do. We may never be able to imagine an alternative kind of life, but I bet we will eventually come across one.

| # Living Worlds

Throughout the continuum as we know it (and a good deal more, as we don't know it) there are cultures that fly and cultures that swim; there are boron folk and fluorine fellowships, cuprocoprophages and (roughly speaking) immaterial life-forms which swim and swirl around each other in space like so many pelagic shards of metaphysics. And some organize into  super-entities like a beehive or a slime-mold so that they live plurally to become singular, and some have even more singular ideas of plurality.

—THEODORE STURGEON, *The Widget, the Wadget, and Boff*

GAIA: IS EARTH ALIVE?

Using a natural philosophy approach, perhaps we can study life's universals without simply projecting visions of our own kind out into the cosmos. Two controversial new fields of thought promise to help lift astrobiology beyond this conundrum of self-reference. These are the Gaia hypothesis and complexity theory.

The Gaia hypothesis, named after the Greek Earth goddess, was first proposed in the mid-1970s by James Lovelock, a British atmospheric scientist and inventor, and American microbiologist Lynn Margulis. Margulis is the mother of serial endosymbiosis theory, which states that major evolutionary innovations occur when more complex forms of life arise out of symbiotic collectives of smaller organisms (as described in chapter 8).

How far can we extend this principle? Gaia scientists regard Earth in its totality as one giant superorganism incorporating many parts of our planet that traditional science sees as nonliving.* The atmosphere and

*Though many older, prescientific traditions have long perceived a living Earth.

oceans are the breath and blood, and the rain forests the lungs of this great global beast. Human culture is, perhaps, its awakening mind. Gaia suggests that life is seen as not merely incidental, but integral to the evolution and functioning of the planet. Gaia is a theory of biology, but also one of Earth history, geochemistry, and climate. The Gaian approach toward the Earth sciences is "geophysiology," studying Earth's functioning and health as a physician might approach a patient.

In 1988, as a graduate student, I attended the first major mainstream scientific conference on the Gaia hypothesis, sponsored by the American Geophysical Union in San Diego. It was fascinating to watch skeptical traditional scientists do battle with those from the new Gaian camp as they attempted to get Earth scientists to take their new approach seriously.* Science still doesn't quite know what to do with the Gaia hypothesis, because it isn't science-as-usual. Yet it is more than just a pretty metaphor. The Gaia hypothesis is guiding the way some biologists model life and its role on Earth, even as other (mostly older) biologists completely dismiss it.

Gaia actually began as an idea about exobiology. James Lovelock, consulting for NASA during the design of the Viking life-detection experiments, was thinking about how to look for life on Mars. He realized that the unusual atmosphere of Earth is by far its most distinctive sign of life. In considering the global properties of life that might be observable from another planet, he started noticing the many ways in which Earth's biosphere behaves like a giant living organism. In 1974, he and Margulis presented the Gaia concept in a paper called "Biological Modulation of the Earth's Atmosphere," published in *ICARUS, International Journal of Solar System Studies.*

The Gaia hypothesis has caused a quiet revolution among Earth scientists, many of whom are now realizing that life participates deeply in the physical evolution and functioning of Earth. A small school of scientists has fully embraced Gaia and dedicated their careers to it directly. A much larger group has been more guardedly receptive to the viewpoint, incorporating it into their work, or at least their worldview.

The Gaia perspective views evolutionary change as a creative interplay between biosphere and Earth—an intricate partner-dance between life and the changing planet in which neither seems to be leading. Life

---

*My friend Dorion Sagan suggested that we try to get everyone to gather one morning and run down the beach naked yelling, "The Earth is alive!" but this idea never caught on.

on Earth is no accidental collection of organisms lucky enough to find a hospitable planetary home. Rather, life has largely created the world we know. Life, we are learning, has altered many of Earth's basic physical properties, investing the air, the rocks, and the water with qualities they would not possess on a dead world. Thus, Gaia has important implications for the kind of relationships that biospheres can have with planets, and this should inform the way that we search for other inhabited worlds.

With a Gaian picture of evolution we see that some properties of the "nonliving" parts of Earth are actually encoded in the DNA of the world's organisms. Are other planets blessed with their own genomes?

It seems beautiful and true, but is it science? Some scientists complain that the hypothesis is more hype than thesis. They say there is no way to test or falsify it. Yet, the Gaia perspective has clearly led to some good science and to a new framework guiding some of the science we were already doing. In my view, it's right on the border of science and natural philosophy.

Gaia scientists have discussed the significance of Earth's unusual atmosphere, which is drastically out of equilibrium. Without the incessant, life-driven chemical cycles that permeate our world, the oxygen and methane would rapidly react, leaving only CO_2 and water, producing a mix of gases that we would find unrecognizable and certainly unbreathable. The strange brew we breathe would never be found on a nonliving world.

Gaia proposes that the cumulative activity of all life on Earth acts to keep conditions here stable and comfortable for life. This happens by the evolution of numerous negative feedbacks in which the growth, death, or evolution of organisms creates environmental changes, which in turn affect the growth and evolution of other organisms. The net effect acts to pull the climate and various chemical balances back toward a certain moderate range if they begin to stray.

Short of comparing and contrasting numerous inhabited worlds, you can't do an experiment to test the idea as a whole, but you can look for active feedbacks on Earth that may be part of such a system. For example, some plankton act, collectively, as an air conditioner for the oceans. When the water gets warm, these guys get frisky and start multiplying. Their growth produces a chemical called dimethylsulfide (DMS), which diffuses up into the atmosphere. DMS is great for seeding clouds. As the amount of DMS rises, clouds build up over the ocean. The ocean surface cools off, which chills out the plankton orgy. The production of DMS then declines. As a result, it doesn't get too cloudy or too sunny for long, and ocean temperature remains in a moderate range favored by life.

The Gaians suggest that such mechanisms have been biologically regulating conditions on Earth for billions of years. This homeostatic self-regulation makes Gaia very much like a living organism, with the atmosphere and oceans behaving like circulatory and respiratory systems. However, obviously Gaia is not like any other organism we know in some important ways. For example, it has not reproduced, although you can't say we aren't trying.

How deeply ingrained is the biosphere in the physical functioning of our planet? We don't know. Gaian science endeavors to find out. It could go very deep indeed. Life clearly has hold of the atmosphere and oceans. Numerous cycles connect the atmosphere with the chemical state of Earth's interior rocks. Can life have actually assumed control of the plate tectonics that controls all terrestrial geology? If so, then the entire thermal evolution of the Earth is controlled by life.

Does it really go that deep? Or might Gaia be a spherical superorganism riding around on a nonliving core? How could you define the boundary dividing creature and core? Perhaps by looking for a level, at some depth within the Earth, where things are exactly as they would have been if life had never come along. When it comes to the deep interior of the Earth, we don't yet know if Gaia is holding the reins or skillfully riding bareback.

The Gaia hypothesis reveals life to be a planetary-scale phenomenon with a cosmological life span. Gaia can help us identify those global qualities that distinguish planets having billions of years of life ingrained in their cyclic chemical activity from those orbs not blessed by this world-altering magic.

COMPLEXITY: LIFE BEYOND WHAT WE KNOW

Complexity theory is the study of self-organization in nature. You've heard of the second law of thermodynamics: things fall apart. Entropy will get us all in the end, leaving nothing but bland disorder. Everything runs inevitably downhill into dull formlessness. Yet look out the window or stare at your hands. What we actually see around us is not a gray and featureless sea of entropy but a living world overflowing with stunning and profligate order. What saves us from the tyranny of entropy is an amazing property of matter: in certain conditions, in the presence of a flow of energy, it self-organizes, forming ordered structures, in seeming defiance of the second law. Complexity theory is the mathematical study of these emergent, spontaneous pockets of order.

The simplest example of emergent complexity is a whirlpool spontaneously forming in a flowing stream. The most complex example may be a living organism, or a society of organisms.

Some scientists are starting to see life as the most refined manifestation of a universal tendency toward self-organization. Given half a chance, order emerges from chaos, and given optimum conditions, matter keeps on self-organizing until it can get up, crawl around, and write poetry. What I find incredibly exciting about this new field is that it seems to be pointing toward a mathematical account of living systems that goes much deeper than merely "reverse engineering" the life here on Earth. Complexity theorists may be groping toward a truly universal understanding of the nature of life.

Like Gaia theory, complexity theory is poised on the shifting border between science and natural philosophy. It cannot be used to make many specific, testable predictions. But, like Gaia, it is an idea that can guide and recontextualize some of our efforts to study the general phenomenon of life.

The tendency of matter, under certain conditions, to self-organize suggests a new picture of evolution. Traditional Darwinian theory has regarded evolution as a "blind watchmaker" where natural selection between random mutations leads to all innovation and adaptation. But complexity theory suggests that evolution may also refine and exploit the nascent emergent properties of matter. Natural selection may be helped along by some spontaneous pattern-forming habits built deep into this universe.

We can begin to study the organizing principles of life that are not necessarily based on any local chemical system. We can simulate evolution in a computer and begin to mathematically address one of our most vexing questions: Is Earth life a fluke—an improbable, singular occurrence—or does it represent the local flowering of a capacity for life inherent everywhere in our universe? Some of us believe that complexity is hinting at the latter.

All the significant developments in our cosmic story can be seen as leaps to new levels of complexity. If the universe tends toward self-organization, and the epitome of self-organization is life, then rather than some accidental occurrence here on an unusual ball of rock, life may be implicit in the laws of nature, a stage of organization that this universe goes through on its journey from atoms to minds. Complexity theory is

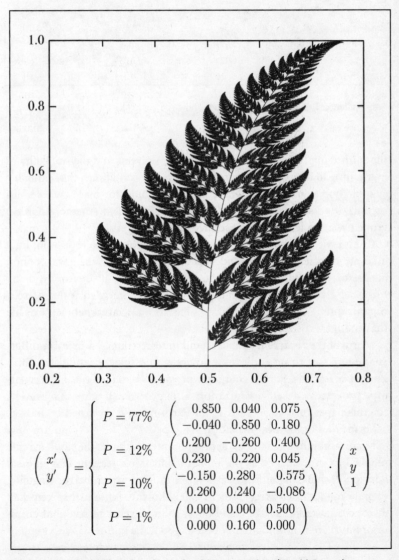

$$\begin{pmatrix} x' \\ y' \end{pmatrix} = \begin{cases} P = 77\% & \begin{pmatrix} 0.850 & 0.040 & 0.075 \\ -0.040 & 0.850 & 0.180 \end{pmatrix} \\ P = 12\% & \begin{pmatrix} 0.200 & -0.260 & 0.400 \\ 0.230 & 0.220 & 0.045 \end{pmatrix} \\ P = 10\% & \begin{pmatrix} -0.150 & 0.280 & 0.575 \\ 0.260 & 0.240 & -0.086 \end{pmatrix} \\ P = 1\% & \begin{pmatrix} 0.000 & 0.000 & 0.500 \\ 0.000 & 0.160 & 0.000 \end{pmatrix} \end{cases} \cdot \begin{pmatrix} x \\ y \\ 1 \end{pmatrix}$$

Life is math: it looks like a fern, but in fact it's an equation. (Harold Cooper)

$z \to z^2 + c$

Self-similarity. The Mandelbrot set (God) and a red algae plant. (Ernst Haeckel)

the connection between the wildness of the universe and the simplicity of math. Instead of a quest for familiar Earth-like conditions, the search for suitable homes for life elsewhere becomes a hunt for places where self-organization is likely to flourish and complexity can emerge. Thus we grope toward a less parochial science of life's universals.

Complexity theory also suggests a new take on an old question, long a staple of science fiction and speculative science: When we do find aliens, or they find us, what will they look like? By revealing many forms of Earth life to be governed by deep geometrical rules of self-organization in nature, complexity suggests a universal geometry of life that should transcend worlds.

Many of the recurrent shapes found in terrestrial creatures are different versions of those endlessly branching and self-repeating structures known as fractals. Simple computer programs designed to simulate natural processes of self-organization also generate fractal shapes. The fractal shapes of living creatures on Earth may be governed by universal principles of pattern formation.

Some computer-generated synthetic fractals have the look of life reduced to its elements, as if Noah's Ark had been caught in a colossal whirlpool, smashing all the beasts of Earth into pieces and recombining them at random, with tree branches blending into antlers, then veins, an insect eye, a sea-horse tail, fish bones, palm leaves, and so on ad infinitum. But there is nothing Earth-bound about the mathematical origin of these shapes; they are just as likely to hint at alien architecture.

If life is fundamentally a process of self-organization, then the fractal shapes of life will be found in other biospheres as well. There is no way to predict precisely what aliens will look like, but the fractal geometry of life gives us reason to believe that when they do finally land on the White House lawn, whatever walks or slithers down the gangplank may look strangely familiar.

LIVING WORLDS

Are there other Gaias? I doubt it. Gaia/Earth is a unique individual. Given its complex, contingent history, it's unlikely that Gaia has an identical twin. The question we could ask, though, is "Are there other worlds that are alive in the same sense that Gaia is?" I call this hypothetical class of Gaia-like entities "living worlds."

How can we combine the insights from Gaia and complexity theory to refine our schemes for finding other living worlds? Complexity theory can, I think, help us become more sophisticated about which worlds are likely to come alive. Gaia theory can make us more sophisticated about the ways in which living worlds can evolve.

According to the Gaia hypothesis, complex feedbacks at all scales have long ago made it impossible to draw a clean line between planet and life. Earth is like a coral reef, where life's environment is built of its ancestors' remains. In the Gaian view, habitability and inhabitation are one and the same, a self-catalyzing property of a living planet. Extrapolated to the wider universe, this suggests that life is not just something that sometimes happens *on* a planet given certain conditions. Rather, life is something that happens *to* a planet. A living world is more than just a planet with water, carbon, energy, and luck. Life is what a planet becomes.*

In this view, conventional notions of habitability (solar distance, size, water) are still relevant to the question of where life can get started. But long-term habitability may depend more on the establishment of a robust and resourceful global living organism than on having a lucky planet and a lucky star.†

---

*The idea that a planet may be alive, not just a home for life but a living thing itself, has long been an exotic, enticing theme in science fiction. The most famous example is Stanislaw Lem's *Solaris,* in which a planetwide ocean is found to be an intelligent but inscrutable living entity.
†Substitute "highly stable system of biogeochemical feedbacks" for "global living organism" here if you find it less offensive.

The question may not be "Why did life evolve here, given certain conditions?" but "How and why did Earth evolve into a living world?" The difference is subtle, but can affect our interpretations of causality in planetary history and suggest features and qualities to pay particular attention to when exploring new planets. This might contribute in a concrete way to our efforts to build the right telescopes and spectrometers and to make the right observations to identify living worlds. It is certainly true (as Lovelock noted thirty years ago) that life has left its brand on Earth—and not in any subtle way. Earth's atmosphere is so unlike that of any imaginable nonliving world that it would not whisper but scream "Life!" to any alien with a spectrometer and a clue.

Complexity theory suggests that life is most likely to emerge in conditions of continuous flow of energy and nutrients, as long as some suitable chemical architecture exists. Carbon in water is a good one. Are there others? Once some kind of living organisms form, riding the chemical flows of a young planet, they will use whatever fluid medium there is (air, water, bkji⅋pqek, whatever) both to extract raw materials and to carry away their wastes. This extracting and dumping will change the composition of these fluids, so organisms in one part of the planet will be altering the environments of organisms elsewhere. Organisms around the planet will evolve to take advantage of each other's wastes. These flows become mutualistic exchanges, then global biogeochemical cycles that bind life to the planet. Once life starts, it is only a matter of time before it develops, in concert with its evolving planet, into a global system of interacting and self-regulating cycles. These cycles will help to maintain the steady flows of energy and moderate environmental conditions needed to facilitate complexity. Within such a rich planetary environment, Darwinian evolution is free to work its magic, all the while getting hints of innovation from the spontaneous formation of ordered structures.

Once global cycles are well established, then a living world is born: the first globalization movement. Each kind of organism starts off in its own little microenvironment and lasts up to the "living world" stage only if luck keeps its surroundings reasonably stable. After that, organisms do not need to be so lucky. Once the global cycles start up, then they all evolve within a larger system that is largely made up of other evolving organisms and all that they leave behind. This interdependence tends to moderate environmental changes: if things change too much (too hot, too cold, too acidic, too salty, etc.), something will start to die off, or multiply, and counter these changes. The net effect is to evolve a global regulatory system. Life is no longer completely subject to the whims of external environmental change. A globally interconnected biosphere—a living world—begins to make its own luck.

What kind of world is needed to maintain such a system for billions of years? Our present efforts to define conditions for life elsewhere are focused on the presence of water. It's worth noting, however, the additional qualities that set Earth apart from most apparently lifeless worlds. One of those qualities, which may be just as vital to a thriving biosphere as water and organic carbon, is our high, steady level of internally driven geological activity.

Earth is always remaking itself, its insides spewing out onto the surface, while other surface areas are sucked back down into the furnace. Like a restless poker player endlessly drawing new cards, Earth's dynamic atmosphere is continually trading in its molecules for new ones from the planet's interior. All of this changeability is directly or indirectly attributable to Earth's size-dependent internal thermal evolution.

Indirectly, Earth's size is also responsible for our climate and weather. Atmospheric changes are driven by the sun, but Earth retains a thick, active, erosive atmosphere and a filled-to-the-brim hydrosphere—unlike, say, the dry, wispy atmosphere of Mars—because our planet is large enough to hang on to its air and water.

Restless tectonics helps maintain active cycling of chemicals between the different "reservoirs" of the planet (oceans, crust, atmosphere, interior). One crucial ingredient of Earth that I think is underrated as a qualifier for life is this great changeability and all the fertile flows of energy and nutrients that it sustains. As a necessary condition for a living world, we should consider vigorous, continuous geological activity, lasting over cosmological timescales, a candidate right up there with water and organic molecules.*

Size matters. There is a critical mass above which planets stay active over the lifetime of their star and below which they die out more quickly. This suggests a lower limit for the size of a living world—somewhere between the sizes of Mars and Earth.

---

*Notice that our remaining hopes for life on Mars are focused on potential hot springs and similar locales where, we infer, remnants of its lost internal heat may still be driving some sort of flow near the surface.

VITALS

How can a "living worlds" view of life's role in planetary evolution help us in our search strategies? Our current efforts and plans for studying extrasolar planets are largely focused on looking for "other Earths"—places with environments, histories, and solar systems as similar as possible to home. You can't blame us for being curious to learn how such alter-Earths have evolved, and how Earth-like they really are in the details. But the search for life might wisely be conducted more broadly. Soon we will start learning what the atmospheres of other rocky planets are made of. So it's an excellent time to be thinking about what we should look for, and how to interpret what we find. Gaia says look for disequilibrium. An excess of oxygen, especially in the presence of a radically unoxidized gas like methane, would be one example. Others would be a huge excess of methane with a little bit of oxygen, or a large amount of sulfur dioxide (SO_2) mixed in with hydrogen sulfide (H_2S) or any other hydrogen-containing molecule. I could go on, but I'll spare you. It may be the disequilibrium of Earth, not the oxygen or water, that is the universal life indicator. As we plan for the first spectral observations of the atmospheres of roughly Earth-size rocky worlds, we should keep Gaia in mind.

We are talking about comparative astrobiology. At present, this is like comparative planetology was before the space age, theorizing with abandon inside a knowledge vacuum. The idea of disequilibrium, of anomalous composition of planetary atmospheres, as an indicator of life is a good one. The idea is testable in principle and may soon be in practice. We need to study many living worlds and their atmospheres. Then we'll be doing science. For now, even if we buy the disequilibrium life test (I do), there is another important question that we can't yet answer: How much disequilibrium?

Disequilibrium isn't an either/or thing. It's a matter of degree. No atmosphere is perfectly in equilibrium. Nonbiological sources of disequilibrium are everywhere in the universe. Some that we know of on nearby planets include gases cooked up by lightning flashes or by solar ultraviolet radiation or squirted out of volcanoes. Note that each of these requires an energy source. Each blows temporary puffs of nonequilibrium gases into an atmosphere, so finding slight disequilibrium would only tell us that *something* is going on there, but not necessarily life. We don't know enough about the varieties of nonbiological atmospheres to be certain what an anomalous signature is. Yet, if Earth is at all typical, life will be marked by *extreme* deviations from equilibrium.

Now, let's think about how complexity theory might help us in defining criteria for living worlds. Self-organizing emergent behavior happens within a flow of energy and/or matter. The steadier the flow, the better. As discussed, puny worlds won't make good candidates, except perhaps for moons in orbit around giant planets, getting tidal heat.

Beyond this size criterion, what *observable* effects go along with vigorous and continued geological activity? For one thing, living worlds are not likely to preserve the signs of ancient bombardment. Planets with active geology will have erased all but their most recent impact craters, so living worlds will not be heavily cratered. If we develop ways to remotely sense the texture of cratered surfaces, we could use this as a probable indicator of a dead world.

We may not be able to make a definitive test for life with remote observations. Yet, we can use such observations to tell us which places are worth investigating up close. Then we'll need to send interstellar probes. Only by sending imaging devices and other sensors will we learn for sure if any place is inhabited.* Even if the results will not come back until long after you and I are dead and gone, we should launch them anyway. We need to send ships to the most promising worlds so that our great-grandkids can find out if they're really living.

We can best explore astrobiologically by roaming widely and keeping a sharp eye out for anomalous order of any kind. This could include strange, nonequilibrium mixtures of gases (or, conversely, *too much* equilibrium in places where other known processes are creating disequilibrium!), strange mechanical shapes and assemblages, or rhythmic environmental changes without any obvious cause. Such anomalous order will indicate either an interesting nonbiological process that we need to learn about† or that we have at last found new life.

I've been talking about the search for life signs on worlds orbiting other stars. But how do we apply this "living worlds" perspective to the other planets in our own solar system? Our current exploration strategy for astrobiology is highly focused on two places that surely have liquid water within: Mars and Europa. But what if we tried to devise an

---

*Or if we receive a signal (see chapter 18).
†Remember Kepler's deduction of lunar cities based on the "anomalous" circular shapes of craters. . . .

exploration strategy based on a more general definition of life as an evolving system of complexity thriving under stable and vigorous conditions of thermodynamic disequilibrium? What if we decided that the main criteria was to look for places where a lot is happening? The two most compelling targets under these alternative criteria would be Io, Jupiter's hypervolcanic moon, and Venus, Earth's "twin," which (as described in chapter 11) seems to be, geologically, very much alive.

Saturn's moon Titan also gets an honorable mention. We don't really know the level of geologic activity, but it does have a thick atmosphere, and a surface pooled with juicy hydrocarbons. If it wasn't so darn cold there, permanently well below the freezing point of water, I'd rate it as the number one candidate for life. As it is, who knows? Reid Thompson (chapter 7) pointed out that intermittent lakes of organic-rich, liquid water appear on Titan for thousands of years whenever an occasional large impact melts a portion of the icy surface. It is conceivable that there could be life on Titan today, using liquid methane/ethane lakes as a fluid medium. I will certainly be paying attention on the morning of January 14, 2005, when the *Huygens* probe, now on its way to Titan, attached to the *Cassini* Saturn orbiter, descends through the hydrocarbon hazes, methane clouds, and thick nitrogen atmosphere to land or maybe even splash down, sniffing the air and sending back pictures all the while. There's a nonnegligible chance that something living will fly, crawl, or float through the Titanian scene.

MARS IS DEAD: LONG LIVE MARS

From a living worlds perspective, the new wave of interest in life on Mars is highly questionable. If Earth's drastically out-of-equilibrium atmosphere is anything close to typical for living worlds, Mars does not qualify. Mars today has a highly equilibrated atmosphere of almost pure CO_2.

If we regard life on Earth as synonymous with Gaia, with the global biosphere that infects and affects all of Earth so deeply, so exuberantly, and if that is what we are looking for on other planets, then we already know the answer about Mars. It is not enough to identify places on another planet with conditions that overlap with those where organisms can live on Earth. Whether or not Mars has little pockets of water, organics, and local energy flows somewhere, in the Gaian sense, it is dead.

Further, large areas of Mars are covered with craters that date back to the earliest days of the solar system. A living worlds perspective suggests that such a surface is incompatible with life. Ironically, the famous "Mars rock," ALH84001, the meteorite with the "fossil" worms, may provide the best evidence that Mars is dead. Radiometric dating suggests that this rock crystallized 4.5 billion years ago, when the planets were newly formed. No rocks on Earth are anywhere near that old. A living world has no 4.5-billion-year-old rocks.

With its rusty surface and stale (equilibrated) atmosphere, Mars seems very unlike a world where life is thriving. If life exists at all it is barely hanging on in isolated outposts, and it hasn't taken over the thermodynamic state of the atmosphere, and the global geochemical cycles, as has life on Earth.

In his writings, Lovelock distinguishes the birth of Gaia from the origin of life. This idea, of an origin of life separate from the birth of a living world, has interesting implications for life elsewhere. It is possible that a planet could develop life, but never become a living world. If self-regulating Gaia is responsible for Earth life's longevity, then we need to find other places where this kind of global organism has evolved, not merely places where the origin of life might once have occurred.

Can a planet be a little bit alive? For a world to be alive, in the sense that Gaia is, life must be deeply ingrained in the physical functioning of the planet. This suggests that life, as a global property, is something that a planet either has or doesn't have, a distinct state of being, just as an animal's body is either dead or alive. An animal can be "barely alive," but not for long. Maybe in the period after the origin of life and before the origin of Gaia (or something like it), a planet is "barely alive," in a fragile state that either achieves Gaia-hood or quickly dies out. During this vulnerable stage, the continuance of life depends on luck, on the environment's not changing too rapidly or extremely, until the expanding web of feedbacks grows to become a global, self-regulating system, like an organism. The image of a dying organism experiencing global systems failure may be more accurate in picturing life's extinction from a planet than that of isolated colonies of intrepid survivors trying to hold out against all odds.

In this view, it is hard to imagine life existing for 4 billion years on a planet without participating thoroughly in planetary evolution. A living worlds perspective suggests that life cannot hang on for billions of years isolated in underground hot springs. Life either thoroughly infests a planet, or it is not there at all. The signs of life on a planet will not be subtle. As on Earth, life will shout out its existence in the air itself. For this reason, I'd be (pleasantly) surprised if we find life on Mars.

Mars may once have evolved living organisms but never become a living world, before environmental changes doomed all life there. Mars did not have as long as Earth did to make this change before it froze over and lost most of its air and water to space. Maybe the difference between Earth and Mars tells us something about how long it takes to transform a planet into a living world.

A new wrinkle in all of this has appeared since Lovelock's work, and Viking's observations, in the 1970s: the discovery of an extensive and deep underground biosphere on Earth. This discovery is partly responsible for the renewed interest in life on Mars and elsewhere in the solar system. Even if conditions on the surface are inhospitable, life may be underground. I don't rule this out, but over long timescales any gases emitted by living things inside a planet will diffuse out into the atmosphere. We should be able to detect underground life from the disequilibrium nature of those gases. Can a planet have internal life but no signs of it on the surface or in the atmosphere? I doubt it. I regard the absence of flagrant disequilibrium in the Martian atmosphere as a likely sign that Mars is dead—not mostly dead or almost dead or just dead on the outside but completely dead. Perished. Deceased. An ex-biosphere.

These considerations reveal a contradiction within current astrobiological thinking. It is widely held that Mars may have life, regardless of the lack of disequilibrium seen in its atmosphere, because there are probably layers deep inside where liquid water exists. Yet the commonly agreed upon signs of life on exoplanets, primarily the detection of atmospheric oxygen, requires that life drastically alter an atmosphere from a Mars-like state.

I can't think of anything I'd rather be wrong about, but I think Mars is a dead world if ever there was one. The Gaia hypothesis suggests that biologically generated gases will permeate the atmospheres of living worlds. Complexity theory suggests that for a planet to support life it must have an active, evolving, recycling surface where self-organization can flourish. Mars fails both tests. The planet's ruddy complexion, easily visible to the naked eye on a clear night, shows that it has literally rusted everywhere, with none of the active chemistry found on the surface and in the air of a living world like Earth.

I'm a skeptic about life on Mars, but also an enthusiastic advocate of

Mars exploration. We'll keep exploring Mars for other good reasons: it is beautiful, mysterious, nearby, and a relatively easy place for human beings to survive with a decent space suit.

In fact, Mars, *because* it is almost surely dead, has certain advantages. On a planet like Earth, nothing lasts long, because a living planet eats its past. Our frozen neighbor planet has wonderfully preserved traces of ancient epochs that have long been erased from Earth. Mars, then, surely holds important clues to our own past. We might even find fossils there, remnants from a brief, early flowering of life before the billion-year winter set in.

The other important advantage of a dead Mars is that we could be free to import life there without violating any strong ethical principles. We could become the Martians, but should we? Only when we are sure there are none already there.

LIFE ON VENUS AND BEYOND

We may need to look beyond our solar system to find another example of a living world, but in our ignorance, we cannot yet rule out some nearby places. Europa is an obvious place to look. As I've mentioned, my favorite underdog places for biology in the solar system are Venus and Io. Both have active chemistry and vigorous flows of energy and matter. And where something's flowing, maybe something's growing.

What draws my attention to Venus is that geologically it is a vibrant world, pulsing with volcanic eruptions, bathed in a chemically fertile disequilibrium atmosphere. I first suggested that some unexplained Venusian phenomena might possibly be signs of life in my book *Venus Revealed*. This book was published in 1997, just before exobiology became astrobiology and such speculations became not only respectable and tolerated but sometimes even encouraged with research grants.* At the time I was consciously floating the idea of life on Venus to see if anyone would bite. It has taken a while, but some of these ideas have recently been picked up and cited in the peer-reviewed literature, and others have been used in science fiction novels.†

Though still not a place where you and I would be comfortable with-

---

*I hurriedly added a mention of the new Martian "fossils" as the book went to press.
†Two SF novels of Venusian life that explicitly credit *Venus Revealed* are *The Quiet Invasion* by Sarah Zettel and *Venus* by Ben Bova.

out a well-designed suit or terrarium, Venus may have some of the essential characteristics of a living world. In fact, some of the very qualities that seem, at first glance, to doom Venus's prospects for life may conceivably help to support a more alien kind of life. That "chemically corrosive" atmosphere would certainly not be kind to organic molecules, but it reveals dynamic interactions between the surface and atmosphere.

If we are looking for a specific kind of life that we are familiar with, we had best look elsewhere. In terms of carbon biochemistry on the surface, Venus is deader than burnt meat. But if we are looking for the kind of chemically charged environment where self-organization could thrive, and the kind of ongoing geologic and atmospheric activity that a Gaia-like planetwide network of organisms could come to participate in, then Venus deserves a closer look.

If there is life on Venus, unless it uses a radically different kind of chemistry than we do, it probably lives in the clouds, thriving on the chemical energy created by absorption of UV light, and deriving nutrients from the active chemical cycles connecting the atmosphere and clouds to the surface and interior.

One of the arguments against cloud life on Venus is that there is no life in the clouds of Earth, except for bugs that are just passing through, blowing in the wind. However, recently the Austrian biologist Birgit Sattler has found evidence of a population of microbes that are reproducing in clouds over the Alps. She is planning further experiments to verify this result. If confirmed, it has important implications for possible life on any cloudy world, not just Venus.

Astrobiologists have discussed the conditions in Venusian clouds with respect to the impressive acid tolerance found in some terrestrial extremophiles. Some Earth bugs would likely be able to live in the clouds of Venus. Comparing conditions there with the comfort range of Earth's extremophiles is a worthwhile exercise, but it may not really constrain the habitability of the Venusian clouds. Anything living there is not going to be an Earth extremophile but a Venusian, and there will be a million ways in which it will be better adapted to its own world than anything that evolved on Earth.

Venus's clouds are complex, stable, global in extent, and populated with a menagerie of unidentified particles and strange, moving patterns of light-absorbing materials. Could these be photosynthetic organisms?

I've thought a lot about the Venusian clouds and published papers about their evolution, construction, materials, particle sizes, climate effects, and chemical sources. I think there *could be* life there. This doesn't mean that I think there *is* life there. I would be stunned to learn that there is, but the supposition certainly doesn't violate any scientific principle that I can think of. If there *could* be life in clouds in one place that we know, then given plenitude and biological opportunism, there *is* cloud life somewhere in the galaxy.

Is it an accident that Earth has the outstanding activity level that it does and also happens to be the one living world we know? I don't think so. The living worlds hypothesis (my theory, which is mine) suggests that life is most likely to be found where vigorous activity is discovered. Not just any activity, but cyclic flows of energy inside and outside a planet, phase changes, and physical flows across surfaces. According to this view, we should explore those places where a lot is happening. Io and Venus are the winners. Europa, Titan, and Mars are all worth a look. Ganymede and Triton are runners-up. Pluto? We'll see.

YO, IO

While we're considering heretical ideas about life in the solar system, let me speak in defense of my other favorite underdog biosphere: Io. If it's active geology that makes a world viable, then have I got one for you. Sure Io has some drawbacks—its position deep within Jupiter's intense radiation belts, an apparent lack of water, and an atmosphere so thin that it would seem like a vacuum to us.

But, Io can barely contain itself. This innermost large moon roils and seethes with such intense volcanic activity that its insides are constantly overflowing, coughing up silicate and sulfuric lava. The surface is an ever shifting collage of green, white, and red plains. Io is the most geologically alive place in the solar system. On a world where the geology changes as fast as the weather does on Earth, who needs an atmosphere? There is plenty of cycling and flow in the incontinent continental crust. Superhot flows of molten rock plow into vast fields of frozen sulfur compounds, violently vaporizing at the margins and sending sulfurous plumes blasting into space, only to snow back down on the volcanic surface. Could there be evolving complexity, perhaps leading to biological evolution at some level within that churning mass? If there is any truth to the living worlds hypothesis, which posits a relationship between geological vigor and biological potential, then Io is one to watch.

When we think about life within the moons of Jupiter, our water fixation keeps us focused on Europa. Nobody talks about life on Io. Yet, energetically and thermodynamically, Io is a lot more promising. The problem with Europa might be a lack of energetic flows to drive biology. If only Io and Europa could join forces, just think what could be accomplished with Europa's watery conditions and Io's heat flow. Actually, it is entirely possible that they *have* joined forces. They do interact in strange ways. Europa's repeated torques on Io's orbit help keep the tidal heat flowing and the Ionian volcanoes pumping. A steady rain of sulfur from Io's volcanoes falls on the frozen surface of Europa and eventually diffuses into the ocean below. For microbial (or other) life, sulfur can be just as good an energy source as oxygen, so there could be a biosphere within Europa powered by volcanoes on Io!

Perhaps Io started out more like Europa, with a watery shell, but became desiccated by the greater energy flow that goes with the territory deeper in Jupiter's tidal hot zone. In much the same way, Venus started life more like Earth, but eventually dried out from living closer to the flame. What happened to the life on these worlds when most of the water went away? Did it disappear or change with the times?

A relentless stream of hyperenergetic charged particles would rip into anything trying to make a living on or near the surface of Io, like bulls in the organic china shop, smashing up the delicate molecules that make up living cells. But could something evolve to harness these reckless bulls, getting them to plow their fields and grow their food? Imagine an organic creature that evolves to secrete or deposit some shell or substance, let's call it "special sauce," that surrounds it and shields it from this radiation. What if this special sauce absorbs the radiation through chemical reactions that turn its energy into food?

Or what about life deep inside Io? A huge flow of energy from Jupiter's tidal heating extends throughout the entire orb. I am not willing to rule out life in such a thermodynamically fertile environment just because we can't think of how it would work. At some depth there is almost surely an "aquifer" of liquid SO_2. Could such a liquid support a biosphere? I have often wondered about sulfur's biochemical potential in the context of Venus and its active sulfur cycle. I think sulfur might have some surprises for us in different environments. If we want to pro-

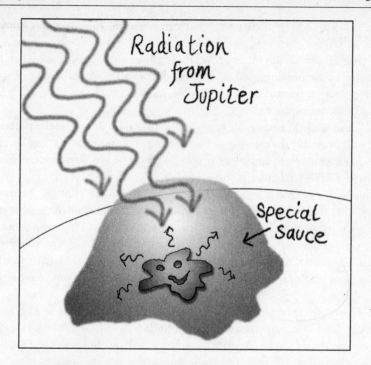

A hypothetical organism on Io secretes nonorganic mystery sauce that it uses to turn energetic particles into food. (Anthony Cooper)

pose a chemical basis for life in hell, how about sulfur, the stuff of brimstone, fool's gold, and rotten eggs? Maybe sulfur is the magic elixir on Venus and/or Io, in the way carbon is here. On Io, sulfur drools, dribbles, flows, explodes, and snows all over the surface. We don't yet know much about the sulfur cycle on Io, but certainly there is one and it is energetic and complex.

In its elemental state (bonded only to itself) sulfur takes on many poorly understood forms in every phase (gas, liquid, and solid). Sulfur reacts in interesting ways with carbon, oxygen, nitrogen, phosphorous, fluorine, and chlorine, all common elements in the universe, all present on Venus and Io. Even in our supreme state of ignorance about how sulfur may act and react in the conditions found on other planets, we do know that sulfur compounds can store large amounts of energy and make complex and unusual structures. In certain conditions sulfur forms polymers, the long chains of repeating structures that give carbon an edge in the life game. What can sulfur do that we don't know about? A lot, I bet.

Admittedly, life on Io, as on Venus, is a long shot. But I think when it comes to astrobiology, we should not discount the long shots until we have good reason to do so, regarding them instead as additional reasons to keep exploring.

These ideas are not quite science, because they do not make precise predictions. But they do suggest an approach to exploration. Look in a wide range of environments for possible signs of life. Any unexplained phenomena—particularly those that involve disequilibrium, physical shapes that seem biological, and unexpected activity confined to narrow thermodynamic conditions—should be considered possible signs of life, and thus worthy targets for further exploration. Life itself should be doubted until the evidence is extremely compelling. Currently, certain unproven assumptions have became axioms that threaten to railroad our exploration program. Yet, a narrow search is not mandated by the state of our knowledge.

These are my current thoughts about the nature of life in the universe, based as much on scientific intuition as on established facts. I have presented these ideas in the science section, but I know that by now I've crossed the line and I'm talking about my own beliefs. Natural philosophy recognizes that these sections ("History," "Science," "Belief") are really one and the same. There *are* no known facts about life in the universe, except that it has happened at least once.

PART III

Belief

18 | SETI: The Sounds of Silence

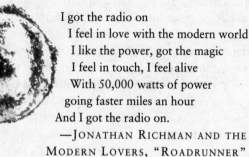

I got the radio on
 I feel in love with the modern world
 I like the power, got the magic
 I feel in touch, I feel alive
 With 50,000 watts of power
 going faster miles an hour
And I got the radio on.
—JONATHAN RICHMAN AND THE
MODERN LOVERS, "ROADRUNNER"

LISTEN UP

I've been waiting for a signal all my life. The news might come any day: a manic phone call late at night or a solemn announcement on NPR as I'm driving to work one morning. I'd swerve and spill coffee down my pants, but it would be worth it. What would your reaction be if you turned on CNN to a head talking about an alien message that had just been picked up by scientists at the Arecibo Radio Observatory? How would the masses respond as the news spread? Pandemonium? Dancing in the streets? Disbelief? Indifference?

Once large masses of people thought there had been such an announcement. It was Halloween, 1938. Orson Welles's fictional radio broadcast of a landing in Grovers Mill, New Jersey, by an invading Martian army was taken seriously by millions of people who didn't hear, or at least didn't heed, the disclaimers.* Panic ensued. People fled

---

*This drama was based on H. G. Wells's *War of the Worlds,* but in the radio version the location of the invasion was moved from London to New Jersey. By the time the movie version was released in 1953, the Martians had apparently learned more about us and figured out that they should really destroy Los Angeles.

their homes, flooded police phone lines, and did their best to prepare for a gas attack from the marauding Martians. In its lead story the following morning, the *New York Times* reported that the broadcast had "disrupted households, interrupted religious services, created traffic jams and clogged communications systems," and that "at least a score of adults required medical treatment for shock and hysteria."

This example can hardly be considered instructive, because first contact won't come in the form of an invading army. If creatures able to travel interstellar distances wanted our planet, it would not resemble a war as much as an extermination or a wildlife relocation program. And why would they even want our Earth? Aliens will surely be better adapted to their own planets. It is highly unlikely that they will be able to breathe our air, infect us, or eat us without a lot of expensive and messy food processing. "War of the worlds," while entertaining (except for those needing treatment for shock and hysteria), is not a likely scenario.

If first contact comes in the form of a radio message, then we may at first know nothing about the senders except for the simple, startling fact that they are out there, broadcasting a signal that seems incontrovertibly the product of some kind of mind. It might be frightening, liberating, uplifting, disturbing, or all of the above, but I say, "Bring it on."

As long as we are still the one "intelligent species," alone in a universe swimming with bugs and scum, we are still the big-brained lords of all we see. Even finding microbes on Mars wouldn't dethrone us, rather it would enlarge our kingdom. The most enticing aspect of such a discovery might be the implication that intelligent life is also relatively close by, because in a universe that is teeming, someone else must be dreaming and scheming.

What we most want to know is whether anyone is out there whom we can talk to and learn from. We want to know what they look like, how they think, what they know, whether they've had the same problems we have, and how they've solved them.

So we search the skies.

FIRST ATTEMPTS

The modern era of SETI (the search for extraterrestrial intelligence) began in September 1959 when the Cornell University astrophysicists Guiseppe Cocconi and Philip Morrison published a seminal paper in *Nature* entitled "Searching for Interstellar Communications."

The idea of trying to communicate with beings on other worlds was not new. In the nineteenth century, several schemes were proposed for contacting inhabitants of the Moon or Mars by drawing huge diagrams on Earth's surface for the aliens to read. German mathematician Carl Friedrich Gauss proposed that massive areas of Siberian forest be clear-cut in the shape of a triangle with adjoining squares, showing the Martians that we know the Pythagorean theorem. Viennese astronomer Joseph von Littrow suggested that a giant circular trench be built in the Sahara desert, filled with kerosene, and ignited when Mars was close to Earth.* In 1909, in the wake of the "Mars mania" caused by Percival Lowell's sensational claims of a canal-building civilization, Harvard astronomer W. H. Pickering advocated the construction of a huge mirror to signal the Martians, and in the 1920s several astronomers imagined flashing huge searchlights in the direction of Mars using a kind of "Morse code" to convey various pictorial images.

Even radio communication had been attempted, decades before modern SETI. In 1899, the Croatian-born American physicist, electricity pioneer (and New Age cult figure) Nikola Tesla declared that he had received electrical communications from extraterrestrials, most likely residents of Mars or Venus. Describing the experience in the February 1901 issue of *Collier's Weekly,* he wrote:

I felt as though I were present at the revelation of a great truth. My first observations terrified me, as there was present in them something mysterious, not to say supernatural, and I was alone in my laboratory at night. . . . The changes I noted were taking place periodically, and with such a clear suggestion of number and order. . . . The feeling is constantly growing on me that I had been the first to hear the greeting of one planet to another.

Twenty years later Italian radio pioneer Guglielmo Marconi proposed that radio could be used to communicate over interplanetary, even interstellar, distances. In a front-page article in the *New York Times* of January 20, 1919, Marconi suggested, "It may someday be possible, and as many of the planets are much older than ours the beings who live there ought to have information for us of enormous value."

---

*Today this sounds like a good idea for the Burning Man festival.

After outfitting his luxury yacht, *Electra,* with radio receivers, Marconi sailed to the middle of the Atlantic Ocean, where he could listen for alien signals free from local interference. He believed he heard them. In response, the *Times* ran an editorial entitled "Let the Stars Alone," warning that through such efforts we might receive "knowledge for which we are unprepared precipitated on us by superior intelligences. . . . If Mars is a million years ahead of us, it is far from impossible that Mars . . . would regard argument with our mundane mathematicians as no more serious an occupation than inciting a kitten to chase its own tail."

Real radio astronomy was born of the invention of radar in World War II. Turned toward the heavens, the powerful new radio techniques developed for spotting enemy aircraft revealed a universe humming with noise and a vast, hidden landscape peppered with pulsars and radiant with radio galaxies. Might there be a deliberate signal, or at least some leaking chatter, hidden among all this radio clatter?

In their 1959 paper, Cocconi and Morrison reasoned persuasively that with the technology of their day (late 1950s) we should be able to detect a radio signal directed at us from elsewhere in our own galaxy. They backed this up with convincing calculations, and they even told us what frequency the aliens would be broadcasting on. Hydrogen, the most abundant element in the universe, hums naturally at 1420 MHz on your radio dial.* This region of the spectrum is generally pretty quiet, so the hydrogen channel is not only useful for studying the structure of the universe, but also a good communication frequency. The aliens would know that any species who had recently discovered radio astronomy would build equipment designed to listen on the hydrogen channel. The *Nature* paper concluded:

> The reader may seek to consign these speculations wholly to the domain of science-fiction. We submit, rather, that the foregoing line of argument demonstrates that the presence of interstellar signals is entirely consistent with all we now know, and that if signals are present the means of detecting them is now at hand. Few will deny the profound importance, practical and philosophical, which the detection of interstellar communications would have. We therefore

---

*MHz means "megahertz" or "million cycles per second." These are exactly the same as the numbers on your FM radio.

feel that a discriminating search for signals deserves a considerable effort. The probability of success is difficult to estimate; but if we never search, the chance of success is zero.

Who could argue with that? After Cocconi and Morrison, it seemed foolish *not* to listen for alien signals.

SWITCHING ON

It was a time of great excitement and optimism about planetary exploration and ET life.* At almost the last possible moment before reality intruded, in the form of data from planetary probes, the golden age planets were still filled with the busy cackle and buzz of life. Indeed, to support their argument for life on the planets of distant stars, Cocconi and Morrison pointed out that it exists on *two* planets in just our own solar system, "Earth and very probably Mars." This, they noted, was a good thing, since a sample of one lacks statistical significance.

My earliest memory, or rather the earliest thing I can remember remembering, is of lying in a crib with several large humanoids peering down at me. These big heads were talking, and I was frustrated, trying to figure out what it was they were saying. That must have been in 1960, around the time when Frank Drake, a brilliant and bold thirty-year-old astronomer at the National Radio Astronomy Observatory in Green Bank, West Virginia, was conducting the first modern search for alien radio signals. He pointed his radio dish at the two nearest Sun-like stars, Tau Ceti and Epsilon Eridani, and listened on the hydrogen frequency suggested by Cocconi and Morrison. (Drake had decided *completely independently* that this was the right frequency to search. So, at least for smart human astronomers in 1960, it really was a universal frequency.) On the first day, Drake detected a strong signal. As he later recalled, "My first thought was 'Could it be this easy?' My second thought was 'What do we do now?' " Fortunately, he did not alert the media, because this first quick success turned out to be a false alarm caused by local interference.

Drake named this first listening effort Project Ozma after the queen

---

*Recently my mother gave me a copy of the *Life* magazine issue published on the day I was born (three months after SETI was born with the Cocconi and Morrison *Nature* paper). Inside is an enticing article headlined "Target Venus: There May Be Life There."

of Oz, "a place very far away, difficult to reach, and populated by exotic beings." Are there whisperings from more aged creatures, waiting for us to grow up and talk to them, coursing through the airwaves above our planetary crib? Ozma detected no messages, but the age of experimental SETI had begun.

Project Ozma immediately provoked a full spectrum of strong reactions. In a 1960 article in *Physics Today*, Drake's boss, astronomer Otto Struve, wrote that Ozma "has divided the astronomers into two camps: those who are all for it and those who regard it as the worst evil of our generation. There are those who pity us for the publicity we have received and those who accuse us of having invented the project for the sake of publicity."

Struve served as chairman of the first American meeting on SETI, a now legendary gathering held in Green Bank during Halloween week, 1961. The ten participants included most of the early prime movers in American SETI. Morrison, Struve, Drake, and Sagan were all there, as were the astronomer Su-Shu Huang (who had invented the concept of "habitable zones" around stars in 1959), the engineer and SETI theorist Barnard Oliver, the biochemist Melvin Calvin (who was awarded a Nobel Prize *during* the Green Bank meeting for his work on photosynthesis), and the dolphin researcher John Lilly. Lilly had just published *Man and Dolphin,* a book arguing that dolphins have a complex language, and that we might learn to communicate with them. That Earth might have evolved not one, but two intelligent communicative species seemed encouraging for the prospect of intelligence evolving elsewhere.

The presence of all of these distinguished scientists discussing the number of intelligent civilizations in the galaxy, and working through the practical challenges of communicating with them, helped to infuse the dawning era of SETI experimentation with a new aura of respectability. As Sagan later described it, "There was such a heady sense in the air that finally we've penetrated the ridicule barrier. . . . It was like a 180-degree flip of this dark secret, this embarrassment. It suddenly became respectable."

The Green Bank participants were so full of camaraderie and hope, and so enthralled by Lilly's reports of success in talking to dolphins, that they formed a club called the Order of the Dolphin, with Sagan as secretary. They even made membership pins with little dolphins on them.

But, the Dolphins never had a second meeting. John Lilly started writing publicly about his lengthy conversations with ethereal, extraterrestrial beings, which occurred while he was tripping his brains out on large doses of psychedelic drugs (acid and ketamine being his favorites for interstellar discourse). The ridicule barrier may have been penetrated, but it was not left far behind. The SETI pioneers decided that the association with Lilly would not help their credibility, and the Order of the Dolphin was unceremoniously dissolved.*

Unfortunately, there are no published proceedings (and no group photos) from the Green Bank meeting. Perhaps no one appreciated the historical significance of the gathering, or they were still too embarrassed to go on record as taking the subject matter seriously. However, SETI had arrived as a science.

FIGURING THE ODDS

Frank Drake began the Green Bank meeting by writing an equation on the blackboard that summarized the major questions encountered when trying to estimate the number of intelligent civilizations in the galaxy. He intended only to suggest an agenda for the meeting, but his formula was destined to become "the Drake Equation," the most famous equation in SETI research. Four decades later, it is still used in nearly every SETI paper, book, and discussion.

The Drake Equation is not really supposed to provide an answer. You can't just plug in all the variables and determine that there are precisely 3,741 civilizations in the Milky Way. What it gives us is a clear way of thinking through the problem and all of the factors that are important in determining the answer.

Before I actually tell you the Drake Equation, I'll give you an analogy: the date equation. Say you are a single person going to a large dance party, and you would like to come away with a date for the following weekend. Arriving in front of the house, you can hear the music pumping and feel the bass rattling your gut. You are excited, but nervous as hell, so you decide to calm yourself with some math. Before going inside, you try to calculate your chances of getting lucky. You

---

*Lilly kept "Order of the Dolphin" listed under "professional associations" on his curriculum vitae until his death in 2001. Shortly after Green Bank Lilly set his dolphins free and focused on self-experimentation with drugs and isolation tanks. The protagonist in the movie *Altered States* is loosely based on Lilly.

start by guessing the total number of people at the party. You notice that people are arriving at a rate of three per minute. We'll call this rate of arrival R. People are leaving at roughly the same rate, but you realize that you can estimate the number of people inside if you know how long they are staying. Let's call this length of stay L. The number of people inside will be roughly R times L. So, if people on average are staying for, say, one hundred minutes, there will be about three hundred inside.

But they are not all potential dates. After all, you have standards and preferences, and some may not be available. So you multiply the total number at the party by several factors, each expressing the probability that the average partygoer will meet one of your requirements. Each of these probability factors will have a value between zero and one. Zero means that nobody measures up to a particular requirement. One means that anyone will do. If half of them are okay, the probability is one-half, or 0.5, and so on.

For instance, you might want to rule out potential dates because they don't fit your sexual preference. We will call this factor f_p (pronounced "f-sub-p") and assume that this is roughly 0.5, meaning that it rules out half of the people there.* Then you are going to multiply that by the fraction that you find yourself attracted to. If you are being picky, we'll say that $f_{at} = 0.1$. In other words, one in ten meets this criterion. Again, it cannot be higher than 1, even if you are drunk or desperate. Now, some people are not going to be available because they are already hooked up and not interested in multiple partners. Let's say optimistically that a quarter of the people (or 25 percent) you are interested in are free. So $f_{av} = 0.25$.

You also have to factor in your own behavior. Some are just so hot, you can't get up the nerve to talk to or dance with them. But all this math is making you feel pretty confident, so we'll say you can deal with approaching three-quarters of them: $f_n = 0.75$. Then we have to multiply again by the fraction who turn out to actually be interested in you. Because you are fascinating and fun to dance with, and because you can talk knowingly and winningly of probability (chicks and cats dig that), no one can refuse you, so $f_i = 1$. Assuming you have not forgotten any

---

*Woody Allen has pointed out that if you are bisexual, it doubles your chances for a date on a Saturday night. Put quantitatively, for these lucky people $f_p = 1$.

important factors, you can now estimate your chances of scoring at the party. The total number of likely candidates, N, will follow the formula

$$N = R \times f_p \times f_{at} \times f_{av} \times f_n \times f_i \times L$$

This is the "date equation." Given the numbers we've estimated, N = $3 \times .5 \times .1 \times .25 \times .75 \times 1 \times 100$. So, N = 2.8: 2.8 people at the party will go out with you next weekend. Jackpot! Although this is just a rough estimate, since we had to estimate the various f factors, your best guesses lead to an N greater than one, so you figure your chances for success are pretty good. Thus emboldened, you check your hair one last time and enter the party.

Now back to the Drake Equation. The equation that Frank Drake wrote on the board to start off the Green Bank meeting looked like this:

$$N = R^* \times f_p \times n_e \times f_l \times f_i \times f_c \times L$$

The Drake Equation is parallel to the date equation, but the party we wish to crash is much larger, more frightening, and more enticing: N is the number of communicating civilizations in the galaxy. Each star in the Milky Way is a potential dance partner for a lucky biosphere, but some won't have the right chemistry. We have to narrow them down. On the right side of the equation are all of the factors that we must estimate in order to size up N.

Here's what they all stand for, going from left to right:

The first three factors are astronomical: R represents the rate, per year, at which stars are being born in the galaxy, f_p is the fraction of stars with orbiting *p*lanets, and n_e is the average number of planets, per star, with *e*nvironments where life can evolve.

The next two factors are biological: f_l is the fraction of suitable planets where *l*ife actually does develop, and f_i is the fraction of these where *i*ntelligent life evolves.

The final two terms in the equation represent cultural or social factors: f_c is the fraction of planetary cultures that are *c*ommunicating over interstellar distances, and L is the average *l*ongevity, or *l*ifetime, of these civilizations, in years.

Multiplied together, all of these terms give us N, our estimate of the number of communicating civilizations in the galaxy. Obviously, most of these numbers are highly uncertain, if not totally unknown. If this were a serious attempt to nail down an answer, it would be laughed out of science. The value of the equation is that it allows us to examine the consequences of our assumptions and explore how changing them changes the answer we get about finding intelligence in the universe. It helps us to identify the most crucial small questions we must answer to crack the big one.

Drake had the Green Bank group take their best shot at estimating each factor. Much of the discussion and literature of SETI in the forty-two years since then has attempted to refine, refute, or improve these estimates. Notice that the terms in the Drake Equation become less certain, and more highly subjective, as we read from left to right. R^*, the rate of star formation, has been pinned down by astronomical observations. The Dolphins knew that between one and ten new stars are born per year.* In contrast, the fraction of stars with orbiting planets was completely unknown in 1961. The Dolphins estimated that half of all suitable stars had planets ($f_p = 0.5$).

They could not agree on a single estimate for n_e, the average number of life-friendly planets per star. Some thought that a conservative estimate was $n_e = 1$. Sagan, believing that several planets in our own solar system probably harbored life, argued for an average of five. They agreed to disagree, settling on a range of n_e from 1 to 5.

Since 1961, we've narrowed down the uncertainties in the astronomical factors ever so slightly. The biological and cultural factors remain pretty much where they were when Kennedy was in the White House, although different beliefs have come in and out of fashion. The Green Bank group reasoned that since life apparently arose quickly here on Earth, any planet that could have life must have life. So they estimated that $f_l = 1$. This view is still quite common in astrobiology. They also estimated that $f_i = 1$, arguing that biological evolution would always lead to intelligence, because of the great survival advantage it conveys. This has proven to be one of their most controversial assumptions.

Finally, there are the "cultural factors." How many intelligent civilizations will build radio telescopes and choose to use them for interstellar communication? In this innocuous term (f_c) are contained a huge host of thorny questions about the nature of intelligence, the universal

---

*In our galaxy, that is. This is not counting the massive stars that burn out in a few million years.

attraction of technology, and the motivations and desires of alien beings. The Green Bank Dolphins decided that one in ten such civilizations would have the required radio technology and the will to use it ($f_c = 0.1$).

The final term in the equation, L, the longevity of technical civilizations, proved to be the hardest to estimate. The best we can do is to ponder our own future, estimate our longevity, and pray for rain. Then we extrapolate recklessly to the rest of the galaxy. The Green Bank group recognized that this number is totally unconstrained.* We might destroy ourselves after less than a hundred years of radio listening, which began in 1960. Or we might learn how to get along with one another, control our technology, live sustainably, and achieve security against natural disasters. (Well, anything is possible.) If we do all that, we might last for billions of years. So there is a factor of 10^7, or *ten million,* separating the optimistic from the pessimistic estimates. At Green Bank, recognizing that we can't really pin down L, they settled on a range of 10^3 to 10^8 (one thousand to 100 million) years for the average lifetime of a communicating civilization.

Multiplying all of the factors together, the Dolphins calculated a huge range of 10^3 to 10^9 for N, the number of civilizations in our galaxy. Take a look at the numbers in the following table and notice a couple of things. First, even in the most pessimistic estimate conceivable to all of these brilliant scientists, the galaxy should contain, at present, a thousand advanced civilizations. This is a modern, quantitative reaffirmation of the ancient principle of plenitude: in a galaxy this vast, with so many stars—each a potential home—we've still got plenty of company, even if the odds are extraordinarily unfavorable to life and intelligence. And that's the pessimistic scenario.

THE DRAKE EQUATION: GREEN BANK ESTIMATES

Symbol:	R*	f_p	n_e	f_l	f_i	f_c	L	N
Meaning:	star formation rate	fraction with planets	number in habitable zone	fraction with life	fraction with Intelligence	fraction communicating	Longevity of civilizations	Number of civilizations in galaxy
Green Bank Estimate	1 -10	.5	1 -5	1	1	.5	10^3-10^8	10^3 - 10^9

*Which is the technical term for not having a freaking clue.

On the other end of the range, the number of civilizations may approach 1 billion, or more than one in every thousand stars. If this is the true state of our galaxy, then several of the few thousand stars you can see with your own eyes on a clear night probably have planets that are home to advanced cultures. In this crowded galaxy, the average distance between civilizations would be only a few tens of light-years, and we might hold two-way interstellar radio conversations within the short span of a human life.

Also, note that the calculated range for N, the number of civilizations, comes out identical to the range estimated for L, their average lifetime in years. That's because in the (possibly somewhat optimistic) Green Bank estimates, all the numbers to the left of L, multiplied together, give a value around 1. Accepting this greatly simplifies the Drake Equation, which becomes

$$N \approx L^*$$

In other words, the number of civilizations in the galaxy is equal to the average lifetime, in years, of a civilization. This may or may not be close to the truth of the matter, but it does highlight the most solid conclusion that we can draw from the Drake Equation: the number of intelligences in the galaxy hinges most crucially on one factor, L, the longevity of communicating civilizations.

If intelligence is self-limiting, if brainy races are often too smart for their own good and survive, on average, only for a century or a millennium, then unless we are lucky (always an annoying possibility when trying to use probability on ourselves), even our nearest communicative neighbors are probably thousands of light-years away. Conversation would be impractical, to say the least, since every answer would take millennia, and most civilizations wouldn't last that long. If, on the other hand, the average high-tech race lasts for millions of years or more, then the galaxy must be full of wise and ancient races. We conclude that the nature of our galaxy, whether it is a quiet desert with an occasional lone cry of consciousness, or a cacophonous jungle of talk radio, hinges on how long technological species can survive.

This realization adds cosmic poignancy to the question of how humans will fare in the long run. If we are in some way representative, then our own prospects contain a hint about the prevalence of conscious

---

*The squiggly equals sign means "is approximately equal to."

souls in our entire universe. Our own nebulous future may be loosely tied to the psychozoic density of the cosmos. The largest unknowns in the equation lead us back to the unknowns of our own nature.

The guys at Green Bank realized all this, and their ideas are still representative of mainstream thought in SETI. Although innumerable caveats and alternate scenarios have been described and proposed, the Drake Equation has survived as the scaffolding of all discussions about intelligent ETs.

BACK IN THE USSR

Meanwhile, in the Soviet Union a small group of astronomers and theorists were discussing the same questions and planning their own alien hunts. Because of the tradition of Cosmism, in Russia mainstream science has long tolerated such far-out thoughts more than it has in the United States. Perhaps because of a Marxist faith in an inevitable progression of historical stages, some of the boldest notions about the evolution of advanced societies came from Soviet scientists.

Certainly, while it lasted, the Soviet Union gave much more official support to SETI than the United States ever has. As I mentioned in chapter 14, in 1962 Iosif Shklovskii published *Universe, Life, Mind,* which included discussions of interstellar radio communication with other civilizations. Shklovskii independently arrived at many of the same conclusions reached by Cocconi, Morrison, and Drake.

The Soviet counterpart to the Green Bank meeting, the First All-Union Conference on Extraterrestrial Civilizations and Interstellar Communication, was held in May 1964, at the Byurakan Astrophysical Observatory in Soviet Armenia. In his opening remarks Academician V. A. Ambartsumyan, head of the Byurakan Observatory, stated:

> We have no doubt whatsoever that life and civilizations exist on a multitude of celestial bodies, but . . . modern technological civilization (on Earth) has its origin no more than two hundred years in the past. And yet, the ages of planets may differ by as much as millions of years. Hence it seems that Earth civilization is not yet past the diapers age, and that there should be enormous disparity with extraterrestrial civilizations.
>
> The problem is therefore essentially a problem of communication between civilizations on entirely different levels of development. . . .

In practical terms, our aim is therefore to obtain rational technical and linguistic solutions for the problem of communication with extraterrestrial civilizations which are much more advanced than terrestrial civilization.

Echoing these comments in the first scientific paper presented at the Soviet meeting, Iosif Shklovskii respectfully critiqued the work of Drake, Cocconi, and Morrison: "It seems to us that Project Ozma was doomed from the very start, for the following reasons: (a) It assumed that civilizations may occupy the nearest stars. (b) Cocconi and Morrison's idea and its realization by Drake assumes that the extraterrestrial civilizations are approximately on the same technological level as terrestrial civilizations. But . . . we are only infants as far as science and technology are concerned."

At Byurakan, Nicolai Kardashev, Shklovskii's young colleague (and former star student) at Moscow State University, pondered what a technological civilization that survives for thousands or even millions of years might look like. Pointing to the exponential increase in energy resources available to human societies in the last couple of centuries, Kardashev proposed that expanding energy use will be a universal hallmark of advanced civilizations.

Starting from this assumption, Kardashev devised a classification system for technological civilizations. Those that have the approximate energy resources of their entire home planet at their disposal he termed Type I. Humanity, he said, will become a Type I civilization sometime in the twenty-first century. Kardashev defined Type II civilizations as those using the power output of their entire home star. Finally, Type III civilizations have access to the resources of an entire galaxy.

Kardashev observed that on Earth a Type I civilization took several billion years to develop. He predicted that the transition to Type II would take, at most, a few thousand years, and that to Type III no more than a few tens of millions. If these numbers are typical, then our level of civilization is a brief stage, and our own kind should be rare. Type II and even III civilizations should be common, since they tend to stick around for a long time.

These considerations, Kardashev pointed out, have practical consequences for our search strategies. If any Type II or Type III civilizations are out there, they might be easier to detect even than Type I civilizations located much closer to us—just as a stadium rock concert miles away is easier to hear than a ukulele across the street. A supercivilization might have constructed a beacon reaching a wide expanse of galactic space, to attract the attention of fledgling species like ours just turning on their radios. Or, their artifacts and internal communications might be visible from vast distances. With a million years of progress under their belt, perhaps they would have undertaken vast "astroengineering" projects, rearranging or reconstructing stars for their own inscrutable purposes. Such cosmic-scale works of civil engineering might be visible across the galaxy, or even from other galaxies.*

Shklovskii and Kardashev concluded that most civilizations out there must be at least Type II. This, they pointed out, suggests an entirely different kind of search strategy from that used for Project Ozma. Drake, in effect, was searching for Type I civilizations who are broadcasting toward the nearest stars likely to have similar planets. This type of targeted search of nearby stars should only be successful if the galaxy is loaded with Type I civilizations. If we think that Type II or III civilizations might be out there, then perhaps we should forget about looking at nearby stars and instead scan large areas of the sky, looking for the Big One. Kardashev's ideas, like Drake's, became part of the lexicon of SETI. Scientists in the field still refer to Type I, II, or III civilizations.

East met West at the First International Conference on Communication with Extraterrestrial Intelligence, held at Byurakan in 1971. Later, Shklovskii wrote, "Never, before or afterward, have I taken part in a more imposing scientific gathering." In addition to astronomers, physicists, and biologists, the organizers made an effort to include representatives of various relevant fields from the humanities, including linguists, philosophers, anthropologists, and historians. All of the Soviet and most of the American SETI pioneers were there. Several Nobel laureates were in attendance, including Francis Crick, who had discovered the structure of DNA. The proceedings make for lively reading because of the fiery clashes between Crick, who thought that we could not really say anything about the probability of intelligence elsewhere, and Sagan, who thought we could.

The assembled polymaths discussed the value of L (the lifetime of civilizations) in the context of the nuclear arms race, population

---

*Type III civilizations always make me think of those talking galaxies in the opening scene of Frank Capra's *It's a Wonderful Life*. In the film they are actually angels, but we might be hard-pressed to tell the difference between angels and wise old type III aliens who have mastered and internalized technology that we cannot even dream of dreaming about.

dynamics, and conflict resolution theory. Discussions spun off into fascinating debates about the universality of mathematics, the meaning of progress, and whether the laws of physics can be changed. These considerations did not, however, really nail down the answer to the Drake Equation.

Near the end of the conference, the historian William McNeill of the University of Chicago concluded, "I must say that in listening to the discussion these last days, I feel I detect what might be called a pseudo or scientific religion. I do not mean this as a condemnatory phrase. Faith and hope and trust have been very important factors in human life and it is not wrong to cling to these and pursue such faith. But I remain, I fear, an agnostic, not only in traditional religion but also in this new one."

WOW!: FALSE ALARMS AND REAL HOPE

As long as I've been aware of anything beyond the street where I lived, I've been aware of SETI. In the 1960s and early 1970s there were many small-scale, independent observing programs, mostly targeting nearby stars. Since no one had ever tried to look before, it seemed plausible that success could come quickly and easily. I first became conscious of SETI during this hopeful period, when the search was young and anything was possible. With my impressionable mind warped by science fiction and rock 'n' roll radio, and with some of the SETI pioneers in my family's social circle, searching for alien signals seemed like a perfectly respectable and smart thing to do.

I remember going to the Boston premiere of *2001: A Space Odyssey* in the spring of 1968 with my family and the Sagans. It was the single most influential, moving, exhilarating, and terrifying moviegoing experience of my life (at least up to now). That night, I couldn't sleep. I kept opening my eyes, half expecting to see a large, black, humming monolith standing sentinel in my bedroom. The movie was powerful, exciting, and frightening because it seemed completely realistic. That all these smart adults took it seriously heightened the aura.* There was nothing

---

*I remember my dad and Carl on the way home discussing all the scientific flaws they had noticed in the film: stuff like dust swirling on the moon, which it wouldn't do in a vacuum, and Dr. Heywood Floyd's head resting in the wrong position when he was sleeping in weightlessness. But *2001* is one of the few movies that even *attempts* to be scientifically realistic and everyone gave it high marks.

in it that couldn't happen in my lifetime. By century's end, we'd be sending humans to the far corners of the solar system. Maybe I'd even go myself. Radio contact with extraterrestrials was bound to happen before too long. Lying awake, I felt that I was rushing through a Stargate, with the future approaching fast.

Iosif Shklovskii later referred to this early period of SETI as the time of "adolescent optimism," a time when we imagined creatures not too different from ourselves pointing their dishes right back at us from planets around nearly every star. This hopeful attitude was reflected in that, back then, it was not called SETI but CETI, for "communication with extraterrestrial intelligence." Shklovskii became much more pessimistic about the prospects for contact in his later years. He came to believe that L (the average lifetime of a civilization) was small, because most technological civilizations were destined to destroy themselves before long. As Shklovskii put it, in his 1991 memoir, published six years after his death at age sixty-nine, this adolescent optimism was "based on faith in human society's unbounded progress and places exaggerated emphasis on the radio-technological prospects for extraterrestrial communication, while ignoring both the humanities and biological aspects." Gradually the term SETI (the *search* for ETI) replaced CETI. With the adoption of this more humble acronym, we admit that we are only playing a game of solitaire until someone else shows up at the table. If and when SETI is successful, then CETI may begin.

In the seventies, NASA started supporting small observing programs to the tune of a few million dollars per year (a couple of pennies from each American). Whereas Ozma had listened in on two stars, the new plan called for the world's largest radio dishes to sample radio waves from a thousand promising suns.

What was it Einstein said about great spirits always receiving violent opposition from mediocre minds? NASA's SETI program was easy prey for politicians who wanted to pose as fiscally responsible. A SETI program might not succeed for decades or centuries or millennia, so it was easy to ridicule and portray as wasteful. In 1978, famously anti-intellectual Senator William Proxmire of Wisconsin awarded one of his notorious Golden Fleece awards for stupid government spending to the NASA SETI program, and in 1981 he succeeded in deleting all funding for radio searches. Frank Drake retaliated by nominating the senator for membership in the Flat Earth Society, and supporters rallied to the SETI cause. The National Science Foundation and the International

Astronomical Union called upon Congress to restore funding. Sagan visited with Senator Proxmire for an hour and reasoned with him. A year after the shutdown, funding was restored and SETI was saved, for the time being. But, in 1993, Nevada senator Richard Bryan introduced a successful amendment to the 1994 NASA appropriations bill that eliminated all funding for SETI. That was it. NASA SETI was DOA.

In a press release celebrating his victory, the senator from Nevada sneered that even after several years of searching, NASA "had failed to bag a single little green fellow." Yet, two years later, his own state government proudly christened state highway 375 "the Extraterrestrial Highway."* Apparently Nevada politicians are not against aliens as long as they generate tourist revenue. In any case, Senator Bryan's amendment dealt a fatal blow to government-supported SETI in the United States. Meanwhile, Soviet SETI had collapsed along with the Soviet Union. So, who on Earth would listen for the whispering of the sky?

Fortunately, a couple of billionaires, as well as many ordinary folks, came to the rescue, and American SETI was privatized and run out of the nonprofit SETI Institute in Mountain View, California. SETI benefited from the 1990s information technology bubble in Silicon Valley, when rich, altruistic visionaries were swarming the South Bay. Much of the funding to privatize SETI was provided by Bill Hewlett, Dave Packard, Gordon Moore (cofounder of Intel), and Paul Allen (cofounder of Microsoft). Each kicked in a million bucks of his pocket change.

The new project, risen from the ashes of government SETI, was appropriately named Project Phoenix. Its director is astrophysicist and SETI veteran Jill Tarter (the field, along with the rest of science, has progressed and is no longer completely male-dominated). Phoenix began observations in February 1995. At present, Phoenix receives 5 percent of the observing time at the world's largest radio telescope—the thousand-foot dish built into a giant natural crater in Arecibo, Puerto Rico.† Phoenix has scanned more than half of its initial target list of a thousand stars, all within two hundred light-years of Earth.

---

*This road passes near "Area 51," where, surely you've heard, the government is hiding and experimenting on the bodies and wrecked saucer of aliens who crashed in 1947 outside Roswell, New Mexico. More on this later, if I'm not silenced by government black ops agents.
†Bizarrely, the enormous size of the Arecibo dish is the result of a large mathematical error. The designers calculated that they needed a *thousand* foot dish to detect radar reflections from the Earth's ionosphere. A *hundred*-foot dish would have sufficed. But, as

To date, the longest-running continuous SETI search was the Big Ear project at Ohio State University. It scanned large areas of the sky near the hydrogen channel for twenty-five years until it was scrapped in 1998 to make room for a golf course.* On the evening of August 15, 1977, Big Ear detected a strong signal from a point within the constellation of Sagittarius. The astronomer on duty, Jerry Ehman, circled the signal on the readout and wrote, in the margin, **"Wow!"** The signal only lasted for seventy-two seconds. Everything about it is consistent with an alien technological source.

The "Wow!" signal has entered the lore of science, and science fiction, as the best candidate yet for an actual alien signal. It is a favorite topic among those who are convinced that aliens have contacted us and the government is suppressing this information.[†]

The Wow! signal winked out quickly and has never reappeared, though hundreds of attempts have been made to find it. In 2001, astronomers used the multiple dishes of the Very Large Array in New Mexico to conduct the most powerful hunt ever for a signal in the Wow! direction. Not a peep. Obviously, Wow! was not a continuous radio beacon, but it might have been a real signal, perhaps a snippet of internal communication between some alien ships. This tantalizing hint has helped SETI scientists maintain their enthusiasm over recent decades.

Today, there are about a dozen active radio search programs. In addition, there is a growing amateur SETI movement, led by electrical engineer (and jovial singer of geeky folk songs) Paul Shuch. His SETI League enlists enthusiasts who want to set up a dish in their backyard. Although the sensitivity of amateur instruments is much lower, there is strength in numbers. Together they can look in more directions at once, so they could detect a strong signal that comes in when the big dishes are pointing elsewhere. The SETI League's Project Argus is trying to link up five thousand radio dishes around the world to continuously monitor the entire sky. As of this writing, they are up to 120.

The power and sensitivity of searches has improved by a factor of

---

Drake wrote, "Fortunately for the history of astronomy, no one discovered the error until construction was well under way and it was too late to change the size." Drake once calculated that the giant Arecibo dish could hold 357 million boxes of cornflakes.

*Way to go, Ohio.

[†]The Wow! signal was even mentioned in an episode of *The X-Files* in 1994, which means that anyone within nine light-years knows about it by now, if they get the Fox network.

more than 100 trillion since the days of Project Ozma. Still, the stars remain silent.

FURTHER

The newest development in SETI is called SETI@home (SETI at home). Anyone with a computer hooked up to the Internet can participate. Just download the free software and then, while you are sleeping or goofing off, the computers do the rest. The mother machine at Berkeley automatically sends your computer a packet of data gathered at Arecibo. When your PC is through searching for the telltale regularities that might indicate an actual signal, it sends back its analysis and waits for the next batch of bytes. The graphic design is almost as smart as the concept. The twenty-third-century Federation-style aesthetic looks much cooler than most software actually used for scientific research.

The project began in May 1999, and by its fourth anniversary more than 4.4 million PC users from 226 countries* had joined, with thousands more volunteering each day. In effect, the SETI@home project has become the world's largest supercomputer. It has already performed nearly a million years of combined processing time and is currently racking up more than a thousand years per day.

Why is this global supercomputer spontaneously self-assembling so quickly? Because SETI is inspiring. It taps into the dreams of a species beginning a new millennium and wondering if there is more to the universe than meets the eye. SETI@home finds fertile ground among the netizens of the rapidly evolving global cybernation. Now that small talk with distant pals we've never actually seen is our daily reality, the idea of a galactic network of distant communicating civilizations does not seem quite so far-fetched.[†] It is fitting that the search for our interstellar neighbors should involve as many residents of Earth as possible. SETI@home represents a new kind of populist science, as multitudes participate in humanity's effort to find the all-important signal in the noise of the universe.

---

*As of May 2003, participants included 59 people from Vanuatu, 433 from Zimbabwe, 37 from Laos, 78 from Iraq, and 59 from the Gaza Strip. The United States, at 1,855,456, has the most participants, and Liberia, with 20, has the least.

[†]No instant messaging, though. That's precluded by the laws of physics. No IM with ET unless we discover new laws (or have them taught to us by ET) or learn to slow ourselves down so that centuries pass like instants.

The search goes on. Bigger instruments, fully dedicated to SETI, will soon come on-line, looking farther and listening to more frequencies with better signal detection equipment. The most ambitious of these comes courtesy of Paul Allen, the Microsoft zillionaire and longtime SETI supporter, who is buying a stairway to heaven.*

Located at the Hat Creek Observatory 290 miles northeast of San Francisco, the Allen Telescope Array (ATA) will be made of 350 individual dishes, each similar to the satellite dishes you see in backyards around the world. Starting in 2004 or 2005, the ATA will expand the volume of space that we can listen to by a factor of a thousand over Project Phoenix. If radio contact with an alien civilization merely awaits our listening a bit farther out into the galaxy, it will happen within the next decade.

SETI has had a bumpy ride. Humans have trouble committing to projects that may require centuries to succeed. It requires a commitment to future generations that is in short supply these days. But, if astrobiology proves to be a sustained movement and not just a breaking wave, maybe SETI can go along for the ride. At the Second Astrobiology Science Conference, at NASA/Ames in April 2002, I heard Jill Tarter declare to an enthusiastic audience that the time was right for the U.S. government to restore generous public funding for SETI.

---

*Allen has contributed $11.5 million out of an estimated development and construction cost of $26 million.

19 | Fermi's Paradox

I know perfectly well that at this moment the whole universe is listening to us, and that every word we say echoes to the remotest star.

—JEAN GIRAUDOUX,
The Madwoman of Chaillot

He'd like to come and meet us but he thinks he'd blow our minds.©

—DAVID BOWIE, "STARMAN"

ABSENCE OF EVIDENCE

In 1952 the composer John Cage produced his most minimalist of all scores, entitled 4′33″, which directs a pianist to play nothing for four minutes and thirty-three seconds. Some regard this as a pretentious intellectual game. Others see a statement about stillness, and the sounds we hear inside our own heads when given space to listen.

What is the significance of the more than forty years of silence heard by SETI? Researchers rightfully point out that the search is just beginning, that we have listened to only a tiny fraction of the stars in our galaxy. Success would mean everything. Failure means little. Or does it?

One of SETI's sayings is "Absence of evidence is not evidence of absence."* In fact, there is no possible evidence of absence. How could we ever prove that the aliens aren't out there? Yet, cumulatively, the silence is a kind of evidence. If the galaxy was thick with signals,

---

*Also recently invoked by Donald Rumsfeld to justify attacking a country for harboring weapons that we cannot be sure are not there.

swarming with radio-noisy species, we would know it by now. We can rule out the most optimistic end of the range of possibilities permitted by the Drake Equation.

Let's face it: the Drake Equation is so loosely constrained that you can conclude whatever you want and prop it up with a mathematical crutch. Sometimes, I think of it as a way for nerds to justify our religion with an equation.

Rationally, I know that it's a big universe, and we've only sniffed around in our front yard. Still, I find myself noticing the four decades of silence and wondering. Oh, I don't doubt that they're out there, but perhaps they are not on the airwaves. The question of the existence of intellectually advanced aliens has, in my mind, become more detached from the question of our achieving radio contact. I am hopeful by constitution. But, my "adolescent optimism" has morphed into a more detached cosmic optimism. I still see the universe evolving toward a state of more fully developed intelligence and self-understanding, but I'm no longer sure the human experiment is a part of that process.

We should keep listening for the next thousand years, message or no message. If we succeed in doing that, whether or not we find anyone else, then we'll be well on the way to bringing to fruition the cosmic intelligence that we seek. If we get our act together to the point where we can commit to anything on such long timescales, then eventually there will be messages blasting loudly through our galaxy. Let's make sure and include some Bob Marley.

So, yes, let's get billionaires to spend millions, or taxpayers to spend pennies, to build huge radio arrays. Let's scan all the stars we can in any way we can think of, for as long as we can. Because you never know. Still, after more than forty years, you do start to wonder:

Where are they?

EVIDENCE OF ABSENCE

In 1943, physicist Enrico Fermi was having lunch with some colleagues at the Los Alamos National Laboratory, and the topic of alien life came up. Fermi, like most scientists with some kind of grasp on what a universe with 100,000,000,000,000,000,000,000 stars implies, was a firm believer in extraterrestrial life. But, he asked, if life and intelligence are likely to evolve on even a fraction of other planets, why aren't the signs more obvious? Why haven't we been visited or contacted? "Where are they?" With this simple question, he encapsulated a major problem for SETI. Today it is known as Fermi's Paradox.

The question might not seem that profound, but Fermi had a deep and subtle point in mind. Thinking like a physicist, he simplified the problem to one of particles starting from a single point, and jumping randomly from place to place. If alien populations moving about the galaxy are pictured as particles spreading through a room, we can ask how long it should be before they fill the room—and calculate an answer.

However conservatively you work the numbers in the Drake Equation, it's hard to avoid the conclusion that we live in a widely inhabited galaxy, even if stars with living worlds are only one in a million. Fermi thought that, by this same logic, we should already have been visited. What if, in addition to developing radio technology for communications, advanced species also develop interstellar travel and decide to explore or migrate to planets around other stars? Then, isn't a search for their presence in our own solar system just as valid as a radio search for their distant messages? How, then, are we to interpret the fact that, as yet, we have found no scientifically accepted evidence for the past or present visitation of intelligent aliens? Can't we conclude that they do not exist and save ourselves the trouble of searching for signals?

This logic was largely ignored during the first two decades of experimental SETI. Fermi's Paradox had been discussed—and quickly dismissed—at the international SETI conference at Byurakan in 1971. The stars, it was decided, are too far away from one another, so interstellar travel does not make sense. The only known way to reach other stars within a human lifetime is to travel near the speed of light and use "relativistic time dilation." According to Einstein's theory of special relativity, as you approach the speed of light, time slows down. If you go fast enough, you could theoretically cross the whole galaxy within your lifetime. But time would not slow down on the planet you left behind, and when you returned home, all your loved ones would be long dead and you would be Rip Van Loser.* Further, the faster you go, the more energy it takes to reach your destination. If you calculate how much energy it takes to travel to the stars at relativistic speeds (including the energy it takes to slow down when you arrive), you will probably decide to stay home.

---

*But, as Dan Hicks sings, "Hell, I'd go!"

Relativity is a real gas guzzler. If you had a superefficient nuclear engine, you would still need to bring something like a billion pounds of fuel for each pound of ship. The energy barrier to interstellar travel was widely accepted within the SETI community, so Fermi's Paradox was not considered a serious worry.

HART'S ANSWER

In the 1980s Fermi's Paradox came back with a vengeance. Several scientists used it as the basis for sophisticated arguments concluding that SETI would not succeed. Astronomer Michael Hart led the charge with his 1975 paper "An Explanation for the Absence of Extraterrestrials on Earth." Hart showed a flaw in the arguments, made by Drake, Sagan, and others, against interstellar travel. Sagan was always keen to point out the ways that our thinking may be limited by our laughably limited experience with life. In his seventies' parlance, he advised that we should always be on the lookout for "chauvinisms." At Byurakan he discussed water chauvinism, liquid chauvinism, planetary chauvinism, and temperature chauvinism. Yet Sagan was guilty, along with the rest, of a kind of *temporal* chauvinism. Why assume all of these trips between the stars have to be made within a human lifetime?

Hart showed that the stars can be reached using much less energy. If you allow for centuries, instead of decades, then the energy needs are much more modest. Traveling at one-hundredth the speed of light, you reach the nearest stars in a few centuries, without a fuel tank a billion times larger than your vehicle. Assuming that interstellar colonies sometimes spawn their own colonies, a migrating species crosses the galaxy in only a few million years.

Now, a few million years sounds like a long time to you and me. Hell, I get bored driving from Tucson to Phoenix. But it's just a fraction of a percent of the 10-billion-year age of our galaxy. Thus, Hart argued, the absence of evidence *is* evidence of absence. Radio SETI was doomed to failure and maybe not worth the investment. Fermi's Paradox, reinvigorated, came to be called, by some, the Fermi-Hart paradox.

The argument is far from trivial. Radio SETI stands a good chance of succeeding only if the galaxy is either (a) richly endowed with advanced civilizations, so that there is a transmitter within a few hundred light-years of Earth, or (b) inhabited by at least a few extremely advanced (Type II or III) civilizations, with beacons that we could see clear across the galaxy. In either case, is it reasonable to think that no one has ever ventured out between the stars? Hart argued that technological intelligence capable of radio communication would also be certain to spread throughout the galaxy on a short timescale. His answer to Fermi's "Where are they?" was "Since they're not here, they're not anywhere."

By the early 1980s, Hart's arguments had become accepted by many, and the SETI debate was polarized. Some argued that intelligence must be spread liberally throughout the galaxy. Others, swayed by Hart's arguments, held that human civilization stands alone, or nearly alone, in our galaxy.

The timing of this new movement was unfortunate, as it coincided with the SETI-gutting efforts of faux-frugal budget-cutters in Congress. Congressional critics easily found respectable scientists to quote, arguing that SETI was a waste of time.* The Hart objection contributed to Congress's pulling the plug in 1981, and again in 1993.

At a meeting entitled "Where Are They: A Symposium on the Implications of Our Failure to Observe Extraterrestrials," held at the University of Maryland in November 1979, Hart forcefully reiterated his views. He did not argue that intelligent, technological species could not evolve elsewhere, only that the evidence suggests they have not yet done so.

Stanford radio astronomer Ronald Bracewell amplified Hart's point with an analogy from Earth history. He discussed the meaning of there being intelligent humans in both Africa and California. Bracewell pointed out that humans can and did walk from Africa to California, arriving perhaps around thirty thousand years ago.† This journey took much less time than it would for intelligent life to independently evolve in California. Comparing the timescales of evolution and interstellar migration, he concluded that intelligent life would be much more likely to travel between stars than to evolve separately on several of them within the same short time interval.

Why, Bracewell asked, are there not multiple, independently evolved, intelligent, technological species on Earth? Not because life, left to its

---

*Indeed, Senator Proxmire, in his attack on SETI, directly referenced a 1981 article in *Physics Today* by physicist Frank Tipler that repeated many of Hart's arguments.
†That is why there are reasonably intelligent creatures, adapted to the African savannas, basking in hot tubs all over California. Most of them, however, are descendants of later invaders who came by wagon, ship, and train and forced out the descendants of the original walkers.

own devices would not evolve technoids again. Rather, once humans appeared, they quickly spread around the Earth and occupied the intelligent, technological niche. By analogy, the galaxy has no species capable of interstellar travel. Otherwise, they would have arrived here and changed a few things. Like Hart, Bracewell concluded that we are the first to reach this stage and it is our destiny to colonize the galaxy.

These are well-argued points, but several alternatives are possible. For instance, perhaps there is only one intelligent race in the galaxy, but it is *not* ours. Maybe an advanced civilization long ago spread throughout the galaxy, but to them we are so clearly not intelligent, and incapable of meaningful conversation, that they don't bother with us. To the truly intelligent species in the galaxy, we may not seem threatening or promising.

I went to my first SETI conference, entitled "The Search for Extraterrestrial Life: Recent Developments," at Boston University in June 1984. I had attended many forums on the same topic at science fiction conventions, but I had never been to a scientific conference devoted entirely to questions about intelligent aliens. For the keynote address, Philip Morrison gave a retrospective of the first twenty-five years of SETI. Participants devoted a lot of time and breath to the likely strategies and timescales of galactic colonization. The resurgent Fermi-Hart paradox led to almost as much discussion of interstellar travel as of radio communication.

Frank Drake presented a new analysis of the Fermi Paradox, including a calculation of the energy required for interstellar travel. His conclusion: intelligent beings do not colonize. He gave two reasons: (1) exponential growth is ultimately self-destructive, and (2) interstellar travel is just too expensive for any rational society to undertake. At the end of his talk, a smiling Frank Drake held up a T-shirt emblazoned with: "Absence of evidence is not evidence of absence."

Various opinions about the strategies and patterns of alien interstellar colonization were defended with analogies from human history: the rate of dispersal of human settlements from island to island by Polynesians in the South Pacific, and the mass emigration of the Irish to North America in the 1800s, were used as examples. These were combined with sophisticated mathematical models borrowed from ecology and physics. The spread of a civilization throughout the galaxy was modeled with the diffusion equation, which predicts, for example, how fast molecules of a gas will spread throughout a room, and has been used to simulate the spread of plant and animal species through a new habitat.

To apply a diffusion model to the spread of species throughout the Milky Way, you have to make some assumptions about long-term rates of population increase, and the willingness of newly arrived colonists to start out on further colonizing expeditions. That's where the examples from human history come in. But when it comes to trying to scientifically model the details of interstellar colonization, of course we're reaching.

When I was a kid, a friend and I had a snow-shoveling business. One winter I calculated that we could make $2,000 if it snowed x number of times and we had y customers and they each paid us z dollars per inch of snow. We'd be rolling in cash and could retire by the time we were twelve. At the end of that winter, I think we had shoveled in all of fifty bucks and a batch of stale homemade cookies from Mrs. Dolan across the street. Even when a problem is framed by a precise, quantitative formulation, it is easy to cook the numbers and reach any conclusion you want.

However, even the slowest models of interstellar migration produce estimates of several hundred million years to populate a galaxy that is more than 10 billion years old. Thus the efforts of those trying to refute Hart actually end up supporting his central argument: if any interstellar colonization has occurred at all, even assuming a halting, aimless movement across space, it should still have had plenty of time to percolate throughout the galaxy. It doesn't matter if *most* societies do not colonize. If only one species, in the 10-billion-year history of the galaxy, had decided to start colonizing, then they could have long ago swept across the entire Milky Way.

Despite Drake's eloquent counterarguments, the SETI camp was forced into a defensive posture. Those who wish to argue against Fermi-Hart must argue that *no interstellar colonization has ever occurred*—a pretty extreme stance to take.

There is no great logical response to Hart's argument against the existence of ET civilizations, no killer retort showing that it must be wrong. The best answers are intuitive. It just seems wrong. It goes against the intuition shared by people over thousands of years, from ancient Greeks to modern geeks, that in such a vast universe we can't possibly be alone. Either argument depends on untestable assumptions. At least the conclusions of the SETI camp can be tested by searching for a message. Long shot or not, as those lottery billboards say, you can't win if you don't play.

WHERE THEY ARE

Those who believe in the standard SETI model of many radio-communicating species but no interstellar colonizers have developed many arguments to explain the lack of an obvious extraterrestrial presence on Earth. These excuses for absentee aliens can be grouped into two categories: physical and sociological.

Physical explanations reason that aliens have never arrived on Earth because of some astronomical, biological, or engineering problem that makes interstellar travel impossible. The most obvious obstacles in this category are the immense distances to the stars and the great, perhaps prohibitive, time and energy it would take to travel between them.

Science fiction writers and speculative scientists have invented numerous ways for future humans to circumvent this limit. Perhaps space travelers could be put in "suspended animation," their metabolisms greatly slowed down until they reach their destination, when machines will awaken them (assuming they haven't developed second thoughts about wanting human company, as HAL9000 did in 2001). Or frozen human zygotes could be brought on the journey, to be raised by "parenting machines" when a new, habitable world is only a few decades off.*

The most plausible option may be to just let the journey take many generations. A staple of SF, and also the subject of some "serious" scientific literature, is the "generation starship" or "world ship": a large spacecraft with a breeding population of humans (and a balanced ecosystem of many other species), launched on a slow cruise toward a distant stellar port. The colonists who eventually arrive will be the remote descendants of those who set out. This is an old idea. Tsiolkovsky discussed it in *The Future of Earth and Mankind,* written in 1928. Robert Heinlein explored it in his novella *Orphans of the Sky* (1941).

Even the notion that slow interstellar travel must take "several generations" is too human-centered. An intelligent sequoia tree would have no problem with a thousand-year journey. We have no reason to believe

---

*This was the mode of travel in Kurt Vonnegut Jr.'s "The Big Space Fuck" (1972).

that intelligent aliens, or even our descendants, will not have life expectancies of thousands of years. For such creatures a voyage of a few hundred years might not be such a big deal, as long as they have plenty of video games and munchies.*

The distinction we draw between individual life span and multigenerational societal life span may lose its meaning for a long-lived species (I'll return to this thought in chapter 23, "The Immortals"). Is it so hard to imagine a sentient being that passes its memories and conscious identity on to its descendants, so that it does not really die? With our stories, books, photos, recordings, and computers we have already invented primitive forms of persistent memory. Maybe our future memories will be passed on directly, in digital form. We may simply continue life's pattern of forming larger units of identity, attaining a more selfless attitude toward future generations, so that a successful arrival of the descendants is seen and felt as a completed journey by those who set out. Some techno prophets believe that immortality for human beings will become a reality in the next century or two. This will, no doubt, cause new problems, but it may solve the dilemma of the stars that we can see but do not have time to reach.

Even if, for some unknown reason, living organisms cannot make the crossing, machines could do it. Soon, we'll be capable of launching robot probes to the stars, and before long we surely will. What if it is machines that colonize the galaxy? Someone could build machines designed to mine the asteroids of distant stars and create more copies of themselves that in turn set off for still more star systems. Such devices are called Von Neumann machines, after the Hungarian mathematician John Von Neumann (1903–57), who worked out the theory of self-reproducing machines. Once launched, it would not be long (in galactic terms) before they reached every star system. We humans will soon be able to build Von Neumann machines, if we choose to. It is hard to imagine that a universe full of high-tech radio buffs will be empty of self-multiplying machine explorers. This presents a further dilemma for radio SETI advocates. If advanced aliens are out there in numbers large or small, might we be more likely to encounter their machine probes in our own solar system than their distant messages on the cosmic airwaves?

---

*And an endless supply of hydroponic herb.

ALIENS AIN'T MISBEHAVIN'

If none of the physical explanations solve the Fermi Paradox, perhaps the answer lies in alien sociology.

One popular sociological explanation is the "self-destruction hypothesis": technological civilizations are inherently suicidal. They always blow themselves up, pollute themselves to death, or otherwise cash themselves in. If true, this solution contains bad news for those who look hopefully to ET civilizations for confirmation that long-term survival with high technology is possible.

The problem with the "self-destruction hypothesis" resolution of the Fermi Paradox is that it must apply to all species, everywhere in the galaxy. Suppose you estimate the chances of the human race surviving our technological adolescence at one in a hundred or even one in a thousand, and you imagine that this is typical of other technological races. If those one-in-a-hundred survivors go on to live for millions or billions of years, then—according to the Drake Equation—there should be a lot of them out there. It is possible to be pessimistic about the human prospect but remain sanguine about the likelihood of a galaxy full of the songs of the survivors.

Another interesting sociological solution is the "contemplation hypothesis." Maybe some combination of spiritual and technological progress removes the desire to physically explore and colonize the entire galaxy. After a certain point, advanced species may adopt a more passive, meditative attitude toward the universe. Such cosmic navelgazers would be more interested in contemplation than colonization.

Maybe the galaxy has not been fully colonized because all wise and ancient civilizations have realized that exponential growth is a dead end.* We can already see here on Earth how a philosophy of unrelenting growth will be self-limiting, and in the global environmental movement, we can even see the seeds of a new ethos developing. The choice is between limits we impose on ourselves and limits nature imposes on us. The same has to be true on an interstellar level.

---

*Even if we expanded our domain *at the speed of light*—a pretty safe theoretical upper limit—and managed to colonize all available stars and planets within a sphere expanding at light speed, then, increasing our population at 2 percent per year, we will still run out of room and perish in our own wastes within the next millennium.

Of course, this contemplative philosophy could be taken too far. A policy of "extreme contemplation" is incompatible with long-term survival. Creatures wishing to insure their survival for millions of years must be space-faring. At a minimum they must learn how to detect and deflect incoming asteroids and comets, lest they go the way of the dinosaurs. To survive unforeseen planetary disasters, they ought to disperse their population to more than one planet, or to space colonies. Finally, if they want to survive for billions of years, they cannot avoid interstellar colonization. They must move on before their sun dies. For a truly long-lived society, interstellar travel is not a luxury.

It is quite reasonable to suppose that surviving civilizations will have developed habits of conservation and expansion that are thoughtful and not reckless. But, what is to prevent a "rogue civilization" from acting in a way that is at odds with the contemplative rationale? It doesn't do much good to have a galactic Kyoto Protocol promoting responsible behavior if one powerful rogue planet decides it has no use for the treaty. This is the weakness of the contemplation hypothesis as a solution to the Fermi Paradox. A "universal sociological explanation" is required—one that applies to the behavior of every species that has ever come along and explored the galaxy. And how likely is that? I can imagine only one way that such a universal code of behavior can apply to all species in the galaxy—if it is codified in laws that are somehow being *enforced*.

Is it possible that an advanced ETI knows about our existence and has decided that we should be left alone? We have seen what happens on Earth when a more technologically advanced, mobile society encounters an indigenous one. It often means the end of the locals. This leads us to the "zoo hypothesis": aliens are out there, perhaps nearby, but they don't want us to know about them, perhaps because of some alien precept of noninterference.* Maybe the galaxy is actively regulated by a powerful, advanced, hopefully benign species, or a "galactic club" of advanced societies that long ago developed a protocol for protecting primitive societies from the shock and dislocation of premature contact.

Perhaps our part of the galaxy is being protected as a "wildlife preserve." Or perhaps they are monitoring us—waiting to see if we will develop in a way that contributes to, rather than threatens, galactic society. If that's the case, they will decide when to make contact with us and

---

*Star Trek's Prime Directive again.

tell us the rules of galactic intercourse. This scenario has been well explored in fiction. At the conclusion of the 1951 film *The Day the Earth Stood Still*, Klaatu, the alien, warned: "It is no concern of ours how you run your planet. But if you threaten to extend your violence, this Earth of yours will be reduced to a burned-out cinder. Your choice is simple: Join us and live in peace or pursue your present course and face obliteration. We shall be waiting for your answer. The decision rests with you." If you were in charge, would you give modern humans the keys to the galaxy?

When we consider such far-out notions as the zoo hypothesis (we are being protected) or directed panspermia (we were seeded by advanced aliens), we must keep this in mind: even before the Earth was born, there likely were technological civilizations living for *billions* of years, possessing capabilities that we cannot even begin to contemplate. Our Sun is not among the oldest stars. Even given the conservative assumption that it always takes 4 billion years for a planet to progress to intelligence (that is, if we assume that all living planets evolve as slowly as ours has), the first civilizations should have begun appearing in our galaxy at least 4 or 5 billion years ago. More likely, there is a spread in the rates of evolution, and on some planets intelligence developed much faster than on ours, so some civilizations should have started appearing 8 or 9 billion years ago, long before our Sun and planets were a gleam in the eye of their parent molecular cloud.* We had just better hope that they have our interests in mind.

Even the zoo hypothesis is human-centered, in that it assumes that we are interesting or important enough to merit a cage in their zoo. A less anthropocentric idea might be the "seedling hypothesis." They may be ignoring our planet because young seedling noospheres like ours are a dime a dozen, carpeting the floor of the galactic forest where a few magnificent trees and flowering vines are much more captivating.

THE GREAT SILENCE

Fermi's Paradox has also sometimes been called the great silence. Although Fermi's original "Where are they?" was meant to refer to

---

*The galaxy is at least 10 billion years old. A few generations of star birth and death were needed to build up enough interesting heavy elements to make life likely. However, the massive stars that do most of this atom-making burn bright and die young, so after a billion years many generations of them had already added their ashes to the stellar compost heap.

alien visitation, it can also be applied to the lack of signals seen, so far, on the cosmic airwaves. It is hard to imagine a purely physical explanation for this great silence. You can't reasonably argue that radio communication between the stars is impossible, since we routinely observe radio phenomena at huge interstellar, and even intergalactic, distances. The Arecibo radio dish in Puerto Rico could exchange messages with an equivalent dish clear across the galaxy.

As SETI advocates are quick to point out, the silence is not all that great. It may be that we simply haven't yet searched the right star at the right frequency. This could all change tomorrow. If not, we can enlist many of the sociological explanations that work for "the great absence" to explain "the great silence." Let's take the zoo hypothesis. Maybe advanced aliens are careful *not* to reveal their presence with radio signals because they know that primitive societies like ours *are* likely to conduct radio searches.

I suppose this is as good a time as any to mention a chilling, but impossible to refute, scenario of cosmic paranoia: What if there is no one out there because they have all been destroyed by something powerful and ruthless? A technically advanced civilization, obsessed with its own safety, might wish to suppress all possible challengers.* Maybe the universe is silent because something is out there that seeks out radio broadcasts and destroys their sources. So, in sending out our first radio messages, we may be like the first monkey who climbs down from the tree, stands upright, and, overcome with pride, shouts out, "Hey, look what I can do!"—awakening the hungry lions who are napping all around. Only species that shut their yaps are the ones who survive. This bleak possibility is, unfortunately, a perfectly logical answer to the Fermi Paradox. Even if such predators do not exist, the fear that they might could be a reason why other civilizations are not shouting their names all over town.

If we regard this as a serious possibility, should we be more careful about drawing attention to ourselves? Should we be banning signals in the direction of nearby stars? Nah. Let's not be wimps hiding in the nursery. You *can* be too careful. I say let's give a shout-out and see

---

*A fleet of Von Neumann machines could lie in wait throughout the galaxy, listening for the first radio signals of fledgling technological races. The deadly probes sweep down on these worlds and obliterate all life. Such killer probes are often called berserkers after a SF novel by Fred Saberhagen that introduced the concept.

who's there. Onward to the stars. Maybe I'm just a sap, but I doubt that alien civilizations would want to harm us. It's much more likely that they won't be interested in us at all while we're still in the "just listening" phase. Maybe the moment a civilization achieves the ability to start broadcasting for real is when they become worth talking to. Like a chat room that discourages "lurkers," maybe galactic communication is done in such a way as to prevent the little seedlings from listening in.

The silence may say more about our own limitations in conceptualizing intelligence than the ability of the universe to produce it. On the days that I don't believe radio SETI will succeed (usually Sundays and alternate Wednesdays), it is not because I doubt that lots of communicating creatures are out there. I'm just not sure they're using radio.

Radio aliens are appealing because they are well-behaved. We know just what we are looking for. Although it makes just as much sense to search vigilantly for strange, "unnatural" objects in our solar system, this idea is problematic. It opens a floodgate that spills out in the supermarket checkout line: pretty soon we are mired in the *National Enquirer* alongside the "face" on Mars, secret alien bases on the Moon, and "Man gives birth to a pickup truck with the face of Elvis!" What do we do when our scientific reasoning leads us perilously close to beliefs that are widely associated with the dreaded pseudoscience?

One alternative to radio that is now being taken seriously is the idea of looking for flashing lights in the sky. Like a ship at sea beaming messages in Morse code, someone out there might try to get our attention with brief and powerful laser flashes. This idea was proposed, and largely ignored, in the early 1960s, when radio SETI was just getting going. Now, with the rapid development of laser technology on Earth, it's easier to conceive of using powerful lasers to flash signals between the stars. Several "optical SETI" search programs have recently come on-line. Most of these are in the United States, but one in Australia covers the southern hemisphere. Frank Drake, the father of radio SETI, is now also doing optical SETI from the Lick Observatory, near San Jose (although he still feels that radio has a better chance of success). Most optical SETI programs are targeted searches, sifting one by one through thousands of nearby stars, but an "all sky" optical SETI program is just getting under way at Harvard. Nobody knows whether this is more or less likely to succeed than a radio search, but rather than second-guess the aliens, we search in any way we can think of.

MESSAGE IN A VIRUS?

One problem with radio or light rays is that they just flash through a system. The message does not stick around, so someone there has to be paying attention. If Earth had been bombarded with a continuous radio broadcast from a nearby star for 100 million years, ending in 1903, we would have completely missed it.

What about a message that could be left on a planet to stick around until someone evolved with the ability to recognize and decode it? It has even been proposed that the genomes of living organisms might have been designed with hidden messages, waiting for us to develop the requisite biotechnology to decode them. There is something poetic in this idea—the essential machinery of life containing a message from one life-form to another.

An early research experience introduced me to this possibility. In the summer of 1978, I was an eighteen-year-old college freshman, working in Sagan's lab at Cornell.* At the time Carl was the editor of *ICARUS: International Journal of Solar System Studies*. As you can imagine, *ICARUS*, a journal run by a flamboyant, telegenic editor with a reputation for publishing speculative papers about exobiology and SETI, received some flaky submissions that did not merit serious attention. Carl asked me to check out a paper entitled "Is Bacteriophage φX174 DNA a Message from an Extraterrestrial Intelligence?" to see if it was legit. The paper, by Japanese biochemists Hiromitsu Yokoo and Tairo Oshima, suggested that a message from extraterrestrials might be hidden in the DNA of a virus that infects the common intestinal bacteria *E. coli*. As outlandish as this sounds, the genetic code contained some bizarre and unexplained patterns that seemed as if they could conceivably represent a message of some kind.

A little background will help here. SETI theorists have devoted a lot of thought to certain mathematical patterns that would indicate an intelligent message. A favorite, along these lines, is to use prime numbers.† Prime numbers are those that cannot be made by multiplying together any other whole numbers: 1, 3, 5, 7, 11, 13 and on up, as high

---

*The same summer I worked with Reid Thompson playing Microcosmic Gods (chapter 7).
†In 1941 Sir James Jeans proposed that we flash the prime numbers toward Mars with powerful searchlights—one of a series of ideas that have been advanced over the years to show the Martians that we are good at math.

as you want to go. No known formula or natural process generates them. If you see primes, you know that mind is not far behind.

A widely accepted idea (on Earth at least) for interstellar message construction is to send digital pulses that repeat with a number that is the *product of two primes* multiplied together. This would suggest that a two-dimensional picture is being sent. For example, if you received a message that was repeating a sequence of 143 pulses, you, or your machine, would say, "Oh, 143 is 11 times 13, and these are both prime numbers. Let's make an 11-by-13 array and see if there's a picture encoded here." This technique—using primes to create easily decoded 2-D images—is a pillar of SETI theory.

Now back to the message in that virus, bacteriophage φX174. It's the first organism for which the entire genome was decoded. A remarkable

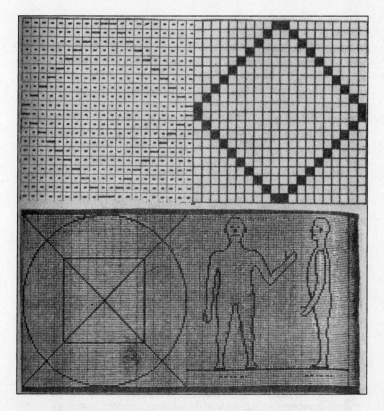

A 1920 design for a message to extraterrestrials. (H. W. & C. W. Neiman)

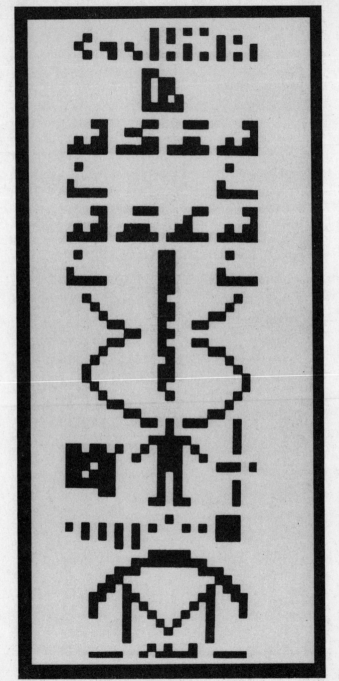

This message was sent by Frank Drake from Arecibo toward the globular cluster M13 in 1974. (Frank Drake)

feature discovered was the presence of "overlapping genes." These are sequences of nucleotides that can be read in two different frames, to encode for two completely different proteins.* In other words, the DNA sequence CAATGGAACAACTCA, can be read as the three-letter "words" CAA TGG AAC AAC TCA, and this will instruct a cell to start building a protein by putting together the five amino acids specified by these triplets. However, starting at a different letter, the same sequence can also be read as ATG GAA CAA CTC, and so on, which will build a completely different protein. It is as if you could write an English sentence criticizing a bickering couple, "CAN YOU TWO NAG," that also contains a message about the ninth inning of a baseball game, "ANY OUT WON," except that to make proteins you would have to continue in this overlapping mode for hundreds of words, with both sentences making complete sense.

But wait, there's more. This organism (ϕX174) contained not one, but three pairs of these overlapping genes. And, if you count the number of letters in these overlapping sequences, you find that they are 121, 91, and 533. Each of these is the product of two prime numbers (11 × 11, 7 × 13, and 13 × 41). How weird is that? Here was the widely accepted signature of an intelligent message, turning up in the strangest of places.

Yokoo and Oshima had done what anyone suspecting a message would do: they tried to make two-dimensional pictures from these sequences. Unfortunately, the pictures looked like random noise. I went back to the original literature on the decoding of the virus to see whether the claims about overlapping genes and prime numbers were accurate. They were. Then, under Carl's instructions, I set about trying to decode the message. I made a series of 2-D pictures from the gene sequence, trying all kinds of tricks, with the hope that a smiling alien face, or the Pythagorean theorem, would pop out at me. With no alien-virus decoding manual, I just tried every system I could think of, but I never did find a message. Being a pack rat, albeit a disorganized one, I still have the "messages" I decoded that summer. Here's what they look like:

---

*Recall that the genetic code consists of a string of nucleotides, similar but nonidentical units, which we represent with the letters A, T, G, and C, and that every three-letter "word" codes for one amino acid in a protein that the organism uses to build itself or run its chemical machinery (chapter 7).

Two dimensional pictures constructed from overlapping triplets of
A and B proteins.

Purine base =
Pyrimidine base =

Picture made using the first base of each triplet

←   "   "   "   "   second   "   "   "   "

"   "   "   "   third   "   "   "   "

My pictures, like those of Yokoo and Oshima, always came out looking like random noise. However, their paper makes two important points: (1) that a biological message might have some inherent advantages over a radio message for interstellar communication, and (2) that structures in this particular genome seem anomalous and could, just conceivably, be a message we had not yet successfully decoded.

Carl accepted the paper and it was published in *ICARUS* in 1979. Since then, the genomes of many organisms have been decoded, including, famously, our own.* Overlapping sequences are found to be common in the genomes of many species. Some of the overlaps seem to be related to prime numbers, but many of them don't.

As physicist George Marx wrote, in defense of the biological interstellar message concept, "How does one send a letter to a faraway planet, a letter that is light enough for easy transportation, that multiplies itself on arrival, that can correct misprints automatically, and that will be read definitely by the intelligent race of the target planet after they have reached scientific maturity?"

Biologically encoded messages actually seem somewhat less farfetched now than they did in the 1970s. We're just beginning to figure out how to read and write in the language of DNA. It is not at all difficult to imagine that less than five hundred years from now, let alone millions of years, we may be able to do intelligent design of just about any kind of organisms we want.

But the idea of a biological message also has large drawbacks. Most importantly, you would have to know an awful lot about the biology of your target planet.† Today this idea strikes me as unlikely, but useful for reminding us that electromagnetic signals, our communication mode du jour, are not the only possibility. It is also useful to think about sending a message through time as well as through space. Huge longevity could give a material message more chance of being found than a radio message, which travels far but doesn't stick around. Once we get used to the idea that a message may well be one-way, rather than part of a dialogue, then it doesn't really matter whether it is sent through space or time.

---

*Or at least, it turns out, that of Celera Genomics CEO Craig Venter, suggesting that the genes for egomania may be overrepresented in "the human genome."
†Unless you also designed their biosphere to begin with, but this thought is "scientifically incorrect."

If you ask any biochemist to explain the prime number products found in the genome of bacteriophage φX174, they will tell you that it is just a funny coincidence. The skeptical mind wants intelligent design, but wants to find it *out there,* not in here. The problem with looking for signs of "anomalous order" is in determining what is anomalous, what is coincidence, and what is simply beyond our limited ability to comprehend natural sources of order. In 1967, astronomers were astonished to discover several strong sources of regularly repeating radio emissions—just the kind of thing that SETI had been looking for. These were first called LGM sources, for "little green men." Today we have discovered over three hundred of these, and we know that the patterns are due to rapidly spinning neutron stars (picture a lighthouse with a strong radio lamp). We call them pulsars, and nobody thinks they were made by advanced aliens.

Any strange enough coincidence can seem like anomalous order, and even on a cosmic scale, they are all over the place. The weirdest and most wonderful cosmic coincidence is that the Sun and the Moon are the same size when viewed from Earth, which makes possible the lovely, spine-tingling spectacle of a total solar eclipse, when the bright disk of the sun is completely, precisely masked by the nearby moon, and the exquisite outer tendrils of the solar corona is briefly visible to starstruck watchers on Earth. Even stranger, since the Moon is slowly receding from Earth, this is also a temporary situation. Total solar eclipses on Earth became possible only in the age when humanity happened along to enjoy them. Evidence for God? A message from some supertechnological creator of worlds? Just a coincidence?

How much coincidence is too much? For any given coincidence this probability can often be calculated, but for all possible coincidences I don't think it can be. So our sense about whether too much weird stuff happens to be explained by "just coincidence" is left as a question of intuition, not probability theory.*

A completely logical resolution to the Fermi Paradox is simply that we're the first ones in the galaxy to make it to the stage of radio technology and space travel. Some have claimed that this is the simplest explanation and therefore the best. The problem is that this explana-

---

*I'm hoping that I'm wrong about this and that, coincidentally, someone who knows the proof will read this and e-mail it to me.

tion requires an extreme violation of the principle of mediocrity, because if we are first, then our location is unique in the entire universe.* It doesn't seem likely. So, it comes down to which unsupported pillar of scientific reasoning you want to violate: Occam's razor (the simplest explanation is correct) or the principle of mediocrity (there is nothing special about our location). Take your pick.

Even if we are not completely unique, or "really alone," we can be "functionally alone." If the nearest communicative species is so far away that we can never have a conversation, then we may forever be isolated. Any messages received will be from extinct societies saying, "We were sentient once, and young."

MAGIC

I have thus far avoided one obvious solution to the Fermi Paradox. Some people believe that there is no paradox, because contact has already been made. UFOs are alien spaceships. Their presence is being covered up by the authorities or simply not taken seriously by narrow-minded scientists. Or, somehow, although individuals have had unmistakable contact experiences such as sightings and abductions, our current conceptions of physical reality are limited in ways that preclude a scientific recognition of the signs of contact.

These views are widely held. Opinion polls consistently demonstrate that more people believe in UFOs and alien abductions than believe in key precepts of modern science. Yet, the mere mention of this solution to the paradox raises the hackles of most skeptical scientists.

The possibility that aliens visited Earth long ago is the least offensive to modern science. My family once took a ten-day raft trip down the Tatsenshini River in Alaska, a wild and beautiful place with no obvious traces of human activity. Our guides were fanatical about insisting that we leave no signs, however trivial, of our presence in that wilderness area. If past alien visitors had a similar preservation ethic, we might not know that they had been here. Or, our solar system today may be littered with alien artifacts, with little chance that we would yet have discovered them, as the solar system is almost completely unexplored. A buried artifact on the Moon would surely still be undetected, let alone

---

*Galileo's persecutors were right!

objects on the surface of, or orbiting, any of the other planets, or within the asteroid belt. It is premature to rule out a past or current alien presence in our solar system.

And then there is "Clarke's Third Law," Sir Arthur's dictum that "any sufficiently advanced technology will be indistinguishable from magic."* Think of how the people of King Arthur's day reacted to the telephones, electric lights, typewriters, and steam engines brought to them by one Connecticut Yankee. It all seemed like pure magic, and they were only separated by thirteen centuries and the Atlantic Ocean. How would we react to technology that comes from someplace beyond us by ten thousand centuries and an ocean of stars?

We can try to imagine the technology of a society millions of years older than our own, but we can't rule anything out. Even things that seem to us like violations of physical laws might not be off-limits to another species who have had millennia to work the problem. Even with technology that is already on the horizon, many capabilities that sound like New Age dreams might become possible. Assuming that they do not lead to our extinction in the near future, nanotechnology, wireless communications, genetic engineering, and much faster microprocessors might allow capabilities indistinguishable from telekinesis, ESP, and immortality. And foreseeable technology is probably a lousy guide to future capabilities.

Advanced technology may often be incompatible with long-term survival. But assuming there are survivors, they may have powers that seem magical, and by nature of their having learned the trick of peaceful survival with these fantastic powers, they will be spiritually advanced creatures. Perhaps they will have solved the riddle of the quantum and will be able to walk through walls. Um, gee, they sound sort of like angels. . . .

Even scientific rationalists like Sagan and Clarke, considering the capabilities of long-lived intelligences, talked themselves into the likely existence of omnipotent, godlike creatures. Yet they recoil with horror

---

*His other laws are (1) when a distinguished but elderly scientist states that something is possible he is almost certainly right; when he states that something is impossible, he is very probably wrong; (2) the only way of discovering the limits of the possible is to venture a little way past them into the impossible. Lesser known is Clarke's Sixty-ninth Law: "Reading computer manuals without the hardware is as frustrating as reading sex manuals without the software."

from people who express a belief in the existence of such creatures but come at it from outside science.

Once we consider the possible technical capabilities of civilizations hundreds of millions of years older than ours, nothing is too magical to be possible.* We are forced to admit that those who believe in angels, spirits, and creatures from other dimensions living among us are not really advocating ideas *inconsistent* with science, only *unverifiable* by science. Certainly, the universe would be a much less frightening place if we could clearly delineate the limits of the possible with scientific reasoning. But to believe that we can is a modern positivist superstition.

How confident can we be that we are not being contacted in very different ways from what we imagine? Might aliens already be here? Given our great ignorance, and the possible, unknown capabilities of advanced alien civilizations, can we really dismiss the possibility that UFOs are real?

In acknowledging Clarke's Third Law, science launches us into the realm of magic. Logic tells me that it is reasonable to look for godlike signs of advanced aliens in the sky. And yet the idea seems ridiculous. It is both logical and absurd. Go figure.

---

*This is maddening for both organized skepticism and organized religion, which have in common that they don't like magic one bit. Amusingly, the Harry Potter books have been condemned by both scientific skeptics and religious fundamentalists as containing too much magic and sorcery for young minds.

20 | Have You Seen the Saucers?

As we staggered out into the Sunday afternoon sunlight, somehow the subject of space flight came up.
Bill Broonzy and his buddies began to sound off
about what the people on the moon would look like.
'Man,' said Bill, 'they gonna be so ugly, if you threw
'em into the Mississippi River, you'd skim ugly for six
months.' 'That's right,' said Sonny Boy Williamson, 'they
got feet comin' out of their ears and eyes comin' out of their toes,
and their mother would cry if she looked at 'em!'
> —ALAN LOMAX, LINER NOTES TO *Blues in the Mississippi Night*

As is your sort of mind,
So is your sort of search:
You will find what you desire.
> —ROBERT BROWNING

THEY ARE EVERYWHERE

You don't have to look very far for evidence of a cultural cacophony
about extraterrestrials. In the minds of many, UFOs constantly zip
through our skies carrying mysterious visitors who take a huge interest
in certain humans. Yet, you do have to be a little careful with opinion
polls. If I was asked, "Do you believe that the universe is full of
extraterrestrial intelligent beings, and do you think it possible that
some of them are now on Earth, or have been in the past?" I think I'd
check the "yes" box. If the question was "Do you believe that reported
UFO sightings are alien spacecraft and that aliens walk among us?" I'd
have to check "no."

I think we sometimes confuse interest with belief and overestimate the extent and danger of public gullibility on this issue. I like reading my daily astrology message but that doesn't mean that I think it "works." Everyone who collects alien books and conspiracies is not necessarily a true believer.

Some people take the profusion of stories itself as evidence pointing to an alien presence. Certainly the intensity and ubiquity of the interest tells us that the subject taps into a deep well, but is this well fed by visitors or by desire?

"Flying saucers" were actually born of journalistic distortion. In June 1947, pilot Kenneth Arnold, flying near the Cascades in Washington State, saw something strange—nine brightly glowing lights, moving in a series of jumps. Arnold described this motion as "like a saucer if you skip it across water." A reporter from the Associated Press picked up on this and described the sightings as "flying saucers." Arnold never said they were saucer-shaped, but ever since then dishware-shaped flying objects have filled our skies. Waves of saucer sightings often correlate with prominent media reports of sightings or human space activities. One of the largest was provoked by *Sputnik*, which got everyone looking up, scanning for new machines.

Among those looking skyward as *Sputnik* orbited overhead in 1957 were "angelheaded hipsters burning for the ancient heavenly connection to the starry dynamo in the machinery of night."* Today's New Age interest in alien friends descends from beatnik mysticism, which morphed into hippie psychedelic cosmic consciousness and the boomer yearning to be members of a special generation who would set themselves free. When the revolution did not materialize as planned and everyone aged past thirty, trust in the coming transcendence was often placed in creatures from beyond this Earth.

The alien symbol—the little green guy with the big black eyes—ubiquitous in the nineties, will soon go the way of other saturated cultural icons like the sixties' peace sign and the seventies' smiley face. Which is to say that it'll never really go away.

UFOs and abductions have become a modern American folk mythology in the tradition of Paul Bunyan; tales that grow taller as they are retold. Some take the stories more seriously than others. As pluralism

---

*As Ginsberg wrote in *Howl*.

Which is the author and which is the alien? (Antony Cooper)

was fertile ground for social commentary in Voltaire's irreverent eigh-teenth-century interplanetary sagas, so today saucer and abduction myths are rich fodder for satire, as seen in the raunchy cartoon *South Park*.* In the pilot episode, "Cartman Gets an Anal Probe," third-grader Eric Cartman is abducted by aliens and given some kind of an implant. Cartman thinks he just had a nightmare, but his pals suspect it really happened, because they have heard about the mysterious "visi-tors" who are probing humans and abducting cattle. Cartman angrily insists that it was just a dream, until a few days later when an eighty-foot radio antenna suddenly sprouts out of his ass.

SCIENCE AND THE SAUCERS

The typical scientific response to reports of UFOs and alien encounters is to assume that there is nothing to any of it. Unquestionably, nearly all UFO reports are due to illusion, delusion, or confusion. Weather balloons, strange reflections, satellites, migrating flocks, and the planet Venus are among the usual suspects. There's a lot up there to misinter-pret. Add to this all the reasons that people *want* to believe—the wish

---

*I do not mean to imply that Trey Parker and Matt Stone are the Voltaires of the twenty-first century.

Cartman suddenly sprouts an alien appendage.

for a cosmic connection, the need for a personal message from the great beyond, and the desire for a protective higher power that takes matters out of human hands. Consider also the close correspondence between alien stories and earlier myths involving mysterious and mischievous little humanoids hiding in the forests, hills, and seas. Now that we ourselves have taken to the sky, we see the little creatures up there, too.

Anyone with her head screwed on more or less in the forward direction will conclude that *most* alien reports have nothing at all to do with aliens. But even supposing that 99 percent of alien reports are false, what about the other 1 percent? One answer is to use inductive reasoning to write these off along with the rest. Why don't you respond enthusiastically every time some stranger calls and tells you he can save you a lot of money on your phone bill? Gotta screen your messages. Astrobiologists are often bombarded with questions and testimonials from believers in alien bodies at Roswell and other government conspiracies. The reflex reaction to the colorful flood of cosmic debris is to dismiss all alien sightings.

Yet, after admitting, in the preceding chapters, that we cannot logically rule out such far-out possibilities as the zoo hypothesis and directed panspermia, and staying mindful of Clarke's Third Law, shouldn't we take another look at the UFO phenomenon? Taking a reflexive, dismissive stance toward all reports of sightings could be as foolish as uncritical acceptance.

A big part of the problem is ufology (You follow, G?) itself. This "discipline" of UFO study has blossomed to feed the endless hunger, nourished on a bottomless dung heap of dubious reports. Anyone with a little imagination and a bone to pick with the government, mainstream science, or with any other authority can hang out a shingle and be a bona fide ufologist (You follow gist?)

Such a clamor of voices and beliefs about UFOs and aliens scream over the pop-culture airwaves that genuine signs of an alien encounter might easily be lost in the noise. Ufology would be an excellent cover for alien activities.

Within ufology, subfields abound. Old-school, no-nonsense ufologists think it's best to focus on proving the existence of lights and spacecraft seen in the sky. They don't approve of groups who push the religious dimensions of the experience or the alleged association with other occult-sounding phenomena such as cattle mutilations, abductions, and crop circles. They worry that these flaky, fringe ideas hurt the chances of UFOs and ufology being taken seriously by mainstream scientists.

Despite the preponderance of obvious misinterpretations, hoaxes, and false alarms, many strange reports have never been adequately explained. In the 1950s and 1960s, the U.S. government, motivated by a fear that the Soviets might have aerospace technology we didn't know about or might try to take advantage of public hysteria caused by UFO reports, commissioned several studies. Many well-known astronomers participated in these efforts, which all concluded that most UFO reports can be explained away as natural or psychological phenomena. But every serious study has always found a smaller number of truly puzzling incidents and sightings.

Take, for example, the case of the "Foo Fighters." These strange balls of light, commonly seen pursuing warplanes, were observed by airmen on both sides in World War II as well as in the Korean War. Each side suspected that the Foo Fighters were some unknown technological innovation of the enemy's. The Scientific Advisory Panel on Unidentified Flying Objects, a secret U.S. government panel convened in 1953, concluded, "Their exact cause or nature was never defined."*

---

*Among the distinguished panelists was Luis Alvarez, who in 1981 was codiscoverer of the giant comet impact on Earth that doomed the dinosaurs.

Strange lights that appear to follow aircraft are not necessarily alien spaceships. I am inclined to think that a more mundane explanation exists, but I cannot prove this, nor could panels of serious scientists who tried their best.

The Scientific Advisory Panel was one of three official U.S. government UFO studies in the fifties and sixties. The longest lasting was the Air Force's Project Blue Book, started in 1952. In 1966, the House Armed Services Committee decided that Blue Book hadn't gotten anywhere, so they handed the files over to the University of Colorado under the direction of physicist Edward Condon. The "Condon Report," published in 1969, did not satisfy anyone on either "side" of the UFO controversy. It concluded that nothing could be concluded and that further study was not justified. Even many UFO skeptics criticized the report as too eager to dismiss some well-documented reports as uninteresting.

IN MY OWN BACKYARD

Eager to do some informal research into ufology, I did end up road-tripping for this book, but I didn't have to drive as far as Roswell. Like Dorothy realizing that her heart's desire was not to be found in some Oz, I found what I was looking for in my own backyard: an alien Mecca in south central Colorado.

Roswell, as an alien destination, is more than a little overexposed—a place where you go to observe the reaction to something that happened long ago. Unfortunately, looking for a real alien culture there today is like going to the corner of Haight and Ashbury to gawk at hippies.

Just like the hippies fleeing the Haight, all the real aliens must have split the Roswell scene as soon as the press picked up on it. Where did the hippies go? Many of them moved to small, forgotten towns in Arizona, New Mexico, and Colorado where housing was cheap and the establishment had not bothered to reestablish much after the mines shut down. The funny thing is, the aliens followed them.

Why is it that New Age meccas are often also places where alien activity is heightened? How are we to interpret this alien attraction to old Western mining towns that became happy hippie artist colonies in the seventies and New Age hangouts in the eighties and nineties? Some would say that New Agers and aliens alike are attracted to those places where Earth power vortices draw in spiritually sensitive beings regardless of their planet of origin. Or is it that the people who gather in such places are more open to the idea of alien visitation and more prone to accept reports of extraordinary experiences? One commonality in various New Age beliefs seems to be a sense of belonging to a chosen population that will, through its own personal breakthroughs in consciousness, collectively help the entire human race to save itself. A willingness to believe in the significance of your own cosmic role helps propagate beliefs and stories in which the aliens are speaking to *you*.

Whatever the reason, it turns out that one of my favorite spots on the planet, Colorado's San Luis Valley, has become a hotbed of UFO activity and lore. I first caught wind of this when, in the mid-nineties, a band I played in—a world-beat funk contraption we called Mom's Instant Hot*—had a series of weekend gigs in Salida, a small central Colorado town of old Victorian houses and art galleries. We'd play two nights at the Victoria Bar on Main Street for a thousand bucks and some seriously skanky little rooms to stay in above the bar. On the day between the gigs, we'd hang out at the nearby Mount Princeton Hot Springs. There, I first began to hear talk about UFO sightings in the area. Everyone seemed to accept that several widely seen and unexplained apparitions had occurred around Salida, and many more in the giant San Luis Valley to the south. It was not unusual to hear locals discussing theories of alien visitation, especially after some new sightings in the summer of '95.

On the morning of August 27, 1995, a number of people witnessed a mysterious daytime UFO, a bright, shimmering object hovering over Salida. Tim Edwards, a longtime local who ran his family's restaurant, the Patio Pancake Place, captured it on six and a half minutes of home video. The whole town, and much of Colorado, was buzzing about this sighting. Edwards was interviewed and profiled on numerous UFO-related TV shows. This experience changed Tim's life. He believes that the occupants of the craft allowed him to videotape it for a reason, and he is now on a mission to let the world know that these creatures are real, and they are here. He transformed the Pancake Place into ET's Landing, an alien-themed restaurant, and he now devotes all his spare time to the cause of ufology.

I've watched Edwards's video, and he definitely captured something

---

* *Westword* magazine "Best of Denver" award "Best World-Beat Band" in 1995.

Great burgers for the weary traveler, no matter where you're from.
(photo by the author)

strange—not any obvious natural or human-made phenomenon. It's a
white, rhythmically pulsing, cigar-shaped thingy hovering way up in the
clear blue sky. I don't believe it's a hoax. You can hear the genuine
amazement and concern in the voices of Edwards and his friends and
family, as they excitedly point out the UFO to others and wonder what
it is. Tim's seven-year-old daughter, Brandy, who first spotted the UFO,
asks repeatedly, "Daddy, can spaceships really grow bigger like that?"
Some "expert" on one of the UFO TV shows concluded that if it was at
eighty thousand feet, then it could be a quarter mile long.*

It might be trivial—a kite or a balloon, an atmospheric or auroral
phenomenon or an optical illusion—but as far as I know, it has not
been satisfactorily identified, so it is a bona fide UFO. I'm more prepared to believe that it was some unusual, rare glowing cloud with a
bizarre physical explanation that we haven't thought of.†

---

*And if his grandmother had wheels, she'd be a bus.
†"Prepared to believe?" Maybe that's my whole problem.

What is it?: A frame from Tim Edwards's video.

DOWN IN THE VALLEY

Just south of Salida, over a mountain pass, lies the San Luis Valley, the location of many UFO sightings and other strange reports. The word *unreal* comes to mind every time I enter the valley. Located eight thousand feet above sea level, it is bounded on the east by the steep fourteen-thousand-foot peaks of the Sangre de Cristo Mountains and on the west by the San Juans, which slope gently up to the Continental Divide. The valley is a hundred miles long (north to south) and seventy miles across at its widest point.

The economy is still largely ranch- and farm-based. Looking into the valley from its rims, you can detect human settlements, but only as tiny dots on a huge landscape that has kept most of its original good looks. Here, you remember how dark the sky can be at night on Earth. If anyone up there is trying to get our attention, or even sloppily trying to avoid it, this would be the place to see them. If the aliens desire a high-altitude, remote, flat, huge, and relatively empty landing site, they couldn't do much better than the San Luis Valley. And you know how they are always drawn to sacred Indian sites, hippie towns, and hot springs.

One third of the way down the valley on the east side lies the perennially sleepy but spiritually restless town of Crestone. Cradled in the forest beneath the dramatic peaks of the Sangres, and blessed by several rushing mountain streams, Crestone is a power spot (as are Taos, New Mexico, and Sedona, Arizona) that attracts aliens, hippies, and New Age entrepreneurs. These days, you can get about a hundred types of massage and a good latte at most power spots—perfect after a long drive through the desert or a journey of many light-years in a cramped saucer.

In the valley, Crestone is UFO central. Alien encounters there are a kind of "rural urban legend" that has been carried mostly by Anglos moving into an isolated valley that already had its share of tales— a new cultural layer imprinting on and mixing with older Christian, Hispanic, and Native American stories more deeply rooted in the land. Throw in a long-standing Western distrust of the federal government. Sprinkle with hippie counterculture, psychedelic drugs, and stir well in a vast, enchanting landscape where perspective plays tricks on the mind.

Crestone has long been considered a sacred place. The Ute Indians knew this, as did the Spanish settlers who came into the valley and threw them off the land. Crestone soon became sacred to those who worship silver and gold, when both were discovered in the surrounding hills, leading to several mining booms in the late 1800s. By the middle of the twentieth century, the mines were all shut down and the town was largely abandoned.

A group of developers tried to turn Crestone into a military retirement community in the 1970s, but the retirees didn't know or didn't care that it was a sacred power vortex. They all moved to Phoenix instead. That was good news for the New Agers and assorted spiritual seekers who have been trickling into the place ever since, buying the relatively cheap land and building yurts, earthships, geodesic domes, meditation centers, healing retreats, and at least one elaborate astronomical observatory.

Crestone is so spiritually awakened that it is damn near insomniac. There are temples, and serious practitioners, of many faiths, including several Buddhist centers, a Carmelite Catholic monastery, an ashram, a stupa, and many other houses of Eastern and Western worship. Local bulletin boards are plastered with advertisements for things like the Gaia Matrix Oracle, and the Karmic Light Ashram and Yogurt Stand. The local paper is full of classified ads for shamanic journeys, pastlife regressions, ancient alien visitors, and serpent wisdom. It is common knowledge, at least among some, that alien craft frequent the valley and take a special interest in the high peaks above Crestone. Indeed, on many mornings giant saucer-shaped clouds can be seen hovering over the Sangres.

Though I snicker at some of this, I am also drawn there by the laid-back hippie friendliness, the spiritual awareness, and the generous spirit of tolerance and respect for others. Also, an aspect of the local culture is quite progressive in exploring alternative energy and sustainable living. Many locals are knowledgeable about new technologies that are enabling these choices.

Knowing that the valley is a focus of UFO sightings and lore, I thought it would be fun to do some informal research during my writing retreat there in August 2001. Much of my data-gathering occurred at several nearby hot springs—great places to leach out stress and soak up local culture. I soon learned that, even among longtime residents with good eyesight, opinions vary widely on what goes on in the sky. The woman who runs the hot springs told me that if you just go outside for ten minutes and look up, you're sure to see alien activity, because they are always up to something. Others who had worked out of doors in the valley their whole lives told me that they'd never seen anything unusual. It is as if two or more intermingled populations are somehow living in the same place under completely different skies.

One place I had to visit was the UFO Watchtower and Campground outside Hooper, in the middle of the valley on the way to Great Sand Dunes National Park.* The UFO Watchtower was created by Judy Messoline, who came to the valley in the mid-1990s to be a cattle rancher. Ranching didn't work out, but in the burgeoning local UFO culture she found another calling and created the Watchtower—a big metal platform raised about fifteen feet off the ground, standing next to a dome-shaped, alien-themed gift shop.

---

*Where Carl Sagan tested the *Viking* cameras on turtles and snakes back in the 1970s, practicing for Mars.

Spot an Alien, win a Frisbee!—at the UFO Watchtower. (photo by the author)

People come in a slow but steady stream to camp out, watch for aliens, swap stories about sightings, and stock up on alien books, key chains, T-shirts, videos, figurines, Frisbees, and blasters. Messoline is warm and welcoming and has all the time in the world to chat. She is eager for visitors to sight aliens from her tower and enthusiastically repeats the stories of those who have. She told me about one girl who had established communication by flashlight with a bright light pulsing over the mountains. If you bring your own tale of alien contact, you can get a free UFO Watchtower Frisbee or bumper sticker by letting Judy tape-record your story and add it to her collection. She hopes to publish a book of these stories.

While I was in the valley, I watched the sky as much as I could, which, along with the hot springs research task, was not too much of a chore. Certainly, a lot was going on up there. I saw two spinning galaxies, one from the inside and one from the outside. I saw eager clusters of bright young stars just leaving their nebular nest and entering galactic life, and tired, old red giant stars taking one last lap around the galaxy, preparing to blow their guts back out into the stellar reincarnation chamber. I saw enormous clouds of dark, obscuring dust drifting down the Milky Way. I saw stars that we now know have planets circling them, and many more that surely must have planets. Closer to home, I saw Jupiter and Mars promenading through the starscape. I also saw several good meteors, including one brilliant, slowly moving fireball that lit up the valley and left behind a lingering glowing trail. And I saw many satellites speeding along and winking out when they fell into Earth's shadow, and occasional high-flying aircraft, each flashing its own distinctive rhythmic stripe across the sky, the silent songs of mechanical birds. Low clouds drifting between the ranges revealed themselves as migrating voids, black on near-black pockets of negative space in an otherwise brilliant starscape.

It was all there, except no UFOs. Ordinarily I wouldn't complain about such a sky, but I was hoping for something else. I had been reading *The Mysterious Valley,* a book by UFO investigator and Crestone resident Christopher O'Brien, which chronicles UFO sightings, cattle mutilations, and other reports of the "high strange" occurring in the San Luis Valley. O'Brien had collected so many stories of glowing and strangely moving lights and bizarre aerial phenomena that I thought I'd have at least a reasonable chance of seeing something. I was determined to watch attentively without preconceptions, expectations, or excessive skepticism. I didn't want to fail to see something extraordinary because I was not prepared to see it. Conversely, I wondered if I could have made myself see something if I was too well prepared.

INVITATION ONLY

After weeks of careful watching and listening, I started to question whether I was observing the same sky as everyone else. One night at the springs some people I met were matter-of-factly discussing UFO sightings and even telepathic contacts with aliens. I told them that I had been trying every night but had not yet seen any highly unusual sky phenomena. I asked for advice. One woman who had experienced many sightings told me that the most important thing is that *you have to invite them.* They will not show up for just anyone, in just any state of mind. I asked if she could tell me anything more specific about how you invite them. She said that you go to a very dark place and focus on inviting them, on receiving them, on wishing them to be with you, and you try to communicate to them that it is safe to come. After a while spent meditating on these themes, sometimes they show up—lights in the sky that hover, pause meaningfully, and seem to communicate with you.

Late the following day I was back on the porch of the Willow Spring Bed and Breakfast, watching the last red rays of sun helping the Sangre de Cristo Mountains live up to their name, when suddenly I saw a formation of seven bright lights flash on simultaneously across a fifty-mile stretch of the mountains. They lit up in unison, and trust me, they were far too bright to be lights on cars, houses, or any conventional terrestrial, nonclassified vehicles. They flickered for about ten minutes and then faded suddenly. Could this be the work of some mysterious intelligence? Possibly. But my mind, always demanding an explanation, got the old hamster spinning and quickly found a plausible scenario. The setting sun far behind my back was just a few degrees above the horizon. When it hit just the right angle, it caused the windows on any building within a certain altitude range in the mountains to bounce direct reflections right at my spot on the valley floor, causing my retina to jump and shout, my optic nerve to tell tall tales, and my mind to wonder.

Maybe I just have the wrong sort of mind. What if instead of inventing a physical explanation for these lights, I had simply invited them in? Would they have accepted? I suppose with my bad attitude it might be impossible to extend an honest invitation. The sad thing is, they are probably telepathic interdimensional mind readers, in which case there is no way to fool them with feigned credulity. They'd see right through me, immediately discovering my impure heart and my weak faith. They are never going to talk to me.

I really tried watching the sky from the perspective of someone who believed many conscious entities were up there (hey, wait, I *do* believe that). What might I see if I could just get into the right frame?

Is it possible to make yourself believe something? I always wonder about people who change religion for reasons of convenience. Can you really just decide like that? I was definitely getting the impression, more than ever, that when it comes to visible signs of intelligence in the sky, "believing is seeing." If part of "trying to see" is "trying to believe," it's easier said than done.

WHERE'S THE BEEF?

Stories of strange, unexplained, and often grotesque livestock deaths in the San Luis Valley go back to the famous case of Snippy the horse. Everyone in the valley knows some version of the story of poor Snippy, whose real name was Lady, who was found dead on September 7, 1967, with strange, fresh wounds. His* skin and flesh had been cleanly cut from his neck all the way back to his shoulders, leaving nothing but a skeleton where the front of the horse should have been. There were no tracks, and no blood on the ground. Strange markings that could have been caused by the landing pads of a flying saucer were found nearby. A neighbor reported that she had seen something pass over the ranch house on the day that Snippy had died. Lady's owner, Mrs. Bertie Lewis, told one local paper, "Flying saucers killed my horse!" And thus a legend was born.

Most of the victims have been cattle. They are usually found with various parts removed with surgical precision—most often the tongue, the brain, other soft tissues in the face and upper body, and the genitals and anus. All of the blood has mysteriously been drained from the bodies, and flies, which usually can't resist a carcass, stay away from these. There is sometimes a medicinal smell like embalming fluid. There is never a trace of the perpetrator, no footprints or tire tracks—almost as if they dropped out of the sky.

The cattle mutilation phenomenon is not unique to the San Luis Valley. Similar mutilation epidemics have been reported in Nebraska, South Dakota, Texas, Oklahoma, Alabama, Kansas, and northern New Mexico. Investigators have also reported mutilations in South America and Europe.

Skeptics, naturally, debunk these stories, chalking them up to hoaxing, credulity, and some faulty reporting and distortion of cattle deaths by lightning strike or predators. The missing soft tissues, they say, have been eaten by scavengers. The debunking arguments often resort to the "why don't these aliens behave as we think aliens should behave" line of reasoning.[†] Much of what is written about the mutilations is sarcastic, and unfortunately some of the ridicule has an obnoxious, elitist twang—"Those dumb farmers are so unsophisticated that they believe in things we ivory-tower academics can easily explain away without ever having to go near a dirt road or a tractor."

The ufology community is divided over whether cattle mutilations have

---

*Yes, as far as I can tell, Lady was a boy horse.
[†]One *Encyclopedia of Skepticism*, asks, "Why would the aliens take only parts of the cattle, and not the whole animal? Why would they leave incriminating evidence behind, or at least disguise their activity better so that nothing appeared out of the ordinary?"

anything to do with aliens. Some believe they result from experiments that the government is conducting in secret collusion with extraterrestrials. They claim that the cattle deaths are often associated with appearances of mysterious helicopters or other less identifiable craft. Others angrily denounce the mutilation theorists for damaging the credibility of real ufology.

Like UFOs, cattle mutilations are in the air in the San Luis Valley. Even people who deny that there's anything to it are aware of it as a local issue or myth, sort of like a famous haunted house or witch. Everyone knows some version of the story of Snippy the horse and can tell you an anecdote or two about the mutilations. A mistrust of government and central authority, common in rural Western areas, has merged with UFO conspiracies in some people's minds. Some ranchers suspect ETs, although others think that's a bunch of manure.

Steeped in the interpretations advocated in skeptics' magazines, I was prepared to be unconvinced. I am much more comfortable with the debunkers' view that mutilation stories all result from mass hysteria, sloppy reporting, and misinterpretation. But since I got out of my armchair, knocked on some farmhouse doors, listened to stories of mutilated animals, and viewed snapshots with the ranchers who took them, I cannot dismiss the impression that something strange, twisted, and hard to understand has actually happened here.

I spoke to a young sheriff who had been first on the scene at one of the mutilations. The cow had fallen on fresh snow and was precisely carved, but there were no tracks and no blood. He'd never seen anything like it and has to this day heard no explanation that makes any sense. I believed that he was simply telling me exactly what he'd seen.*

One of the most interesting people I met is a seventy-two-year-old rancher named Virginia, who, with her daughter, runs a large cattle ranch in the valley. Virginia is sharp as a tack, opinionated, and warm once she decides she likes you. She is not obsessed with or enthralled by cattle mutilations—clearly, she would rather that the topic had never entered her life. She is more concerned with running her ranch and her work with the local historical museum. But when the subject comes up,

---

*Then he proceeded to regale me with stories about all the times he's found couples having sex in cars at night around Crestone. He found this topic very entertaining and it was definitely less unpleasant to picture than cattle mutilations.

she doesn't shy away from it. Virginia has the dubious distinction of owning the ranch where one of the most famous and well-documented cattle mutilations occurred, in 1980.

She invited me into her kitchen and made some coffee, lit a cigarette, and we got acquainted. The second time I visited her she confessed to liking me more than she expected to, based on our initial phone conversation when I called and asked if I could come and meet her. I don't know what she expected, but I definitely wasn't pumping her for information for an article in either *Enquirer, National* or *Skeptical*. She did tell me that she is tired of BBC film crews showing up and wanting to interview her about her damn bull. When we got around to the topic of cattle mutilations, she showed me her snapshots of the dead animal. My eyes were unavoidably drawn to the smooth incisions where the poor beast's missing genitals had once been. It was carved up in a way that looked careful and deliberate—and sick.

Virginia has heard all the debunking arguments and thinks they're all bunk. She knows cattle better than most people know people. This rancher knows what dead cattle look like in various states of decay. She knows what a lightning burn looks like. She knows how a carcass appears after it's been gnawed by predators. She knows her bulls and she knew that this particular one was alive two days before it was discovered. Nothing will keep flies away from a dead bull, but Virginia assures me that even the flies wouldn't touch this carcass for days.

We didn't dwell on the topic of the mutilations. I didn't need to make her go through all the gory details. I had satisfied myself that she and her story were for real, and that was enough. I did ask her if she had ever seen a UFO in all her years working out of doors in the valley. Not a one. She has, however, seen many mirages and strange reflections, and she speculated at length about optical and psychological phenomena, sounding like one of the sharper writers for the *Skeptical Enquirer.*

I left Virginia's ranch completely convinced that there have been mysterious, unexplained cattle deaths. The actual number of animals affected is no doubt much smaller than that reported by the investigators who have made a cottage industry out of cattle mutilations and other anomalous phenomena. When people start to write books and build careers out of this sort of thing, the numbers tend to swell as new reports, whatever their origin or details or degree of similarity or validity, get added to the mass of confirming evidence. No doubt a lot of rotten meat is heaped together in the cattle mutilation bin, but there is also something strange and unexplained.

Virginia doesn't know what killed her bull and would welcome a new idea that made sense. Some of her best photos, however, were "borrowed" by a well-known investigator who distorted her story in his popular book, changing many details to fit his alien conspiracy theory. She doesn't seriously entertain the notion that aliens were involved. The closest she can come to a possible explanation is some sort of perverse cult ritual, but she cannot explain the complete lack of vehicle tracks or footprints, or several other bizarre details. Mostly she just chalks it up as a genuine mystery.

After talking to the sheriff and Virginia and a few other folks, it seems to me that these events are not as easy to explain away as the debunkers would like to believe. The skeptics may even be doing something they often accuse their "opponents" of: avoiding the truth out of fear of the unknown. You can come up with a rational explanation for anything. This doesn't mean that your theory is correct, but its mere existence can be comforting. In the case of the cattle mutilation phenomenon I am not convinced by any of the "rational explanations." Yet I see no reason at all to link cattle mutilations with extraterrestrial life.

There are some mysteries. Are we being unpatriotic to the flag of science if we admit there are some mysteries?

WHO GOT THE BUNK?

Some ufologists do attempt to apply rigorous methods to documenting strange phenomena seen in the sky. But they are up against an awful lot, including the rest of ufology, and we scientists don't usually talk to them. Many of us have had bad experiences with aggressive ufologists accusing us of all kinds of nasty things for not taking their ideas seriously.

If we are being honest, then our scientific attempts to debunk UFOs must contain caveats. This doesn't mean that debunking false reports is not worthwhile. Indeed, it is essential if we are ever going to be able to recognize the real thing. But sometimes we forget that we don't really know much about aliens. Unfortunately, the skeptics' attitude toward UFOs often has a moralizing tone, justified by a concern that the masses will turn back to medieval darkness if we don't wake them up by shining the spotlight of science right in their faces.

One of our defenses against dealing with UFO claims is to lump them in with all the newage* drowning modern culture, seeping from the ground in places like Boulder and Sedona. I used to keep a file of especially flaky New Age literature. My favorite was an article in a Tucson rag by some guy waxing rhapsodic about the universal vibrations in electricity, which vibrates everywhere with a frequency of sixty cycles per second, demonstrating an important global harmonic something-or-other. I found it hilarious that this fellow was finding cosmic significance in a frequency that is "universal" *because of an engineering convention,* and one that is not in fact globally applied. When it comes to the New Age, it's all too easy to set up straw men and blow them down. Dismissing or ridiculing *all* beliefs and perceptions associated with a New Age spiritual consciousness on the basis of the most silly examples is a common practice among skeptics, and one that reveals some loose thinking on the other side of the crystal mantra.

Scientists don't like it when we are portrayed as unidimensional, robotic, spiritually insensitive, tight-assed, honky geeks.[†] Straw men can be used, just as inappropriately, to knock down our philosophy. You could quote biologist Richard Dawkins saying that faith is an evil that should be stamped out just like smallpox. You could point to the aid and comfort some American skeptics have given to the Chinese government's brutal, murderous campaign against Falun Gong. All houses of philosophy have rooms built of straw.

The most extreme UFO believers and debunkers are caught in a feedback loop in which each side validates the other's existence. Overzealous efforts to discredit UFO reports help to reinforce the wide perception of scientific skepticism as intolerant and narrow-minded. Believers accuse debunkers of being in on a conspiracy, which leads to more hysterical debunking, and so on.

Debunking, unfortunately, must be done case by case. Unfortunately, there are a lot of cases. If the Reverend Sun Myung Moon can marry five thousand people at once in Madison Square Garden, why can't we do a mass debunking? Some attempts have been made, but many of the arguments are quite weak. They generally amount to "If they are real aliens, why don't they act as we would expect aliens to act?" and rest on assumptions about the nature of alien technology, society, motivations, and so on. In some cases, the skeptics, just like the believers, have

---

*Short for "New Age." Rhymes with *sewage.*
[†]We're not all honkies.

clearly made up their minds in advance. Each community is quite sure it is saving the world from the other.

The whole debunking concept, when applied to other people's belief systems, as opposed to specific reports of events or phenomena, is antiproductive. It doesn't lead to greater understanding. The term itself does not connote an effort to win over or convince those who don't agree with you. It's meant to show them to be the idiots they really are. "Your beliefs are bunk" is a trifle condescending.

Some people who see themselves on the science side of a science/anti-science divide develop their own strain of credulity, accepting anything published with the stamp of scientific approval as automatically authentic. Organized skepticism is always in danger of an ironic slide into its own form of dogmatism. The UFO debunkers know what they expect to find. This makes me think twice about debunking reports. After I think twice, I usually agree with them. Much debunking of specific claims has been done carefully, thoroughly, and convincingly. But many scientists do have a strong ideological commitment to keeping UFO reports within a certain class of phenomenon. If a UFO report turned out to be evidence of actual alien spacecraft, and aliens who do not follow our rules, a frightening tear would be rent in the fabric of our worldview.

THERE'S A HOLE IN MY PHILOSOPHY

I grew up hearing a lot about UFOs from my parents and their friends, and from reading Asimov and Clarke and communing with the science fiction crowd—all people who loved to think about alien life and space travel, and who would have welcomed real alien contact more enthusiastically than anyone else. Yet the dominant view was that UFO believers were generally quite deluded.

That is not a controversial statement even among UFO believers, most of whom seem eager to distance themselves from those *other* UFO believers, whom they regard as *really* flaky. But are all of them deluded? Sagan and Asimov and my dad thought the answer was yes.

They were my authorities and one of their commandments was to question authority. So I had to question them, even while questioning whether this was always a good idea. At various times I've forced myself to rethink my stance on UFOs. Not wanting to just form my opinion based on authority, or received knowledge, I've had to ask myself if we might all somehow be deceived on this issue.

Science says, "Without objectively verifiable evidence, assume that it doesn't exist." But it is more accurate to say, "Without such evidence, we can't say whether it exists." We must be careful not to become lazy and let our skeptical mind-set become a closed one.

We have a certain view of how aliens will and will not behave and manifest their presence here. We get huffy when these imagined rules of interplanetary etiquette (of necessity based on projections of ourselves) are not followed. Skeptics complain that the aliens reported by UFO enthusiasts don't act like real aliens. Real aliens would not spend that kind of money on space fuel (energy is money). They'd stay home and improve things in their own systems. Real aliens wouldn't be interested in kidnapping humans and examining us or stealing sperm and eggs. We can't think of any good reason for them to behave like that. Real aliens would surely leave some spare parts or trash or footprints behind for us to study. Don't you know *anything* about aliens?

Yet, science faces some special challenges in applying itself to the question of intelligent aliens. Our methodology and philosophy assume that nature doesn't care about and isn't aware of our experiments. (Some ufologists assume the opposite.) We don't really know how to study something that knows it is being studied or might not want to be studied, or that *might even be studying us.* All our standards of evidence and proof—repeatability, multiple witnesses, material evidence, and so on—might fail with something that is actively messing with our minds, aware of us, and being careful *not* to be of interest to mainstream science.

Imagine for a moment that aliens were aware of our scientific method and were careful not to reveal themselves, perhaps out of compassion. You could envision their rules for avoiding our scrutiny:

Memo to All Space Brothers: Remember that human contact is to be avoided whenever possible. They are stuck in the "science" phase we went through eons before we went intergalactic. We can use this to predict their reactions and avoid suspicion. Under no circumstances leave any physical evidence that could be used to scientifically deduce our existence and extraterrestrial origin. It's inevitable that humans will occasionally detect our activities, and this is acceptable as long as they don't have what they consider to be a "scientific" case. So if you are detected, make absolutely sure that the observation is not replicable, and keep your spectral scramblers on. Such occasional cases are puzzling to them and help maintain our secrecy by sowing doubt about all sightings.

Science has given us criteria for distinguishing the physical from the metaphysical. But if a conscious entity is studying us, which box does it go in? If advanced technology is indistinguishable from magic, the boundary between the physical and the metaphysical vanishes again, as if science never happened.

Are SETI aliens, fashioned from logical scientific extrapolations, more likely to be realistic than UFO aliens who don't follow our rules? Not necessarily. Despite the undeniable truth of Clarke's Third Law, in our debunking of alien stories we insist that aliens must conform to our current notions of evolution, our current understanding of the laws of physics, and some extrapolation of our own technological capabilities. Because we must extrapolate from the known, and because we cannot consider to be real any phenomenon for which there is no scientifically acceptable evidence, we are not open to magic. So scientists may not be any better qualified than anyone else to predict what aliens will be like.

Here's what we don't always cop to: Our scientific arguments against "the extraterrestrial hypothesis" for UFOs depend on a framework of assumptions. These are the pesky metaphysical leaks and leaps in our airtight worldview—the things we feel we know to be true, but cannot prove.

It wouldn't hurt our credibility to acknowledge that science has its own superstitions. We assume the existence of an objective reality that is independent from our consciousness. We assume that our minds do not create or affect what we observe. We also assume nature is consistent and repeatable, and therefore knowable. In all of this I could replace "we assume" with "I believe." I don't doubt any of this.* This set of regulations for nature seems so obvious and reasonable to me that it almost seems absurd to question it. But if you dig down deep beneath our solid tower of reason, deduction, and provisional truth, you see that the whole thing is planted in loose sand, supported by received, or intuitively perceived, knowledge.

I'm a believer because this is the way the world seems. Further, I think that most everyone knows that this is the way it is. You can spin intellectual counterarguments to your heart's content, or you can medi-

---

*Except, I do believe that our minds strongly influence what we see. I'll come back to this.

tate your way clear out of the galactic disk, but on your way back home tonight notice how your every move, breath, and thought is steeped in a solid world of consistent phenomena. If this is an illusion, I don't think we can shake it. Maybe John Lilly did, but we can't all swim with the extraterrestrial mind-fishes. Even the Dalai Lama has to sit on the can like the rest of us. No matter what you believe, reality is something that we directly perceive, and we all operate on the experiential understanding that the world has external, material solidity.

Much as we "real alien researchers" would like the UFO phenomenon to just go away, we can't dismiss all UFO reports out of hand. We might miss something important. Further, we alienate a large segment of the public when we appear to be closed-minded, snotty, and overconfident.

In general it doesn't really bother me what people believe. I care more about how people behave toward one another, and some of the nicest people I've met have also seemed to have had some of the wackiest ideas. UFO believers and SETI scientists reject each other's philosophy, but both rely on the same core argument from plenitude. It's still the best justification for the existence of aliens: With so many stars and planets, there just has to be other intelligent life. Why should we be the only ones? You will hear this exact same logic and sentiment trumpeted from the stage at conferences of both ufology and astrobiology.

I've found something else that scientists and ufologists have in common, something wonderful that is widespread among diverse communities with vastly different approaches toward alien life: a sense of humor. Certainly, some take themselves and their beliefs too seriously, but there is wide recognition, on all sides, of the absurdity of the subject matter, and an ability to laugh about it. This could be a good starting place for scientists and ufologists to meet. If I ever ran a joint SETI/UFO conference, inviting a constructive dialogue between skeptics and believers, I would make the first and last session of every day a comedy session.

GO FISH

The last day of my third writing retreat in the San Luis Valley coincided (not by coincidence) with the Leonid meteor shower, which was predicted to be spectacular that year. I went out to a dark place for an all-night meteor picnic. Occasional low clouds wafted through all night, but the sky was mostly clear, not in the way it is sometimes "clear" in Denver or anywhere on the Eastern seaboard where I grew up, but so intensely dark, bright, and varied that it seemed close, deep, and enveloping. The meteors began bursting at a whole new level—behind the slowly drifting cloud curtains but far in front of the piercing stellar backdrop. All of this activity above me at different levels increased the feeling of depth, reminding me that the sky is also a landscape that we live within.

The only sounds were the crackle of my little fire, and occasional packs of coyotes yipping across the valley. I lay back and watched as the thin remnants of an ancient comet swiped across Earth's orbit, and little bits of its dusty debris trail slammed into our atmosphere, showering sparks across the sky. Each bright streak revealed the unreal speed with which things fall from space. Seventeen thousand miles per hour is the absolute minimum, and these suckers were going a great deal faster. A meteor shower is the only time you can actually see anything move that fast, getting a visual, visceral sense of orbital velocities in our solar system. This is how the carbon in our flesh originally found its way to Earth 4.5 billion years ago—as cometary dirt and dust flashing down through the skies of a world ripe for life.

I grilled a steak and watched all night. The sky show was great but I still didn't catch any UFOs. No aliens, but at least the quest for aliens got me out there. It was beautiful down there in the valley, taking a nice cold meteor shower out under the wheeling galaxy. In this way I suppose ufology is like fishing, worthwhile even if you don't catch a thing.

21 | Cons, Piracies, Conspiracies

Help me believe in anything
I want to be someone who believes.
— COUNTING CROWS

Space may be the final frontier, but it's made in a
Hollywood basement.
— RED HOT CHILI PEPPERS

THE CASE OF THE FACE

It's important to remember that conspiracy theories are not always
wrong. Things that sound crazy sometimes aren't. A good friend of
mine grew up in Argentina. When he was a teenager there during the
1970s, some swore that people were disappearing without a trace. For
a while his family, and most everyone they knew, dismissed this as
wacky paranoia. Now we know about Argentina's Dirty War when the
fascist government "disappeared" thousands of liberals, intellectuals,
and suspected or potential dissidents. This really happened, but some
who first tried to call attention to it were dismissed as crazies. From
this we should not conclude that every bizarre theory is true, but we
should at least briefly consider the possible truth of things that sound
crazy if a lot of people believe they are happening.*

Some of the alien conspiracy theories get pretty bizarre, and some of
them are quite comical. There's a sort of Zippy the Pinhead appeal to

*It's also a good time for us to remind ourselves of what can happen when protecting
security over freedom becomes your government's prime imperative.

the alien head symbol and the alien conspiracy story, a joke that is funny in part because some people don't get it.

I assume you've all heard about the face on Mars. You know, this guy:

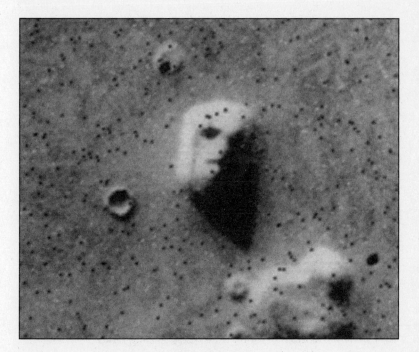

(NASA)

You do have to admire this face. It's been braving the elements much longer than Mount Rushmore, the Easter Island heads, or even the Old Man of the Mountain in Franconia Notch, New Hampshire, which survived for 30,000 years before crumbling into rubble on May 2, 2002.

Of course, erosion is much slower on Mars than in New Hampshire. It's so dry that if you were a face on Mars, you could sit for billions of years, staring at the sky and thinking slow rock thoughts while your wrinkles were gradually scoured away by the thin, dusty gales. Your major enemy would be boredom. That and secret NASA plans to blow you up.

No scientist I know has ever regarded the face as anything but a random rock formation that looks vaguely facelike in a distant photograph with a certain angle of light. But many don't see it this way. In books, magazine articles, Web forums, conferences, and television exposés, proponents of the face-on-Mars conspiracy present detailed analyses showing that the face and surrounding "monuments" in the Cydonia region of Mars—eroded pyramids and the like—are constructed in ways that show clear signs of intelligent design.*

The supreme test of the face on Mars, for believers and skeptics alike, became possible in April 1998 when *Mars Global Surveyor,* which had begun mapping the Red Planet, had an opportunity to take the first close-up images of the Cydonia region. Within NASA some debated whether we should shoot the face (with *cameras,* I mean!), because this might be seen as legitimizing the question of an artificial construction. Proponents of the face would portray the mere rephotographing as a victory. Of course we should shoot the face. Better to show that Mars has nothing to hide, and neither do we. Besides, whatever we found there would either completely embarrass the face people and shut them up for good, or else we'd uncover the greatest mystery of all time. Either would be a positive result.

Here's what the *MGS* pictures showed:

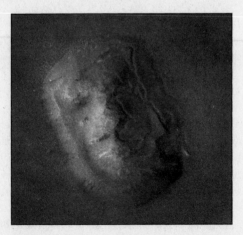

(NASA)

---

*Ironically, the preeminent "researcher" and popularizer of the face is Richard Hoagland, the man credited by Carl Sagan with the idea of including an interstellar message on *Pioneer 10* and also one of the first people to suggest there might be life in Europa's ocean. The borderline between science and pseudoscience can be porous.

There you have it. It's a jumbled, eroded mountain. Not a face. Case closed, and we can move on to other things, right?

Wrong.

The hard-core face community was ready to "save the phenomenon" at all costs. If these features were not in the pictures, then something had to be wrong with the pictures. The face people spun elaborate theories, implicating NASA in faking and altering the pictures, or even sending up a nuclear bomb to destroy the face before *MGS* got there.

Nothing is inherently absurd about the basic premise of finding some kind of monuments on Mars. As I've already discussed, there is no good logical reason why we shouldn't find alien artifacts anywhere in the solar system. On the contrary, there are good arguments for the possibility of such artifacts. Remember, all it takes is *one* rapidly spreading galactic civilization, or *one* successful launch of a fleet of self-copying interstellar Von Neumann machines, to have swept through our entire galaxy long ago turning asteroids into more probes. If they were ever here, they might have left something behind. So, yes, of course we should look for the unexpected, the seemingly engineered.

"But why a human face?" is a good skeptical question, but one that is not hard to answer with a pulp-fiction imagination. Maybe the monument builders were themselves humanoid or wished to communicate something to humans once we were ready to find it.

Yet the face on Mars is a classic example of pseudoscience: Proponents are so committed to a theory that they don't really care about new data or find a way to twist the data to support the answer they've already chosen. They are willing to seize on the slightest hint of monuments in an area of eroded rock formations, and to uncover "contradictions" in "official statements" that seem in some convoluted way to support their theories. At the same time, they have a remarkable ability to ignore a wealth of evidence that completely rules out their claims.

The presentation is also classic pseudoscience. Their tracts are full of equations and complex-sounding phrases that appear scientific, until you actually work your way through them and you find either complete nonsense (doesn't say anything) or complete bullshit (intended to deceive). I think that some of the purveyors know exactly what is going on. They are peddling theories they don't believe themselves, and that is the definition of hucksterism.

Conspiracy theorists are convinced that Dr. Ed Weiler, NASA Associate Administrator for Space Sciences, is personally hiding information about life on Mars. This actually came up at an SSES meeting* at NASA head-quarters, when Weiler (whom the committee reports to) was briefing us on the status of the Mars Exploration Program. He mentioned that he was being besieged with messages and accusations about the face and NASA's cover-up of it.

Why should NASA want to hide a discovery of life on Mars? It would be the most astounding success NASA could ever hope for, the ultimate validation of its reason to be, and a sure solution to our funding problems. Even the notion that NASA *could* hide something that important is silly. I've worked at a NASA research center, and I can tell you exactly how long an order to maintain silence about something on Mars would last, if you tell me how long it takes to type it out and hit "send."

What these people don't understand is that, more than anything, I wish they were right. It is my secret fantasy that when, in the interest of fairness, I spend a few moments on their Web sites, I will find myself saying, "Now, wait a minute! There really is something there. Why haven't I seen this pyramid before?" I would be on the phone in seconds calling my friends on the Mars Orbiter Camera team. If I encountered any obstruction whatsoever, I would call the press immediately and report that government scientists were hiding important findings from the American people. Hundreds like me would do just the same.

Science has, in the past, missed important discoveries by dismissing ideas that seemed too far out there, or too good to be true (such as continental drift, for instance). Reading history encourages us to be open to thinking outside the box, to at least consider frameworks within which fringe ideas could be true. These face people and their ilk only encourage us to ignore the fringes and, with the little time we have before we become food for worms, focus on that which seems potentially fruitful. They say they want to force NASA to open up and face the truth. In reality they make us clamp down our filters more tightly, which could someday actually make us miss something important.

CIRCLEMAKERS

There is a bit of a parallel here with the crop circles that some people think are created by aliens. You know what I'm talking about: the

---

*NASA's Solar System Exploration Subcommittee, which I served on from 1998 to 2002 (chapter 15).

A fractal pattern that appeared mysteriously in a wheat field in England.
(Lucy Pringle)

strange, often quite lovely, geometrical patterns that show up overnight in English wheat fields (and sometimes elsewhere as well).

The crop circle mystery turned out to be an ingenious hoax. Well, not a hoax exactly, but an art project in which the perpetrators, believers, debunkers, and journalists all played their roles and continue to do so.

The artists Doug Bower and Dave Chorley, of Southampton, England, came out in 1991 and confessed to making the crop circles. They demonstrated how they had done it, cleverly wielding boards and rope, and how others could join the fun. Their many worthy successors, who call themselves the circlemakers, continue to create beautiful, increasingly intricate patterns in nocturnal excursions into fields of wheat and other crops.

Naturally, some believers just became more adamant that crop circles had to be the work of extraterrestrials. The hoaxers, they said, were themselves a hoax, perhaps employed by the government. They might even be alien agents. The crop circles are real and contain important messages from other intelligences.

The remarkable performance goes on. Sometimes I worry that the debunkers who don't get the joke might be too successful. Only after they succeed in entirely wiping out the irrational scourge of crop circles will they realize that they have helped kill off an art form. On the other "side" are those who believe that the circles are alien signals of an approaching earth energy transformation in which we'll, um, you know, all vibrate together to a higher plane and join the ascended masters with their serpent knowledge. You don't want that side to be too successful either, or some cereologist cult might take over the world and force us to worship wheat. I suppose the current dynamic between strident debunkers and earnest circle believers is healthy. It gives the circlemakers a reason to carry on, bless them, and provides fine family entertainment for the rest of us. Long live crop circles.

DISCLOSURE

When I heard that the highly publicized start of the Campaign for Disclosure tour in June 2001 was going to be held at the University of Colorado, in Boulder, I decided to attend as a receptive fly on the wall, rather than as a skeptical scientist. I'm not saying that I left all of my critical faculties in the trunk, but part of me was thinking, "Do the experiment honestly. Don't write it off without listening. Maybe you'll be surprised."

In the lobby outside the auditorium a phalanx of eager and earnest Disclosure Project volunteers gave out literature and sold books, videos, and CDs containing massive amounts of witness testimony about government UFO secrets. As the audience jammed the auditorium, I let myself get caught up in the palpable excitement in the air. Here I was in a room packed with people all enthralled by some of the same questions that most intrigue me. Like them, I wanted to learn something new and inspiring about alien contact.

The last time I had been in that same lecture hall (Chem 140), it was for a public panel discussion about SETI, with scientists from the SETI Institute and the NASA Astrobiology Institute. Although attendance was higher at the Disclosure meeting, the audience looked similar. It was a classic Boulder crowd: well-heeled hippies with carefully matted dreadlocks falling over designer tie-dyes, bespectacled academics toting tattered notebooks, and smatterings of spandex, bike helmets, laptops, dogs, beards and peasant dresses (not necessarily on the same person but not necessarily not), the occasional whiff of patchouli oil or pot (but absolutely no tobacco smoking, under pain of death).*

---

*Boulder is a bubble town nestled against the mountains thirty miles northwest of Denver. It's sort of like the city in *Logan's Run,* a pleasant place, and anybody who is unhappy or unattractive or too old or unwealthy is recycled, and used to grow organic, free-range fruits and vegetables.

The Disclosure audience seemed more engaged and excited than the folks at the SETI symposium. Something stronger was calling them on. Although the auditorium was sweltering on that sunny afternoon in June, every seat was taken and people were crammed in the aisles and doorways.

The Disclosure Project is an offshoot of CSETI, the Center for the Study of Extraterrestrial Intelligence, which has no official relationship to the SETI Institute. None whatsoever. CSETI was founded in 1990 by Dr. Steven Greer, a former emergency-room physician from Virginia, who now devotes all his time to uncovering and spreading the truth about secret government involvement with aliens.

As we waited for the program to begin, slides looped on a giant screen, depicting grainy saucer- and cigar-shaped spacecraft photographed against scenic backdrops. The event officially began with "an exclusive showing of a two-hour video of fifty government and military witnesses to UFO and extraterrestrial events and projects." The video was a seemingly endless parade of sincere people earnestly testifying about the incredible things they had witnessed and government efforts to keep them quiet. They had sighted craft, witnessed crashed saucers and alien bodies, and even personally encountered alien beings. The witnesses included retired military personnel, retired intelligence officers, retired commercial airline pilots, academics, flight controllers, FAA officials, and NASA people whom I had never heard of, except for one astronaut who is a well-known supporter of numerous esoteric and mystical causes and who seems to me to long ago have left his powers of discrimination in orbit.*

We were told matter-of-factly about numerous saucer landings and crashes and top-secret branches of the military that have known about this for decades. A daytime landing at Edwards Air Force Base in the 1950s was filmed, but the film is hidden or lost somewhere at the Pentagon. An Air Force sergeant told us about giant artificial structures discovered on the back side of the Moon, including large geometric shapes, towers, and spherical buildings. Someone has been on the Moon before us.

At this point, suspending my disbelief became more difficult, because I know the Moon. Assuming the soundness of my basic framework of reality, which I do assume, the giant structures on the Moon are not there. My attempted open-mindedness began to crumble and my critical inner voice began to crack jokes against my will.

---

*I did, later on, do Web searches on several of the witnesses, and the ones I checked seemed to be more or less who they say they are.

One witness was repeatedly identified in captions as a "McDonald Douglass Engineer," which made me wonder whether he worked in aerospace or fast food. Then it all clicked: of course, McDonald's, dead cows, aliens. It's the military/industrial/fast-food/aerospace/alien complex. That kinky humanoid Ronald McDonald has been sent here to lure our children into eating mutilated alien beef.

They had even managed to videotape some guy with a *cosmic* top-secret clearance. Only about twenty-five people have one. Not even the president has this level of authorization. (So who the heck decides who gets it—the alien overlords?)

After the two-hour video of witness testimony, Dr. Greer himself, flanked by a few acolytes, appeared to deafening applause. Greer is charismatic. You like him. Before he said a thing, I found myself wanting to be receptive to his message. He explained that he is just a country doctor from Virginia who has been swept up in incredible events. And he did look like a doctor on rounds. Handsome and bespectacled, with a receding hairline, conservative tie, and a pen in his pocket, you would trust him with your family's health. The guy is built, too. His shirtsleeves were rolled up, revealing impressive Popeye forearms.

A natural motivational speaker, he is relaxed, sympathetic, passionate, self-deprecating, and witty. Confident, cadenced phrases hang on important words in a way that gently commands your continued attention. He sat casually on the lab desk at the front of the room, legs dangling, and smoothly, without notes, addressed the hushed room.

Dr. Greer began by identifying his message with the good ol' principle of plenitude: In such a huge universe, how could we possibly be alone? Then he said solemnly that what he was going to tell us about was nothing less than the greatest secret of the twentieth century. He asked us to imagine the realistic possibility of achieving a world without poverty, pollution, conflict, or energy shortages, in our lifetime.

Sounds good to me.

All this was possible, he said, if only we could convince the powers that be to release the secret knowledge that could make it happen.

SECRETS AND LIES

Then Greer connected the dots for us, putting all that witness testimony in the context of a massive, decades-long conspiracy involving rogue elements of the U.S. government working in secret with extraterrestrial technology. Several technologies that sound futuristic to us, it turns out, have been in the hands of the military for decades. For example, we mastered antigravity in the 1950s. Secret military groups have "quantum zero-point energy power," an infinite, completely free, and nonpolluting source of energy. All this and much more they learned by studying numerous crashed saucers, starting with the one in Roswell in 1947. They have "reverse-engineered" these wondrous alien inventions—meaning they took them apart and figured out how they worked.

According to Greer, the alien knowledge is confined to organizations within our own government that most people know nothing about. These groups, blacker than black ops, are known as cosmic ops, and they are privy to high-security information so sensitive that the president and the head of the CIA have both been denied access. (This is unconstitutional and Greer doesn't think that we citizens should stand for it any longer.) Greer cited numerous specific witnesses with names, dates, and impressive titles. For example, a brigadier general told Greer that three to five alien bodies were discovered at Roswell, and many artifacts were retrieved.

Greer also told us about a top-secret U.S. Army Extraterrestrial Retrieval Team, also known as Operation Blue Fly, in charge of recovering objects of unknown origin that fell to Earth. This team had retrieved material from several crashes as far back as the 1940s and early 1950s.* These UFOs are particularly interested in our military sites, especially nuclear weapons facilities, and they have repeatedly shown that they can render our ICBMs inert.

According to Greer, the presence of aliens was initially kept secret for benign reasons—to avoid an eruption of public panic. But after the Moon landings and decades of space exploration, it's obvious that the public could handle this knowledge. Now, the secrecy is being maintained to preserve the present economic order. Alien energy and propulsion technologies, if shared with the world, would eliminate poverty, save the environment, and end much human strife and suffering. This would be bad for business. No more fossil fuel industry. No more nuclear power industry. No internal combustion vehicles. The most powerful industries of our time will be history as soon as Disclosure is

---

*Why are these aliens such bad pilots?

achieved. Among those companies that are in cahoots and have black operations working on alien technology are SAIC, Lockheed Martin* Northrop Grumman, Raytheon, and EG&G.

These evil, rich, polluting, militaristic alien-hiders will resort to any means necessary to maintain their control. And that leads us to the true motivation for space-based weapons.

How are they going to maintain control? By provoking an interstellar war, once the secret weapons have progressed to the point where we can hold our own against the aliens. All other rationales for space-based weapons are just excuses: Reagan's Strategic Defense Initiative was needed against the Russkies. The current system is needed to guard against "rogue nations." That's what we've been told. Someday soon, there will be a hoax involving an alien invasion—a Gulf of Tonkin–type incident to build support for space weapons and the interplanetary war effort. They will provoke a war with an alien civilization that will so scare the populace that people will allow them to install a military government over the whole Earth. Ultimately that's what it's all about. The largest fascist plot ever, backed by weapons and technology of unimaginable power.

There is an urgency to this threat. According to Greer, our government's classified military projects have now achieved virtual parity with alien abilities. The interplanetary war could start at any time. Soon an army of alien ships will fill our skies.

ON THE SIDE OF THE ANGELS

Greer knows how to push some buttons with his populist, antiauthority message. Big cheers and loud applause filled the room when he said we must not militarize space. His presentation skillfully tapped into humanitarian concern for the poor, antiauthoritarian sentiment and fears and hopes about our future.

Despite my horror and amusement, I also felt stirred by the spirit of the crowd. I related to the collective concern and anger about corporate disregard for the health of the Earth. When an articulate and passionate speaker says something you strongly agree with, and everyone around you is clapping and cheering, it's hard not to feel somewhat roused. To

---

*Lockheed has had operational antigravity craft since the sixties. Greer can prove it.

be simultaneously shocked, repulsed, and genuinely moved by a speaker was, as we say, something of a mind fuck.

The loudest cheers came near the end of the speech, when Dr. Greer issued a call to action. We, he said, are the last generation that has a chance to change course. With genuine pain in his voice, he reminded us of the unbelievable poverty in the world, and he reasserted that it could be changed only if the suppressed information is released. The good news is that several dozen civilizations are ready to welcome us into a family of planets, if we are ready to live in peace.

"What can you do to help?" He told us: Give money (all this educational, investigative, and legal work is expensive). Proselytize (if each of you here today holds a meeting in your home, and if each of these generates ten more meetings, then the Disclosure movement will grow like wildfire).

Greer concluded on an emotional peak: We have a beautiful planet, which we're cannibalizing. We are working for "the good future." We're on the eve of a civilization that can endure in peace for thousands of years. The entire cosmos can be open to us. "You are on the side of the angels. Join us. Thank you very much."

A long, roaring standing ovation followed, with me scratching my head.

EXTRAORDINARY EVIDENCE

It really was impressive to see the faces and stories of the Disclosure witnesses endlessly roll by. Indeed, that's the point. Greer doesn't even try to present physical evidence. His evidence consists of the sheer number of people with some kind of credentials who are willing to talk about these experiences. He wants us to believe that the weight of numbers transforms these stories, collectively, into something more than anecdotal evidence. And it could. A large amount of anecdotal evidence is always worth paying attention to. The fact is, if I heard a lot of people *I knew* talking about alien encounter experiences, I'd take it seriously.

Oh, I know people who have seen inexplicable things, and I've seen some myself. But the social universe I live in is entirely disconnected from the one in which these sightings, crashes, and cover-ups occurred. There are no six degrees of separation here. The separation is total. We are supposed to think, "All these people couldn't possibly be wrong," but I couldn't help thinking, "All these people can't possibly be right." Otherwise, I would have heard about some of this more directly.

My disbelief in this conspiracy is certainly not based on faith in the absolute integrity of our government and military. I would not be surprised to learn of secret government operations that skirt the Constitution. It wouldn't be the first time. You don't need alien bodies to believe that some people within our government are cynically risking world peace to protect the wealthy.

Of course there are secret military aircraft and secret projects in space. Strange sightings are far more likely to be signs of "military intelligence" than of alien intelligence. In fact, many UFO legends arose from sightings of classified craft, which were initially denied by the military and then admitted to years later. By increasing the level of distrust in official stories about UFOs, these actual cover-ups have contributed to the conspiracy culture. In this way, the U.S. and Soviet militaries helped to create ufology—another bizarre by-product of the Cold War.

During the Cold War, the military didn't try too hard to dispel these misinterpretations and sometimes actively encouraged them. Ufology provided convenient cover. The most famous of these incidents is Roswell itself. The air force initially claimed that the "flying saucer" found on the Foster Ranch outside Roswell on June 14, 1947 (ten days before Kenneth Arnold's first sighting of a "flying saucer" in Washington State) was merely a crashed weather balloon. Ufologists uncovered many inconsistencies in the government's story, which helped fuel the UFO myth and suspicions of the official line on alien contacts. In 1994, the Air Force released a credible report showing that the crash debris came from Project Mogul, a top-secret air force program to develop stratospheric balloon-borne equipment for detecting Soviet nuclear explosions.

The air force also admitted to another experiment that explains many of the reports of alien bodies. In the early 1950s, they dropped numerous "anthropomorphic dummies" from altitudes as high as ninety-eight thousand feet, to test escape mechanisms for astronauts or pilots forced to eject at high altitudes. Several of these dummies landed near Roswell. Over 150 of them were dropped over Wright-Patterson Air Force Base, which may account for alien sightings at that location.

The truth is that both the skeptics and the believers were mistaken about Roswell. For years, skeptics wrongly insisted that there was no cover-up. Believers accepted that the bodies that "looked strangely like plastic dummies" were proof of the existence of extraterrestrial humanoids.

In 1997, a declassified CIA report stated, "Over half of all UFO reports from the late 1950's through the 1960's were accounted for by manned reconnaissance flights. This led the Air Force to make misleading and deceptive statements to the public in order to allay public fears and to protect an extraordinarily sensitive national security project."

So, yes, the military has secrets and lies and some of these have involved UFOs, but the whole Disclosure conspiracy just doesn't hang together. Once I got over my abortive attempt to be open-minded, the whole thing seemed incredibly dumb. While amusing, the gullibility of the crowd was also a tad scary. The audience *wants* to believe, responds to the charismatic leader, the sense of community in the room, the shared pursuit of a cause.

Greer seems authentic to me. I don't put him in the same class as the "face on Mars" hucksters—people who probably know better who have made big bucks on books promoting a ridiculous conspiracy. The Disclosure stuff, all said with a straight face, evident sincerity, and I would say genuine concern for humanity, is basically a case of extreme credulity.

The Disclosure Project ultimately struck me in the same way that much New Age thought does: I often find myself sympathetic with the quest for new solutions and new frameworks, but repulsed by the flaky, uncritical thinking. Can't we subvert the dominant paradigm without abandoning reason altogether?

HOLLYWOOD BASEMENT

What makes me so sure? It comes down to this: the whole thing sounds too much like a cheesy movie plot. If it sounded like a *good* movie plot, then I'd have to consider it more carefully. But I can't believe that if and when alien contact happens, it will seem just like a comic book or a B-grade movie.

Admittedly, the Disclosure guys know exactly as much about ET intelligence as the SETI folks: nothing at all. Both carry assumptions about aliens based, in different ways, on extrapolations of human traits. But the Disclosure scheme, like many ufological books, appears to borrow heavily from the pulp science fiction of the forties, fifties, and sixties. These people need to watch less television and get out more.

Anyone who drew their ideas about alien civilizations from the deep well of science fiction literature, their spiritual sense from the great religious texts, their ideas about the wider universe from the insights of modern science, and their sense of evolution's possibilities from the direct experience of nature (rather than Jerry Falwell, *Babylon 5,* and *Wild Kingdom*) would not create such cliché aliens and plots.

The idea that alien technology could be captured, held secret, and reverse-engineered, that aliens could interface with humans in any way that would seem like a "war," is just as ludicrous as the idea that our computer disks will fit perfectly in alien drives and our viruses will be able to crash their computers as in the blockbuster movie *Independence Day.* These fantasies carry an unrealistic expectation that our interactions with aliens will be safely within the realm of our previous experiences, including, especially, our TV watching experiences.

I cannot prove my "non-B-grade universe" hypothesis. Like Occam's razor, the Immaculate Conception, or the tenets of any faith, this is an unprovable belief about the way things are. Maybe reality does resemble a B-grade movie, and I am deluding myself into thinking we live in a cosmos that is more original or finely crafted. Maybe God is a hack director, banished to our universe because she couldn't get good work on the Coast.

Anyway, once I satisfied myself that the Disclosure claims could not possibly be true, I lost interest in learning any more of the details of their conspiracy. It's more interesting to research the real world. Disclosure is, ultimately, a little sad, a little silly, a little scary, and more than a little stupid. But, why should I care what other people believe, if what they are doing is harmless?

Because, I don't think that the Disclosure Project is completely harmless. In June 2002 Greer jumped on the post-9/11 "it's a shame that thousands of people were killed but at least I can use it to promote my cause" bandwagon in a new paper he circulated on the Internet stating, "One of the few silver linings to these recent tragedies is that maybe—just maybe—people will take seriously, however far-fetched it may seem at first, the prospect that a shadowy, para-governmental and transnational entity exists that has kept UFOs secret—and is planning a deception and tragedy that will dwarf the events of 9/11."

At the very least, these guys are destructive to our society in the way that a useless mutation in an organism is maladaptive—simply because it uses up energy and resources that are needed elsewhere. On the way out of the Campaign for Disclosure meeting, we passed by donation boxes where you could make a tax-deductible contribution to the cause, and people were stuffing them full of bills. Maybe the Disclosure Project is really a vast right-wing conspiracy to sap the life out of the environmentalist and pacifist movements by getting people to believe in this crap.

All in all, it was one of the weirder days I can remember.* Believe it or not (some of my relatives won't), I had gone to temple that morning, for a bar mitzvah. The alien revival meeting had less singing but more clapping and cheering than the service at the temple. It had better special effects, too, but in the end I liked the bar mitzvah a lot more, because it was all about real hope for the future and bringing a new generation into the spiritual community. It was uplifting, and it didn't give me the creeps. Just gimme that old-time religion.

---

*Although I do wonder about some of the ones I can't remember . . .

22 | Believing Is Seeing

I'm breaking through
I'm bending spoons
I'm keeping flowers in full bloom
I'm looking for answers from the great beyond.
—R.E.M.

When one admits that nothing is certain one must,
I think, also add that some things are more nearly
certain than others.

—BERTRAND RUSSELL

ARE YOU EXPERIENCED?

Once, a few years ago, I was lying on a beach north of Santa Cruz
when a strange dronelike machine zoomed overhead, heading south
down the coast. It looked like a pilotless airplane with the front sheered
off. I don't think I was too wasted at the time, but this was California
so I just thought, "Whoa, dude, that's too weird!" and assumed it was
some sort of experimental military aircraft. It didn't hover over me
with an intense blue light or suddenly move away at incredible speed or
anything like that.

Years earlier, driving up the East Coast on I-95, I saw a glowing,
shifting, unnaturally bright purple patch in a clear sky. It moved
around coherently, as though it were being directed or piloted.

When I was a kid, I had a recurring dream of being taken on an alien
spaceship. I still remember vividly the mixture of excitement and fear I
felt, the diffuse green light, the gridlike, patterned floor, and the gravity
that always seemed slightly off-kilter, making it awkward to stand up.

Most amazing of all was the view of the rapidly receding Earth through the porthole. I dreamed of this same ship enough times that I remember thinking, "Here we go again." The aliens were mysterious, their intentions hard to discern. Though I never saw them, I knew they were there.

My interpretation, if dreams need interpreting, is this: What do you expect from a kid obsessed with science fiction and surrounded by adults who talked about spaceflight all of the time?

If I were the type prone to believe in UFOs and alien abductions, all these memories could form a pattern of evidence convincing me that I had in fact been contacted. Over a lifetime, we all see and experience plenty of weirdness. If you are a person who thinks, "A lot of strange things are happening, and it all fits together to reveal a pattern of alien presence," then anything unusual you see and hear will help validate your scenario.

Our beliefs and expectations mold our interpretations of experience. This affects what we see and how we remember it. After a while you have a pattern of memories that confirm your beliefs. This can work both ways, of course. If you think, "We can find a rational explanation for anything," then you always will.

INTERRUPTED JOURNEY

You know those "audience brush with greatness" segments they used to have on *Letterman* where people stood up to tell about the time Henry Kissinger stepped on their foot, or how their sister fixed Venus Williams's carburetor? Well, *I* once saw Gene Simmons from Kiss eating a hot dog at the Denver Airport.* And my dad roomed with John Mack at medical school.

John Mack is the shining intellectual light of the UFO abductee movement. They prefer to be called experiencers, rather than abductees. Mack has published two books on his work with experiencers. In therapeutic sessions he helps them to remember and come to grips with their (often traumatic) histories of alien contact.

He has also elaborated a philosophical framework to explain the abductions, who the abductors might be, and what it all implies about who *we* really are. Mack believes this phenomenon has crucial implica-

---

*No sign of the tongue.

tions for human history, evolution, ecology, and the relationship of our consciousness to a deeper reality.

John Mack is a distinguished, widely published, Pulitzer Prize–winning, tenured physician and professor at Harvard. Among those who are completely convinced of the reality and importance of the abduction phenomenon, Mack is uniquely mainstream and respectable.

Or was.

I guess it's slightly strange that my parents' circle of friends included both Sagan, the SETI pioneer, and Mack, who has become the leading voice of the abductee movement. The beliefs and philosophies of these two men were once quite similar, back in the day. But over time they developed radically different conceptions of alien intelligence, which required two entirely different worldviews. Sagan was a passionately rationalist astronomer fascinated by the possibility of alien contact, and Mack, a psychiatrist, came to believe that contact with extraterrestrials required nothing more than a prepared mind. Somehow, they both ended up in our swimming pool.

John and my father, Lester, met in 1951, the year they both started at Harvard Medical School. They shared interests in literature and culture that went far beyond the medical curriculum. They became roommates and attended each other's wedding. In some ways they have had parallel careers. They graduated from Harvard together in 1955. Both went into psychiatry and then psychoanalysis, and both eventually became professors in the Department of Psychiatry at Harvard Medical School. John became a celebrity in some circles after his 1976 psychohistory of T. E. Lawrence won the Pulitzer Prize (the year before Sagan won his Pulitzer). Lester became famous in some circles and infamous in others for his books advocating sanity in drug policy.

They continue to socialize, though not as much in recent years. This friendship, as you might imagine, has been strained by the abductee movement. Though John's involvement began over a decade ago, it still seems like a surprising new development to his old friends.

The topic of alien abductions was not entirely new to our household. We had an early brush with the concept. The first abduction case to gain wide attention, the archetypal case, was that of Betty and Barney Hill, a couple who, under hypnosis, recalled being kidnapped by the occupants of a UFO, taken on board, and subjected to strange experiments, all while driving in rural New Hampshire one night in 1961. *The Interrupted Journey*, a 1966 book about the Hills' abduction experience, became a best-seller. In the 1975 made-for-TV movie Barney was played by James Earl Jones.

At first the Hills experienced only "missing time" and had no memory of what had happened to them. Their memories came out in hypnosis sessions with a Boston psychiatrist, Benjamin Simon. Betty and Barney seemed to independently recall the same details of the abduction. Betty drew star maps that the aliens had shown her.

My father knew Simon and borrowed the tapes of the hypnosis sessions. One evening, after the kids were in bed, Lester listened to the tapes along with Sagan and James McDonald of the University of Arizona, a meteorologist who was one of the few mainstream scientists taking UFOs seriously.*

After this listening session, Lester proposed that the Hills were suffering from a psychiatric condition known as a folie à deux—a joint delusional syndrome first identified in mid-nineteenth-century France.†It generally occurs between two people who live together and are closely related. One is dominant and the other passive. The dominant companion first develops the delusional system and the passive partner is gradually drawn in.

Betty Hill was a strong-willed, outgoing, successful woman with high standing in her community. Barney was by all accounts dominated by his wife. Betty was effusive about her abduction experience. As my father described it in a recent e-mail to me, "She talked about it as though it were the defining event of her life. One might say that Barney could scarcely afford to express any skepticism, or, 'if you can't beat them, join them.'"

Lester's folie à deux hypothesis became a widely accepted explanation for the Hills' "recovered memories," at least among skeptics. So the concept of alien abductions was not unknown to Lester and Carl when John began publicly espousing his theories. Nevertheless, John's acceptance of these experiences at face value was somewhat shocking.

Not completely shocking. John always seemed to be searching. In the seventies he got deeply involved in EST and became close to Werner Erhard. In the eighties he became increasingly interested in altered states of consciousness. He was especially enamored of psychedelic psy-

---

*I was upstairs having strange dreams. . . .
†In the original case from which the name stems, two women remembered time-traveling to eighteenth-century France.

chiatrist Stanislav Grof's "holotropic breathing" technique. It is a measure of the high esteem with which he was held by his old friends in the Harvard academic community that he was able to convince the archskeptics Carl and Lester to join him in a session where they were all "breathed" by a woman trained in Grof's technique. Mack had visions, hallucinations, and transcendent insights, but Lester and Carl just got a little dizzy. Eventually John's interests soared far beyond interpersonal or even transpersonal communication and he launched into interspecies communication with other humanoid intelligences—no radio telescopes required. To many of his old friends and colleagues he seemed to be heading for Lilly-land.

EXPERIENCERS

I'm not going to delve deeply into abductions. Many books and articles have been written by believers and debunkers. The subject has been debated on a *NOVA* documentary and discussed ad infinitum on daytime TV, late-night radio, and nonstop coast-to-coast-to-coast Web chatter. As my wife Tory said tonight at dinner, "You don't need to go on and on about that. People can watch that on *Oprah*."

In fact John did go on *Oprah*. She introduced him by saying that although ordinarily she would not invite a guest who made such bizarre claims, she was intrigued that a respected Harvard professor and eminent psychiatrist takes abductions seriously. And that is one role that he has played—helping to create a mainstream venue for stories of extraordinary experiences.

It is troubling but fascinating to see people you know head down such different roads of belief. All the time you hear about beliefs and ideas that seem like complete tabloid nonsense: UFO cloning cults in France, miracle cures from the dust of the Grand Wazoo, and authentic crystal balls. Most of this cosmic debris is easy to write off either because you don't know these people and can just assume they are nuts, or because you *do* know them and *know* they are nuts. But although the ideas espoused by the abductee movement definitely seem pretty loopy, I don't believe John Mack is crazy—which, for me, presents a puzzle.

The experiencers and their supporters are certain their recovered memories of alien encounters are real. They believe that the messages imparted to them are important, not just for themselves but for all of us. Although the experiences are solitary, personal, and dreamlike, there are many recurring elements that give the abduction narrative a certain solidity.

These common elements include being woken up by small humanoid creatures in your bedroom, being transported to what seems like an operating room on a "ship," being confined to a table, and having experiments done by alien "doctors" who seem particularly interested in human reproductive organs and who sometimes take samples of genetic material. The alien creatures also transmit various messages—often warnings about ecological disasters to come if we do not stop defiling the Earth.

Mack believes that the remarkable consistency of the stories establishes the reality of the experience. He and his experiencers are convinced that these are not just dreams or hallucinations, that they have actually been taken away to another place and encountered other beings.

When I read the experiencers' individual stories, I'm struck by how frightening and emotionally powerful the memories are. However, experiencers and their supporters lose me when they start interpreting the aliens' behavior. This is when they descend toward the grade-B zone. The aliens are trying to alter us genetically. They want to create a race of hybrid alien/humans. Some investigators feel this is to be feared and resisted, because the aliens are creating an invasion force, and when enough of them are among us, they'll just take over. But Mack and his followers think the aliens are benevolent. They are here to help us survive by assisting in a necessary transformation of consciousness, and human/alien hybrids will help us through the difficult decades ahead.

Meanwhile back in reality, the debunkers sometimes get a little bit grade B themselves. They argue that aliens wouldn't behave in ways that seem irrational to us, and that our encounters with real aliens would not seem like dreams. Why would the aliens come all this way from the gamma quadrant only to steal our sperm and eggs? How could they possibly breed with us? It doesn't compute. Real aliens would have more respect for basic principles of physics and biology.

In a harsh critique of Mack in *The New Republic,* entitled "The Doctor's Plot," James Gleick wrote "how infinitely unlikely it is that our corner of the universe should be receiving alien visitors in such strikingly near-human form at just the eyeblink of history when we have discovered space travel."

In his book *The Demon Haunted World,* Carl Sagan asked why aliens haven't set off burglar alarms or been captured on security videos. Debunkers are especially vexed by the aliens' failure to leave any footprints or garbage. Sagan noted, "There is certainly no retrieval of cunning machinery far beyond current technology. No abductee has filched a page from the captain's logbook, or an examining instrument, or taken an authentic photograph of the interior of the ship, or come back with detailed and verifiable scientific information not hitherto available on Earth. Why not? These failures must tell us something."

Sagan stated that he would believe in the aliens if they left behind unknown alloys, or materials with extraordinary properties.

Those were, of course, his own fantasy aliens. He'd have been convinced if their technology was impressive in ways that he could imagine, and if they left us samples. But they'd have to play fair, respect the laws, and manifest themselves physically in a repeatable way that we can verify. Mistrust and verify.

As I've pointed out, any aliens who came to Earth would probably be thousands, if not millions, of years more technologically mature than we. Who are we to tell them the rules? This provides a pretty good loophole against any debunking based on the seeming illogic of alien behavior and capabilities.

Other, more promising critical arguments invoke the similarities between the abduction stories and pop-culture images of aliens. The experiencer aliens have a definite physical resemblance to movie aliens such as those portrayed in *Close Encounters of the Third Kind*—the skinny little guys with big heads and almond-shaped eyes. Abduction critics also theorize about different kinds of hallucinations or sleep disorders, the power of suggestion, and therapists who are often eager to help experiencers remember their cosmic adventures. The skeptical mind wonders if the wide propagation of science fiction imagery, popular books containing the basics of the abduction story, extensive media coverage, and conscious or unconscious suggestions from the therapists are not more than adequate to explain the consistent imagery reported by experiencers.

Critics also love to make titillating hay out of the fact that many of the memories involve lurid interspecies sex acts or violations of the abductees' private parts. You have to admit that these details do give it a pulp fiction edge.

If numerous people who had never been exposed to the alien imagery pervading our culture or the abduction myth itself had independently remembered the same experiences in detail, then it would point to something much more interesting, although not necessarily extraterrestrials. Mack claims to have demonstrated just such a multiple independent origin for his patients' stories. Yet critics point out that he uses unorthodox therapeutic methods such as hypnotic regression and becomes highly involved in evoking and even "cocreating" the memories.

One thing is for sure: John is not a fraud or a huckster. His decision to go public with these beliefs must have been difficult, because he surely knew that a lot of his old friends would think he'd gone completely off the deep end. Folks who don't know him and don't accept the legitimacy of the abductions might assume that he is after money and fame. In reality, he has made major sacrifices, I think because his curiosity and conscience demand it.

Although he must have known what he was getting into, the wrath of the skeptics has sometimes been intense and personal, as if John has been a traitor to science. James Gleick, writing in *The New Republic*, wondered if "the whole thing isn't just a calculated scam." He described Mack as a "mark" and a "gull" and his behavior as "sleazy," "slippery," and even "sickeningly corrupt." Attacking the messenger provides a convenient solution to the puzzling juxtaposition of impressive scientific credentials and wacky beliefs.

Maybe Mack has lost his grip on our particular consensus reality. Maybe he has adopted a worldview that doesn't make any sense to you and me (or at least to James Gleick and me). But he is not stupid or sleazy. I know him to be a smart, competent, and genuinely compassionate person. He is not out to make a quick buck. He really believes.

MIND MATTERS

To explain the abduction phenomenon, John has synthesized a worldview based on very different premises from those at the core of our scientific worldview. I think the essential difference comes down to different assumptions about the relationship between mind and matter.

For Enlightenment-guided science it was quite the revelation that our mental states arise from processes described by physics, chemistry, and cellular biology.

Thought is chemistry? Wow!

This has encouraged us to think of consciousness as something that arises out of a preexisting physical universe. We believe that matter exists with or without mind.

On the other side of the great divide lies a different assumption: that mind is primary and that the entire realm of the physical is somehow created by consciousness. This belief underlies a lot of New Age thought. If you don't like it, then try to prove it wrong. I dare you.

Mack's interpretation of the abduction phenomenon is thick with talk of invisible realms and other-dimensional realities. He refuses to be pinned down on whether these domains are "real," focusing instead on the inadequacy of that term.

"What about the spaceships?" you ask. "If these aliens are just materializing from another dimension, why do they need to travel in ships? Do you believe the ships are physically real, or are they just a

The world inside our heads. (Borin Van Loon)

metaphor?" If you ask such questions, be prepared to be told that the difference between "real" and "metaphorical" is not necessarily absolute, since our mythical imaginations collectively help cocreate our reality.

Science can do nothing to disprove the existence of subtle realms that are *defined* as realms science can do nothing to disprove. Science can only declare that it has no use for them. If you start with the assumption that consciousness is primary, then the truths that science holds to be self-evident are no longer so. So there is a choice here. Do we have any data to assist us in choosing the correct worldview, or should we flip a coin?

I am reminded of my experiences in the San Luis Valley that convinced me that different people can live in the same place and somehow see very different skies above them, depending on what they expect to observe. Either these objects are "really there" and some people somehow can't see them, or they are not "really there" and some people are able to enter mental states where they believe they see them, and where desires can mold memories. Either possibility is unsettling.

Remember what I learned about the secret of contacting UFOs: "You have to invite them." They won't show themselves for just anybody. They know if you've been naughty or nice, or at least if you are radiating the right vibes. This is not very different from "the magic will only work if you believe." Scientists have a visceral negative reaction against such participatory magic: It seems like cheating. We suspect fraud. It's not fair if the universe is somehow playing with our heads—it's supposed to just lie there, ignore us, and let us figure it out.

When you probe,* you find that many UFO believers are not bothered by physical arguments against the plausibility of sighting or encounter reports, because we cocreate our reality, and those objections are too narrow-minded and Western. They suspect that our thoughts and feelings can have cosmic consequences and that certain humans are of great interest to extraterrestrials.

The idea that mind is behind everything is not limited or original to UFO groups. Personally, I have no problem with consciousness being a major part of the architecture, the story, and perhaps even the point of the universe. Much philosophy, religion, physics, and cosmology finds an essential role for mind. In fact, Buddhism, the religion that makes

---

*Probe in this context meaning to simply ask questions and listen . . .

the most sense to me of all the majors, seems to hold the view that, fundamentally, it's all in our heads.

Although I've never found a religion that seems like a perfect fit, I love what I know of the teachings of Buddhism. Its most important principle seems to be compassion. If there is a perfect spiritual principle, I would vote for this. Then there is the "self is an illusion" part, the "death is an illusion" part, and the "oneness with the entire universe" part. They're speaking my language. These all seem true and important, and entirely consistent with the picture of the universe that science paints for me.*

Buddhist texts urge compassion, not just for all humans, but for "all sentient beings." The idea of nonhuman intelligent creatures is right there in the language they use. For a while I've been trying to figure out the Buddhist view of extraterrestrial life. The answer I most often get is "Of course there are many other worlds with many other sentient beings. We've always assumed that." And sometimes they add, "But so what?"

During one writing retreat in the San Luis Valley, I met a number of people who were in Crestone for a large Buddhist meditation retreat with Tsoknyi Rinpoche, a visiting Tibetan master.† One of the participants told me that the question of extraterrestrial life was entirely meaningless and uninteresting. Everything, he said with complete confidence, is the creation of mind, so the idea that stars are many light-years away means nothing. Other beings are right here, intermingling in our space. Life on other planets should be of no concern to us, because there is no difference at all between out there and down here.

It was an interesting concept and it made me think. Later I wished I had thought faster. I would have asked him for the keys to the new BMW he had pulled up in, since he could have just materialized another one for himself.

Is the Buddhist belief in the centrality of consciousness the same as the New Age primacy of mind, or the idea that you could manifest real lights in the sky by achieving the right mental state? It's easy, in a place like Crestone, to lump Buddhism in with all those other New Age beliefs that are so popular. Except, of course, Buddhism is hardly new.

---

*Another thing Buddhism has going in its favor is that if you don't believe and don't follow all instructions, you still don't go to hell or anything. At worst you just come back as a cockroach or a lawyer for the NRA, and you get another chance.
†Richard Gere was even there, I swear to God.

Another part of Buddhism that I love is the idea of ego death, or ego denial: seeing your individual self as an ephemeral, illusory whirlpool in a larger sea of consciousness. A lot of New Age belief seems to be about heightening the importance of one's own mind in the cosmic scheme. I'm not sure this jibes well with ego death. There is a difference between "consciousness helps create reality" and "the universe is all in my head." There is a difference between "it's all mind" and "it's all in mine." There is more ego serving than ego death in the notion that UFOs are *interested in me personally.*

Obviously, we do not fully understand the role of consciousness in the universe, nor the limits of its influence over "purely physical" phenomena. But I don't believe that the distances to the stars are immaterial because the stars are just in our eyes. I do think that there is a real, solid world existing outside all of our heads. I believe in a universe to be discovered, explored, and explained, that the stars really are light-years away and that this distance could present an immigration barrier for some kinds of beings, but not for others.

Yet, even though I believe in an external, material reality, that doesn't mean that our perceptions and conceptions will ever be more than glimpses. Just think about the complexity of the simple act of observing a light in the sky—the optics, the electromagnetics, the neurophysiology, the cognitive processes. Even a "direct observation" is far from direct. It is not so surprising that two people of different philosophies would see different things in the same sky.

We are always filtering sensory input. We screen out *most* of the information coming from our environment most of the time. Otherwise our circuits would be on constant overload. We are always interpreting and perceiving events through filters shaped by experience and belief. Even if you "saw it with your own eyes," it might not be the same thing I would have seen with mine. This is equally true for things that you feel with your own hands. It's still electrons, nerve firings, neural networks, and pattern recognition all the way down. No observation or perception is completely accurate or completely captures the essence of any phenomenon. So, yes, to some unknown extent, for all of us, believing is seeing.

This boils down to questions that are quite different from a simple disagreement over whether saucers or lights are in the sky, or whether memories of strange encounters are real. It's about the relationship between matter and consciousness, which I find much more interesting than grade-B saucer-shaped UFOs powered with some kind of super-duper quantum antigraviton engines.

Skeptics bristle angrily at the suggestion that our materialist, scientifically sanctioned, "Western" view of reality is too rigid—that there may be other realms and phenomena that are "real" but do not answer to our objective standards of proof. If we cannot agree on a basic framework of reality, then how can we hope to agree on a common interpretation, or even common facts, of our history, or any moral and ethical code? This, they say, is the dangerous path toward faith healing, Holocaust denial, and *Heaven's Gate.*

The idea that anyone's version of reality is as valid as anyone else's can lead us to some scary territory. I don't like the idea one bit. However, in science we are taught not to reject ideas just because we don't like them.

John Mack is definitely not a Holocaust denier. And he's not crazy. If you spoke with him about anything other than the abduction phenomenon, you would think you were talking to a bright, articulate, thoughtful, knowledgeable person, and you'd be right. I've had great conversations with John about science, philosophy, and spirituality. I find that we agree on many things. But we reach an impasse when we get to the nature of the entities encountered by his patients. He believes that someone, be they extraterrestrial or extradimensional, is actually interacting and communicating with the experiencers.

I just cannot accept that these beings and experiences are real, so I search for rational hooks on which to hang this conviction. I grasp at Occam's razor: "There are simpler explanations." It is easier to believe that many people have similar, vivid, and disturbing hallucinations than that our entire conception of reality is flawed. But this doesn't prove anything, and as I discussed in chapter 7, you can always ask, "Who elected Occam?" The doctrine of simplicity seems true to scientists but nobody knows why.

Still, I don't buy the reality of the abduction phenomenon. Why? In short, because it feels all wrong. I reject it because it does not fit my worldview. This is the best I can do.

I like to think it is, at least, an informed judgment—informed by the fact that science, making the assumptions it does, has been so successful. And I don't mean making all these toys, conquering diseases, or even touching the Moon, all of which are pretty cool. I mean that in science, we've found a framework of ideas that works so well in describing so much of the universe we see, in a coherent, self-consistent, and profound way. To me that suggests that all of those things we don't admit we are assuming, or taking on faith, are really true. The success of science is a confirmation of our faith in an objectively describable material reality.

John doesn't seem to care too much if these guys came from another star or if they just materialized from another dimension that is somehow interwoven with ours. This is somewhat infuriating to those of us for whom "Where did you come from?" would be among the top few questions we'd ask an alien. However, as I've already admitted, it's wrong to say that science can rule out anything when it comes to the capabilities and motivations of other sentient creatures. So there is a door of credulity, if you choose to open it or just to float right through, that admits all kinds of creatures and experiences.

I don't want to think about that too much, though, because what would happen if I started to believe John Mack? How would I ever explain it to my parents?

PRINCE OF OUR DISORDER

My father regrets not being closer to his old friend John and finds it hard to understand the direction he's taken. To the extent that Lester feels he understands the UFO abduction phenomenon, he sees it in religious terms: here is a new world religion, possibly a major one, beginning—as they all do—among a small, committed band of outcasts. In fact, when John first told him of his interest in UFO abductions, Lester thought John meant that he wanted to study it as a psychosocial phenomenon and an emerging belief system. He told John that it sounded like an interesting project.

Most believers would agree that theirs is a religious experience, but they would obviously disagree with my father on its nature and importance. John's Pulitzer Prize–winning psychohistory of Lawrence of Arabia was entitled *A Prince of Our Disorder*. It occurs to me that this title describes John himself as he is today. What we don't all agree on is whether the disorder is in our heads, in Western society, or in John's worldview. Am I allowed to check "all of the above"?

There is truth in the descriptions of humanity's self-inflicted predicament heard in the communiqués coming through the experiencers. Whatever mind generates these stories is not entirely ignorant of the human condition. These aliens, if they really existed, would be right to warn us that we are acting in ways that threaten our future. Guilty as charged. I even agree with the impulse that we need a global spiritual response to this situation, to learn to change our behavior species-wide.

These aliens sure are perceptive little buggers. I don't believe that there really are ships and little guys. I think those messages of global concern are coming from John and the experiencers themselves. These are warnings we need to hear, but why do we need alien intermediaries? As Priscilla said in *The Courtship of Miles Standish,* "Speak for yourself, John!"

23 | The Immortals

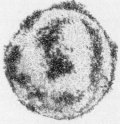

Oh yes, there is hope—infinite hope. But not for us.
—FRANZ KAFKA

Do not fear the universe!
The ghosts of the alien
dead will beckon thee
into the future!
Put on your spacesuits, and
dance; dance into the future . . .
—SIFL & OLLY

WHEN WE ARE THE ALIENS

Stuck here in our tiny little patch of time, we do what we can to unearth the past and peer into the future. Stranded for now on our little patch of universe, we do our best to understand the rest of it. Can we predict the behavior of alien civilizations? Many people have tried, using projections of our own future, often based on lessons from human history.

The cultural shock experienced on both sides when European explorers encountered less technically advanced societies is used as a cautionary tale to dampen our enthusiasm about alien contact. This warning is often refuted with the argument that the aliens could not possibly be as savage as we, or they would not have survived to become an advanced technical civilization. Others insist that only an aggressive society will survive the evolutionary process, and therefore aliens must be ruthless colonizers.

Looking back over a much longer timescale, we've seen how narrow interpretations of life's history on Earth can lead to questionable conclusions about life in the galaxy. The flawed Rare Earth Hypothesis (chapter 9) shows how easy it is to read too much into one planet's history.

Now, turn 180 degrees (on an axis perpendicular to time) and look into the future. What do you make out in your Earth-future crystal ball? Even if the picture were clear, would that really tell us about the fate of intelligence elsewhere?

We cannot help but see a close connection between L, the average lifetime of civilizations, and our expectations for the survival of our own species. Many discussions of L in the literature explicitly tie the two together, concluding that if we are likely to destroy ourselves within the next century, then L must be short. Yet, our chances of survival may be irrelevant to the value of L in the universe. It's understandable that we try to use our own projected future as data, since one fuzzy data point is better than none. Except, in this case, I think we commit a serious error when we assume that we are likely to be typical of worlds that might become long-lived broadcasting civilizations. We are much more likely to be highly atypical.

In biological systems, the most common outcomes are often not the most important ones. Consider the hundreds or thousands of sea-horse ponies born in every litter. For any individual, the overwhelmingly likely outcome is to quickly become fish food. But it's the one-in-a-thousand survivors that matter. At this point, we are one of the ponies. If we base our views of intelligence elsewhere on expectations for our own future, then we are committing a logical fallacy and perhaps selling the universe short.

ARE WE INTELLIGENT?

Well, I don't know about *you,* but . . . I can count to 1,023 on my fingers in binary, or 1,048,575 if I also use my toes, which makes me at least as good as a primitive microchip, albeit much slower. Seriously, though, are we humans an intelligent species? When we hunt for intelligence elsewhere in the galaxy, are we looking out there for something that already exists down here? What, exactly, are we seeking?

Throughout this book, I have weaseled around defining "intelligence" because it's like trying to define "life": though we think we know what it is, we can't quite put it into words.

When we debate how long it will take for intelligence to arise on some random planet, we always start with the history of Earth and deduce that it typically takes 4.5 billion years. We use ourselves as the benchmark and assume that intelligence, on a level that is relevant for discussions of Cosmic Evolution, is something that has now arrived on Earth, in human form.

But, what is it? We can come up with some properties and abilities of intelligence, even while we cannot come up with a perfect definition that everyone, even on this little planet, will accept. If we were going to list some of these qualities in a Lonely Planets personal ad, it might read:

> LONELY: SBF* species seeks a special friend. You must be capable of abstract thought and symbolic language, must have the ability to learn from experience, to pass on learned knowledge by teaching others, to purposefully modify your environment, to anticipate the future and act accordingly, to make tools, dream, and play the drums. If this sounds like you, let's meet for conversation and possibly something more.

Who on Earth could answer such an ad? Some other animals have some of these abilities, but none have them all (dolphins can't play the drums).

More pragmatic definitions have been proposed for specific scientific tasks. Among those who study artificial intelligence, a common definition is this: a machine is intelligent if it can pass the "Turing test," named after the tortured British mathematical genius Alan Turing, who first devised it in 1950.† By this definition, if a machine can mimic human intelligence so accurately that we can't tell the difference, then it is, in fact, intelligent. This is reminiscent of the empirical definition of life mentioned in chapter 7: "We'll know it when we see it."‡

---

*Sentient but fragile.
†Turing invented computer science, excelled at mathematics and philosophy, and was an early innovator in what is now called "chaos theory" and "complexity science." Two years after publishing his landmark paper discussing artificial intelligence and introducing the "Turing Test," he was arrested in Manchester for homosexuality. At his trial, he proudly refused to deny that he was gay—a courageous and dangerous stance in the 1950s. His sentence included hormone injections that were supposed to "cure" him. Two years later, he committed suicide.
‡Also similar to "community standards" definitions of obscenity.

SETI theorists use their own narrow and practical definition of intelligence. They usually avoid the tricky subtleties by simply defining intelligence as the ability to build a radio telescope. That this definition was conceived by radio astronomers is always good for a few yuks. (A woodwind player might define intelligence as the ability to play modal scales on the clarinet.) The advantage of this simple definition is that it relates intelligence explicitly to an observable phenomenon, thereby making it accessible to scientific experimentation. It tells us exactly what to look for.

The "radio definition" of intelligence implicitly includes humans among the galactic intelligentsia. However, when it comes to radio signals, we only listen, we don't send. So, even if the galaxy were overrun with "civilizations" manifesting our level of "intelligence," there would be no messages on the air. The Fermis of countless worlds could all be asking, "Where are they?"

There is a large asymmetry in galactic radio discourse. It's much easier to listen in than it is to broadcast. But there is a kind of broadcasting that someone out there *must* be doing for SETI to succeed, and *we can't do it yet*. What is required is not just a level of technology or transmitter power, but a long-term commitment. If you do the math (with the Drake Equation), you find that for SETI to be viable, for us to have a reasonable chance of finding a signal, there must be many civilizations broadcasting continuously for thousands of years.* We are not even close to being able to become one of these serious broadcasters.

We've sent out some symbolic broadcasts—scribbled a few simple messages and tossed them out there in leaky electromagnetic bottles. We've attached notes to our four spacecraft (so far) that are leaving the solar system, just in case they wash up somewhere. And of course, if anyone is really hunting for the likes of us, our presence is not a well-kept secret: for decades, we have been leaking our sitcoms, talk shows, and ebullient commercials for Jesus, minivans, and beer.† A spherical shell of radio sig-

---

*In the lingo of the Drake Equation, unless L, the average lifetime of a communicating civilization, is many thousands of years, then N, the number of communicating civilizations, is so low that no one is around for thousands of light-years.

†When aliens come to Earth, if their preconceived notions of us are conditioned by TV broadcasts, their first report home might read, "The females have smaller mammary glands and the males less well developed musculature than we expected. There are a greater variety of types of humans than we thought based on their messages to us, but no sign of other humanoid species. We can find no trace of the ones with the funny forehead ridges."

nals is expanding outward from Earth, its diameter increasing at twice the speed of light. As I write, this sphere forms a ball of news, entertainment, psychobabble, and advertising 166 light-years in diameter.* In the time it took you to read this sentence, it grew by another million miles. The nearest stars are only about four light-years away.

These indiscretions might have tipped off some of our closest neighbors that something is up on the third stone from the Sun. Would they conclude from these transmissions that we are intelligent, or merely that some nutcases have stumbled upon primitive radio technology?

When we focus on the technical aspect of how to become a "radio communicating" civilization, there are no great hurdles. We can make bigger dishes, increase our listening sensitivity, and broadcast with ever-increasing power. From this perspective, we seem almost there. We calculate and speculate about finding others that are slightly spiffed-up versions of ourselves and take it as an article of faith that such a stage will arise soon after the one that we are in now.†

But it takes more than technology to be a broadcasting society. It requires that you survive with high technology for many thousands of years and commit to projects that last, at the minimum, for millennia. Your standard SETI aliens have science and technology that are similar to ours, but they must have solved many of the great social, political, and spiritual problems we now face. The abilities that will enable a species to participate in interstellar communication may be part of a qualitatively different phenomenon than what we self-referentially (and self-aggrandizingly) call intelligence.

Our discussions of L have an implicit focus on how long an intelligence "like ours" might last. We tend to ask, what is the distance to the nearest planet with someone like us? But we should also be asking, what else might intelligence become? What can it grow into that it hasn't yet become here (and may not), and how long might *that* last?

CONFEDERACY OF DUNCES

As Doris Lessing wrote in the first volume of her autobiography, "Forgive me for the banality of this observation, but there is something

---

*The first commercial radio stations with regularly scheduled broadcasts were heard in 1920. Television broadcasts started in 1947.
†Shklovskii's "adolescent optimism."

very wrong with the human race." We might refer to what we humans have achieved so far as "proto-intelligence." Let me briefly summarize what it is about us that seems so strictly proto.

Think of the characteristics listed above in the personal ad for an intelligent soul mate: the ability to learn from mistakes, anticipate dangers, to think your way out of a paper bag, and so on. We humans certainly do have these abilities as individuals, at least on our better days. Individuals can alter their environment and change their behavior to aid in survival. Occasionally, communities of humans manifest these qualities as well.

But consider the behavior of the human race collectively. We are dumber in numbers. As a species, as a global entity, we aren't able to respond to information and make intelligent decisions. An individual, behaving as we do, would seem dumber than a dodo (and look what happened to them). With human intellect, the whole seems to be less than the sum of the parts.*

When we think of aliens sending an interstellar radio signal, we usually picture them as representing their entire species, and imagine them attempting to communicate with humanity as a whole. This makes sense, given that the likely timescale of any conversation, where each reply might take centuries, requires a group effort. Individuals cannot talk to the aliens by radio. For our species to achieve the level of maturity that allows for—indeed that may be defined by—interstellar travel or communication, we'll have to learn to act collectively.

Without such an ability, we are vulnerable to many extinction threats, including several of our own making. I worry most about these "unnatural disasters." In our headlong, blind rush toward new technology, we may be cooking up dangers to ourselves that will leave us nostalgic for the quaint threat of nuclear self-annihilation.

TOUCH OF GRAY

In April 2000, Bill Joy, the cofounder and chief scientist at Sun Microsystems, published a powerful article in *Wired* about the growing threat of what he calls GNR technologies (genetics, nanotechnology, and robotics). Joy described a scenario that is frightening, in large part because it seems quite credible. He looks at nanotechnology (the ability

---

*I guess this is the opposite of an emergent phenomenon. Or is it "emergent stupidity"?

to build submicroscopic machines by manipulating matter on molecular and atomic levels), combined with genetic engineering and increasing computer speed and miniaturization, and asks, "Where is all this going?" He concludes that we may soon have the ability to design self-reproducing agents with unprecedented power to remake our planet. What Joy is most concerned about is "the power of destructive self-replication."

One nightmare scenario is commonly referred to by the nanotechnology elite as "the gray goo problem." If a genetically engineered, robotically enhanced nanobacteria that can outcompete naturally evolved microorganisms escapes from a lab, it could replicate like mad and destroy the entire biosphere, turning our pale blue dot into a gray goo glob.

Bill Joy has worked all his life to create better software and microprocessors, believing that his work was helping to create a wonderful future for all of humanity. Only recently has it occurred to him that he may have been helping to build the tools of human extinction. He points out that unlike twentieth-century weapons of mass destruction, which generally require rare materials, highly specialized training, or large institutions to construct, the new GNR weapons of mass destruction might soon be easily created by any individual with a little bit of technical knowledge. The march of new technology is moving in a direction that may empower individuals to do massive harm.

Undeniably, these same technologies also carry great potential for problem solving and liberation from hunger and suffering. I wish I had more confidence that the institutions implementing and watching over genetic engineering and nanotechnology experiments were doing so with the right values and proper acknowledgment of the risks.

Let's accept for the sake of argument my paraphrasing of Bill Joy's point: that soon any damn fool who was pissed off might be able to destroy the world. In a world where a couple of teenage video-game geeks decide to blow away their classmates, where people drop "smart bombs" on cities and crash planes into buildings, it's impossible to believe that nobody would do it.*

Do we have a plan for dealing with these huge new threats to our survival? Of course not. The human race has no plans. Individuals and nations do, but not proto-intelligent humanity. Our lack of collective

---

*Say, could you pass the Xanax?

identity becomes more of a threat to our survival as certain kinds of technology become more advanced. No one would deny our cleverness, but can we find wisdom, and how long can we last with one and not the other?

As I described in chapter 19, some people have argued that the "self-destruction hypothesis" is the solution to Fermi's Paradox. Where are they? Simple: they're all dead, having designed powerful technology without the wisdom to control it. Maybe it's just natural selection on a cosmic scale. All those other living worlds out there that don't produce idiots savants like us might end up surviving, thus ensuring that the self-destructive races are not the ones who inherit the galaxy.

Perhaps our kind of proto-intelligence truly is an unstable development and does not usually survive. But, even if it happens rarely, I believe that *sometimes* proto-intelligence evolves into something else: true intelligence.

TRUE INTELLIGENCE

When a reporter asked Gandhi what he thought of Western civilization, he replied, "I think it would be a wonderful idea." I would say the same about human intelligence.

If we are not intelligent, then what is? Here's my crack at a definition: A truly intelligent species must have the ability to behave, collectively, in ways that ensure long-term survival. It must have learned to avoid self-destruction, anticipate and avoid natural disasters, intentionally and thoughtfully alter its environment and live sustainably within it.

What we have seems pretty special, just as Earth once seemed central. But our consciousness may be just a faint spark, an inchoate stirring of what may someday, somewhere, lead to true intelligence. After all, why should this be the pinnacle? Given our complete inexperience on the cosmic stage, I would argue that what we have is most likely just some vague foreshadowing of what would be called true consciousness by the cosmiscenti. We are not the center of the universe, and our level of awareness is not the apotheosis of evolution.

If you consider the continuum of increasing consciousness—say from a rock to a cabbage to a rabbit to an orangutan to you—why should we assume that it stops with us? Is it so hard to imagine that there are higher levels on this path and that, for all we know, on the whole spectrum, we are more like cabbages than kings?

In chapters 5 through 9 I recounted the major turning points in Cosmic Evolution. Each new stage has involved changing the mechanisms by which matter has organized itself into ever more complex and stable gatherings: protons, atoms, and molecules. Then a molecular dance marathon in which the last ones left standing won the prize of survival. Then some learned how to copy themselves and the rest was heredity. We went cellular, became organisms, and the whole game since then has been getting together in new kinds of groups (cells, colonies, individuals, communities) and repeatedly fusing identity.

At some time in the last million years, some of us cell throngs started to talk to one another and make big plans.

We have no good reason to believe that this history of forming new groupings, each with unforeseen capabilities and a new sense of collective self, has to stop with individual organisms. Is there a way to learn a group sense of identity strong enough to manifest collective intelligence, without becoming Nazis or Stalinists or the Borg? Can we maintain the individual freedom that makes life worth living while gaining a new kind of freedom: the collective cognitive abilities that will allow us to survive?*

At present there are some faint glimmerings of a collective, planetary consciousness growing on Earth. We see them in the bright side of the World Wide Web, in the growth of SETI@home, in some global nongovernmental organizations, in the pulsing rhythms of world music that are collectively evolving into something new and wonderful as they echo around the planet, and in the global, transnational perspective brought back from space by astronauts and cosmonauts.

Perhaps, like life, true intelligence will not be a trait of individual organisms, but something new that will happen to a planet as a whole. Recall the noosphere of Tielhard de Chardin and Vladimir Vernadsky—the planetary development of a thinking realm, a zone of intellectual activity that arose out of the biosphere to become its organ of consciousness. The new-o-sphere. What might it become?

Perhaps here it won't become much of anything. Nietzsche said that

---

*The current debate about globalization needs to be seen in the context of these questions. There is no sense in being "antiglobalization." Globalization must happen if we are to survive long term with high technology. The question is how we do it and what values dominate. Will globalization simply empower massive corporations to control Earth's resources in a short-term orgy of profit, the future be damned? Or will humanistic and multigenerational values prevail?

the human race was an intermediate step between ape and superman. It's interesting to think of *Homo "sapiens"* as the missing link between apes and truly intelligent creatures, between clever animals and wise beings.

How long could such a species survive? How long have you got? I believe that a civilization that achieves true intelligence can survive for the rest of time. As Shklovskii said at Byurakan in 1964, "There is, however, a possibility that some civilizations, having reached a highly advanced level, will find themselves past the inevitable crises and internal contradictions which plague the younger civilizations. The evolutionary timescale of these quiescent civilizations may be considerably larger, approaching the cosmogonic scale."*

"Approaching the cosmogonic scale" means approaching the age of the universe. For the purposes of the present discussion, true intelligence has, by definition, achieved effective immortality.

Humanity's semismart transition state cannot last. This fragile, proto-intelligent phase may be one that most species don't make it through. An immature noosphere like ours could be a risk that a biosphere must endure, hoping to come out newly empowered. It's a high-stakes gamble. Either we get snuffed rather quickly, or we emerge immortal. A great valley of stability is out there, but the journey is fraught with dangers, and we don't have a map.

In this view we are near either the beginning or the end of the human adventure. I believe that if we survive a tight bottleneck we have now entered, we will emerge as one of the immortals. I don't know if this bottleneck lasts a half century or a millennium, but either way it is a trivial interval compared to life's long history. Certainly in a few thousand years—a "blink and you missed it" moment in Cosmic Evolution—we will be through it one way or another.

It seems almost inevitable that other sentient beings will have in common with us this "race between education and catastrophe."† Will some learn to use technology in a way that ensures survival rather than destruction? That will determine whether the universe is lively or lonely. If even a small fraction choose life, then life will still dominate. This is a good reason to have some hope for the universe (if not for us).

---

*This was three decades before he dismissed the early rosy views that he once held, along with his coauthor Sagan and most of the SETI community, as "adolescent optimism."
†Quoting H. G. Wells.

We don't know the odds, but this is the game we're in. The problem of survival is not fundamentally technological. It is spiritual and moral. It is evolutionary. Technical solutions may provide temporary Band-Aids, but they do not save us from our nature. If we want to be one of the survivors, we must create a global society where curiosity is tightly bonded to compassion, and where (this is hardest to picture) not a lot of people want to do violence to others. You're probably not going to like this next thought, but one solution would be to just surrender to the machines.

WELCOME TO THE MACHINE

Arthur C. Clarke once speculated that "all really high intelligences will be machines. Unless they're beyond the machine. But biological intelligence is a lower form of intelligence, almost inevitably. We're in an early stage in the evolution of intelligence, but a late stage in the evolution of life. Real intelligence won't be living."

Many futurologists have predicted that we will evolve into machine-human hybrids with our consciousness intact or even enhanced. Some feel that this transition might come within the next century. Variations on this theme range from Bill Joy–type doomsday scenarios to utopian visions of "uploading" our memories and thoughts into an immortal, pain-free machine state and building for ourselves any bodies we choose. The biological stage may be a mere precursor to what technologist Ray Kurzweil calls "the age of spiritual machines."

In many ways, we already are human-machine hybrids. I sit thwacking away at my computer all day, my thought processes, memories, and communications increasingly dependent on it. While I work, I am often connected to several different computers in different cities. Eventually, I leave work and head home (don't worry, there are computers there, too). As I turn into my driveway, I'm simultaneously cranking the steering wheel, stepping on the brake, hanging up the phone, tuning the stereo, and pushing the garage door opener. Every evening I do all this as effortlessly as Homer at the beginning of each episode of *The Simpsons*. We've constructed an elaborate high-tech matrix within which we are merely the organic, semi-intelligent component. Already, machines are us, and we are them. Goo goo ga joob.

If we receive an interstellar message we may never know if it was sent by machines or biological organisms. Perhaps it will come from sentient organisms who have evolved radio dishes for sensory and communicative organs and computers for memories and minds, as we ourselves may now be doing. Even if we meet the aliens in person, will we be able to tell if they are machine or organism? Will we be able to differentiate "individuals" from tightly knit, machine-enabled communities? These puzzles give us reason to question our cozy categories. Notice that when I've written about "the immortals," I haven't said whether I think it is civilizations, species, machines, or individuals who will evolve to live forever. I've intentionally blurred these lines because I think that for the immortals such distinctions may have become meaningless.

According to one theory, a kind of mineral life may have existed on Earth before carbon-based life. Now, carbon has so remade our world that if this former life ever existed, all vestiges of it have long since been erased. Will our silicon machines one day erase all vestiges of carbon life from their world?* Our carbon-based egos recoil in horror at the thought, but from the point of view of the machines, this may only be the beginning of something magnificent that we can scarcely envision. Rejecting the value and sanctity of machine sentience may someday be regarded as just another form of ignorance, racism, or bigotry. The new machine-human hybrids may keep us on for a time as useful organs in their silicon structures. Then, someday, they may leave our fragile, ephemeral bodies behind altogether and take to the stars. Five billion years hence, as Earth is roasted dry by our bloated, red, dying star, our descendants may briefly pause to remember us as they ride off out of the sunset, seeking other green worlds or the company of like-minded spiritual machines.

IMMORTAL FOR A WHILE

The idea of immortality, I've noticed, is troubling. People are quick to reject the notion. A group of academics at a "philosophy of astrobiology" discussion group at the University of Colorado once gave me a hard time about this. The topic was SETI. When I argued that we should consider a definition of intelligence qualitatively different from our own—one that might indeed be immortal—they practically shouted me

---

*This could be the ultimate Microsoft marketing strategy!

down. They only wanted to discuss intelligence as a variation on what we have on Earth. What good is it to talk of hypotheticals?*

To me, it seems inevitable that in our vast universe at least a tiny fraction of species will escape self-destruction, attain great understanding of nature, and learn how to avoid natural disasters. The combination will make them immortal.

A common response is to point out that no species has ever attained this. Yet, the same could have been said to a cell contemplating animalhood a scant billion years ago. It had never been done.

When we reflexively dismiss immortality as a pipe dream, we are being unduly influenced by our limited experience, and the narrow species-level picture of evolution we were taught in high school. It's true that species come and go. But, forget about species. On the molecular level, the immortality of Earth's DNA is a fact of our existence. And what about the life of Gaia? Earth's biosphere may well last forever, or at least as long as the Sun keeps nuking along, and maybe longer if we play our cards right. Indeed we, the noosphere, might be the biosphere's ticket out of here, its vehicle for extending its lifetime *beyond* that of the Sun. If we think of ourselves as just another species, our odds don't seem great. But if we are the noosphere, why shouldn't we become immortal, like the biosphere that birthed us? And even if we blow it, wouldn't you think that some biosphere somewhere has produced an immortal noosphere?

If some fraction of sentient species might achieve immortality, then that changes everything. True intelligence is not an easy gang to join, but once you're in, you're in for life—the life of the universe. They're not going anywhere, so immortal species would just accumulate as the universe cooks along.

Why discuss a hypothetical like the immortals? Because their existence is a reasonable supposition when we drop the pretense that we are the supreme beings. And because if they do exist, it leads to a different picture of our universe and the cosmic role of conscious awareness.

Frank Drake has said that it is the immortals whom we are most likely to hear from with SETI. As fantastic as this sounds, the more you think about the timescales of Cosmic Evolution and the inconceivable

---

*One professional philosopher there told me I was using the wrong definition of *definition*. Whatever.

power of technology in the service of a spiritually advanced society, the idea of the immortals starts to make more sense.

However rare it may be, the birth of an immortal is, by definition, a one-way, irreversible transition. If immortality can be achieved, then the immortals must exist in ever-increasing numbers. I do not know what the average distance is between civilizations. No one here does. But I believe that it is *decreasing*.

The universe is progressing in a direction toward greater intelligence, conscious awareness, and self-understanding. The dark universe becomes gradually more lit up with consciousness. The wise ones, the immortals, are accumulating, or have already accumulated to a density that, in their wisdom, they deemed enough.

If we focus on our sorry-ass excuse for intelligence, rather than the potential for true intelligence, of which we've seen only hints, then we miss the forest for the seedlings and, perhaps, fail to see the true significance of our own existence.

To accept this picture, you do not have to assume that the transition to immortality is likely, or that *we* will ever achieve this state. You must only admit the possibility that immortality is occasionally the outcome of an evolutionary process to conclude that the universe must be heading in this direction. When I do the math, and plug in reasonable numbers for the unknown parameters, the answers I get convince me that this picture is correct.

We need a new equation if we want to calculate the number of immortal civilizations in the galaxy. Instead of the Drake equation, we'll use a *rate equation,* such as we would use to calculate the buildup of a chemical in an ongoing reaction. We can ask, "*At what rate* is the universe becoming sentient?" and try to model this process. To determine the rate, we need to estimate the probability that a planetary biosphere will achieve immortality within the lifetime of its star.

Once again, we are forced to estimate a quantity that we can't really know anything about. But, as with the Drake Equation, we can explore the consequences of different assumptions. We can ask, quantitatively, given a certain average likelihood of achieving immortality, what are the implications for the nature of our universe, and the role of consciousness within it?

Our new rate equation will somewhat resemble the Drake Equation and will include most of the same variables. But there will be no L, because we are talking about true civilizations, which have no finite lifetime.* There will also be a new "f factor," f_{IC}, representing the fraction of civilizations that make the transition to immortality.† Recall that the Drake Equation looks like this:‡

$$N = R \times f_p \times n_e \times f_l \times f_i \times f_c \times L$$

Our new rate equation looks like this:

$$R_{IC} = R \times f_p \times n_e \times f_l \times f_i \times f_{IC}$$

R_{IC} is the rate of formation of new *immortal* *civilizations* (per year) in the galaxy.

For all of the factors, we'll use the most pessimistic numbers that the Order of the Dolphin used at Green Bank. Then, if we assume that $f_{IC} = .001$, meaning that one in a thousand proto-intelligent species (like us) become immortal, then we calculate that in our galaxy, a new immortal civilization is born every two thousand years. If $f_{IC} = .00001$, meaning that only one in a hundred thousand of our kind makes it to immortality, then the galaxy gets a new immortal civilization every 200,000 years. Put more simply, if one in a hundred suns gets a planet with life, and if one in ten thousand of these develops true intelligence before their sun dies, then in the galaxy there would be a new immortal civilization born every million years. Using these numbers, we conclude that in the 10 billion years leading up to now, our galaxy would have given birth to between ten thousand and five million immortal civilizations.

This "back of the envelope" rate equation is actually quite conservative. It assumes that no one is messing with the equation. It assumes that none of the factors on the right side of the equals sign are changed when the number of true civilizations gets large. This is only true if the immortals are leaving the universe alone. If any immortals decide to sow the seeds of life, or to teach immortality to others, the factors in the equa-

---

*Or at least no finite lifetime shorter than the rest of time.
†We are not going to worry about a seemingly immortal civilization occasionally suffering some disaster and going extinct. This can be handled in our equation by slightly adjusting the value of f_{IC}.
‡All of the factors were defined on pages 295–301. From left to right they are rate of star formation, fraction of stars with planetary systems, number of habitable planets per system, fraction with life, fraction with intelligence, and fraction that communicate.

tion will change over time. Then the problem becomes highly nonlinear, and the above equation will not suffice. The more civilizations that exist, the faster new ones will be made. This results in a nonlinear equation, and the number of immortals will increase exponentially.*

The galaxy will become saturated with them, up to a density level they deem healthy. They may want their space and not crowd things too much. After all, they didn't get to be immortals by ignoring natural limits. Their cultures may have merged into a pangalactic civilization— the galactic club. When will we be allowed to apply for membership?

This might all sound like a flight of pure fancy, but it's not. It's a flight of fancy reinforced with math and logic. The weakest link is the assumption that any species ever evolves to a state where it has the ability to use its intelligence to survive. If this does happen, then the rest of this picture follows almost inevitably.

If the transition probability (to immortal status) is large enough, or becomes large enough, then the universe as a whole may experience a sudden, irreversible transition from barrenness to universal consciousness. If that is the case, then it is not a matter of a rate, but a wait. Then the question is, how long does it take for the fire of universal consciousness to spark and flare up around the cosmos?

ASTROENGINEERING

As I scanned the table of contents of volume 1, number 1, of the new *International Journal of Astrobiology,* published in January 2002, the title of the final paper reached out and caught my eye: "A Search for 'Frozen Optical Messages' from Extraterrestrial Civilizations." Say what?

In this paper, Austrian astronomers Ronald Weinberger and Herbert Hartl reported on a twenty-five-year search for "unnatural looking" objects in the sky. They reasoned that an aged and powerful civilization

---

*Nerd alert: Do not read the following unless you are a certified nerd, and don't say I didn't warn you. Suppose the fraction of living worlds that develop immortal noospheres increases as the inverse of the nearest-neighbor distance to an immortal (the closer a good role model, teacher, or guardian angel). Then F_{IC} will be proportional to 1 over λ (where λ is the mean free path between immortals). So F_{IC} is proportional to the total number of civilizations. Therefore dN/dt is proportional to N, so the equation for the number of immortals as a function of time (t) looks like $N = Ce^{kt}$. They increase exponentially.

might create large, bright, or anomalously shaped artifacts visible at great distances, possibly even across the entire universe. They might even move stars around or alter the form of galaxies. Aliens may undertake these astroengineering projects to satisfy their own internal needs—living space, energy, public art installations, or whatever it is they're into—or they could be making deliberate attempts to communicate with faraway creatures by creating highly visible structures that are clearly not natural. For example, if we saw a grouping of identical stars orbiting around some central massive object, with equal spacings between the stars, we might conclude that this was a massive work of engineering, rather than a random configuration.

Such "frozen optical messages" would avoid some of the disadvantages of radio messages. For one, the message would stay put until someone came along to recognize it.* Maybe this cosmic graffiti tagging would just be meant to say, "Yes, we're here."

The idea that we might discover alien civilizations by searching for the signs of astroengineering was suggested in the 1930s by the British cosmic philosopher and science fiction writer Olaf Stapledon. It has subsequently been explored by numerous visionaries, including Freeman Dyson and Arthur C. Clarke. Just as human consciousness, whether we like it or not, is remaking the Earth, so cosmic consciousness may one day remake the stars and galaxies.

When I read these papers about searching for astroengineering projects, it is hard not to chuckle. Recently I found myself almost reflexively making light of the Weinberger and Hartl paper about searching for frozen optical messages. I had left the inaugural issue of *International Journal of Astrobiology* out on a table in our institute library. Perhaps protecting my reputation from being tainted by my association with astrobiology, I heard myself making some disparaging, eye-rolling remark about this work to a colleague.

Afterward I realized that I had not been honest about my beliefs. It reminded me of a couple of times back in sixth grade when, hungry for peer approval, I expressed popular opinions that I did not really hold or made fun of some "dorky" kid who I did not really think was dorky, in hopes of avoiding the dreaded dorky label for myself, and later felt like a jerk about it.

---

*Also an advantage of biological messages, as discussed in chapter 19.

Why do we laugh? Percival Lowell's wish-fulfillment "discoveries," which embarrassed planetary astronomy, were of giant engineering works on Mars. Searching directly for alien artifacts, whether they be probes (Von Neumann machines) in our own solar system or distant, giant works of astroengineering, has a logical justification that arguably is no more or less secure than the rationale for radio searches, yet it has a much higher yuk factor (or a more impenetrable ridicule barrier). Acknowledging our technological infancy, how can we deny that we might, by looking long and far, see the handiwork of the immortals somewhere out among the galaxies and stars?

The Weinberger and Hartl paper, describing the results of their twenty-five-year search of the sky for astronomical objects that might be artificial in origin, concludes, "A number of very peculiar objects were indeed found, but none of these appeared likely to be the product of alien masterminds. We may conclude that at least within about 10,000–20,000 light-years around the Solar System no highly advanced extraterrestrial civilizations intend to reveal themselves through such objects."

HELP!

Can we communicate with the immortals? Can an amoeba communicate with us? It can give us dysentery. If the immortals do take our communicative desires and abilities seriously enough to tell us anything, it will not be a conversation between two equals. When you read the speculative literature—in science fiction and scholarly discussions—about the possible impact on humanity of receiving a genuine alien message, opinions are all over the map. Some believe it would be our salvation, the best thing that could ever happen to us, our induction into a new, higher state of existence. Others believe that it would spell disaster.

One of the benefits of an alien message might be to help foster a collective terrestrial identity through the knowledge of "the other." The idea that the immortals might help fledgling races get through the technological juvenile-delinquent stage has been explored in fiction, notably in Arthur C. Clarke's *Childhood's End* and *2001: A Space Odyssey*. Frank Drake believes that the immortals will broadcast information that will help others to become immortal. Carl Sagan suggested that an alien message might contain information that would help us to avoid a nuclear holocaust.* A deep consideration of the possibilities of highly advanced technological societies has led some hard-core, no-nonsense, skeptical scientists to the quasi-religious conclusion that there are immortal, infinitely wise creatures out there, ready to deliver us from darkness.

---

*Knowing about Carl's concern that international conflict might doom the human race, and also his belief that an alien signal might help to unite humans, I once asked him if he had ever considered trying to fake an alien message, for the purpose of helping humanity. His answer surprised me. He said that, yes, he had considered it and decided that it was not a good idea because too many smart people could figure out that it was a hoax.

Astrotheology

On yet another world, intelligence had been born and was escaping from its planetary cradle. An ancient experiment was about to reach its climax.

—ARTHUR C. CLARKE, *2001*

searching for a lighthouse
in the breakwaters of our uncertainty,
an electronic murmur,
a bright, fragile *I am* . . .
—DIANE ACKERMAN, FROM
Jaguar of Sweet Laughter

You may say I'm a dreamer. . . .
—JOHN LENNON

SOUL SURVIVORS

Today many of us invest hope in extraterrestrials who may come and save us from ourselves. The idea of contact with advanced ETs is as close to a scientifically palatable miracle as anything we can imagine.

Might contact have a transformative effect on the human race, helping us enter a future in which we, too, become broadcasters and galactic travelers? This question drives both SETI and ufology, each of which has at times been described as a new religion in the making. Perhaps our belief in advanced extraterrestrials is both a proto-religion and a proto-science. Natural philosophy once mingled our religious and spiritual quests together with what we now think of as the scientific quest. Our common wonder and desire involving alien life creates a place, out on thin ice, where science and spirituality can meet, become reacquainted, and perhaps practice working together. The success of this marriage, I believe, will ultimately determine the longevity and fate of the human race.

Science may be a candle in the dark but it is also a lit fuse, and our future depends on an ability to grasp a truth that comes from somewhere beyond science: that if we don't do a much better job of loving one another here on this Earth, then we are going to miss the galactic party.

It really bothered me when Bob Dylan came out with that song "License to Kill," in which he sang:

> Man thinks 'cause he rules the Earth he can do with it as he please
> And if things don't change soon, he will.
> Oh, man has invented his doom,
> First step was touching the moon.

These lines put right in my face the dissonance between my dreams of liberation through space exploration and Bob's dystopian view of space travel.

Hadn't Dylan heard that the times are a-changin'? What about the utopian visions of Tsiolkovsky, which echoed through my youthful space activism?—the dream that the move into space will spark a new liberation of the human spirit, a natural continuation of our walk out of Africa and around the globe, an assumption of our true humanity, the inevitable next step in the rise of cosmic consciousness . . . Something is happening here, and you don't know what it is, do you, Mr. Bob?

Which do I believe? Tsiolkovsky's beautiful visions, or the view expressed by Dylan in that wretched song?* Will space technology be part of our beginning or our end?

One of the criticisms of our space visions has always been that they are escapist, that we need to solve our problems here on Earth First. Taken to an extreme, this attitude deadens all exploration and outward movement. I say, our movement toward our future in space must be

---

*I love that song.

part of a general, painfully slow expansion of human consciousness that will ultimately help us clean up Earth and get along down here.

Space technology has made possible powerful new forms of spiritual experience and communion. Apollo astronaut Russell (Rusty) Schweickart expressed better than anyone else the feeling of seeing the Earth from space. From his description you can feel what it was like to be up there floating in a tin can far above the world, with only a fishbowl between your head and the Earth rushing silently by at twenty-five thousand miles per hour. In this way, he spoke for Earth and helped us to realize the significance of our cosmic moment.

In 1974, describing the identity shift that comes with seeing Earth from space, Rusty said:

> When you go around it in an hour and a half, you begin to recognize that your identity is with that whole thing. And that makes a change. You look down there and you can't imagine how many borders and boundaries you crossed again and again and again. . . . You know there are hundreds of people killing each other over some imaginary line that you can't see. From where you see it, the thing is a whole, and it's so beautiful. And you wish you could take one from each side in hand and say, "Look at it from this perspective. Look at that. What's important?"

Of the Earth viewed from the Moon, he said:

> It becomes so small and fragile, and such a precious little spot in that universe, that you can block it out with your thumb, and you realize that on that small spot, that little blue-and-white thing, is everything that means anything to you. All of history and music and poetry and art and war and death and birth and love, tears, joy, games, all of it is on that little spot out there that you can cover with your thumb.

You ask yourself, he said:

> Have you earned this in some way? . . . You know the answer to that is no. . . . You know very well, at that moment, and it comes through to you so powerfully, that you're the sensing element for man. . . . I've used the word *you* because it's not me, it's you, it's

us, it's we, it's life. We've had that experience. And it's not just my problem to integrate, not my challenge to integrate, my joy to integrate—it's yours, it's everybody's.

He's right—we need to integrate this experience. I don't know of anyone who has gone into space and come down more filled with hatred and misunderstanding of others, or gazed from a spacecraft window and thought, "Hey, let's pave those rain forests and put up a parking lot." The view from space seems universally to invoke a feeling of oneness with humanity and life, and reverence for our planetary home. Technology has provided us with the perspective that triggers this unforeseen spiritual reflex. By entering space, we begin the transition into *Homo cosmicus.*

Meanwhile, back on Earth, humanity seems to be in somewhat of a pickle, at least partially of science's making. The age of nuclear weapons, environmental poisoning, and modern ethnic and religious warfare has sapped our hope and confidence, made us fearful and uncertain of the future, even afraid of the air we breathe and the water we drink.

Science has been regarded as our saving grace, the skill that will open up a glorious, safe, long, equitable future for humanity. It has also been cast as our doom. Which will it be? That may be determined by our success at the reunification of science and spirituality.

In some scientific circles religion has a pretty bad name. We get hung up on the specifics, on the stories that, taken literally, are incompatible with the beautiful truths science has allowed us to uncover. The differences in the story from religion to religion make them all seem arbitrary, just as the sniping between different groups of avid UFO believers casts doubt on all of their theories. Many of us also associate religion with murderous crusades, persecution of insightful thinkers, and—especially these days—suicide bombings.

Often, scientists are suspicious of spirituality in general. Ironically, this attitude helps fuel those belief systems that bother us the most. We often complain about New Age irrationality, but through an overreaction against religion, we have contributed to a gaping spiritual void in our culture. The result is a hunger for beliefs, any beliefs, and this need has helped create the New Age. Yet science, too, can fulfill spiritual needs. We blame spreading irrationality on scientific illiteracy. Yet, in my opinion, it is alienation from science, not science illiteracy, that is the root problem (one breeds the other). If we want the world to see us as wizards, not muggles, then we can't sell our services to the highest bidder, and we need to spread the magical (and spiritually evocative) story of Cosmic Evolution.

What is spirituality, anyway? How should I know? Do I look like the Dalai Lama? But what I mean by spirituality is the religious impulse stripped of religion. Spirituality is what's left when you peel away all of the inconsistencies in the lessons and stories.

Teilhard de Chardin believed in love as a cosmic principle. Even though he represented this as a Christian concept, and monotheism has never been my cup of tea, I find his writings meaningful, intelligent, and inspiring. I believe the phenomenon of humanity on Earth is a local example of a trend toward higher consciousness and spiritual enlightenment that transpires all over this universe.

Emerging complexity is the pattern linking together all of Cosmic Evolution. *Complexity* is a clinical-sounding term, but the principle allows us to see where the structure, the beauty, and perhaps even the self-awareness of the universe comes from. I see spirituality as an intuitive awareness of the internal and external forces impelling us to realize our place within this complexifying, unfolding, self-seeking, beauty-reeking cosmos of ours. Emergent complexity is all about the power to manifest surprising changes at higher levels through forming new connections. Spirituality is embodied in the connections we make with one another and the potential for wider group identification, for human love and unity.

Natural selection can act on a much larger scale than we're used to thinking about. Imagine one hundred thousand worlds, all with some form of "intelligent life" at roughly our current level. Certain qualities will aid survival, and these will be selected for. Eventually some worlds will live to reproductive age and spread life to other worlds, as Gaia may soon start to do.* Technical advancement without spiritual progress creates a dangerous and unstable condition that will be selected against. Natural selection on a galactic level will favor those living worlds where technical and spiritual advancement proceed together. Cosmic spiritual advancement by Darwinian natural selection!

For me the "living worlds" idea I described in chapter 17 expresses a viewpoint that is at once spiritual and scientific. I can get away with

---

*As discussed in Dorion Sagan's 1990 book *Biospheres*.

that, since it is not supposed to be science, but natural philosophy. Like the Gaia hypothesis from which it derives, this view integrates a scientific, mechanistic way of thinking about planets (in terms of self-regulating global patterns of cycling fluids, evolving organisms, and exchanges of energy and matter) with a spiritual intuition of our world's basic wholeness and aliveness. An awareness of Earth's essential aliveness seems to have appeared independently in so many cultures that I have to wonder if it was not part of the spiritual system of the first humans. Science, in moving forward, has caught up with some ancient wisdom: an appreciation that we are cells in something larger.

I don't know if that is what anyone else means by spirituality. I'm just a planetary scientist, what do I know about it? But this view—spirituality as an intuitive internalization of the universe's urge toward emergence—makes me see humanity's current predicament—a world rich with technical know-how but rife with inequity, scarcity, and violent conflicts—as fundamentally a spiritual crisis. Paradoxically, and sadly, contemporary versions of ancient religions seem to have mostly inflamed it. Some of my best friends and relatives are monotheists, but given current world events, it appears that monotheism is failing us when it comes to the all-important goal of world peace and unity.

Where else can we look for new solutions? One answer is to search the skies. If we're lucky, we may actually learn of another technological species. This could happen in a few years with the Allen Telescope Array coming on-line. I don't know *how* that would change us, but I definitely think it would. The discovery of other intelligent life visiting or residing within our own solar system would, of course, carry the most potential for changing our existence, but even radio contact might precipitate a sea change in human actions and values. I don't really buy anyone's predictions of what this would be like, but it seems obvious that we do need some kind of transformation, and I'd love to see the experiment play out. Maybe I've read too much SF, but it could be pretty damn freaky. Still, you don't want to miss it, do you? If it's going to happen someday, I'd like it to be on my watch. Might I suggest this coming Thursday?

SETI is a long-term, multigenerational, transnational quest. Thus even in the absence of a signal it serves as an example of the kind of effort we need to make in order to survive to take a place on the galactic stage.

In SETI literature a "more advanced" civilization implies one with more powerful technology. Sometimes it seems that the ultimate goal of SETI is to commune with other galactic nerds and compare notes about our machines. Ironically, this obsession with hardware may limit our longevity as a species, decreasing the chance that we will ever make contact. But SETI is also a spiritual quest that, by keeping us aware of the possibility of other sentient species, promotes a perspective on our role within the cosmos that can only help us.

We are at a curious and frustrating stage of our evolution. We can conceive of a truly intelligent, sustainable, communicating society. But we don't know if we can become one. So we search the skies for confirmation of a hopeful image of ourselves. Any aliens sending us signals have most likely been technological for thousands, if not millions, of years. They survived the moment we seem stuck in, and they may be immortal. If we only knew that somebody else had survived this stage, even if we didn't know anything else about them, it would serve as a "proof of concept," giving us reason to hope that we might become that which we seek.

If we hear from them, it confirms the possibility of the "good future." Could it be ours? I mean a future in which we are not just listening for signals, but proudly sending them, comfortable with technology, at peace with ourselves, secure against the capriciousness of a well-known cosmos, rich enough to send commercial-free broadcasts around the galaxy, confident enough to reveal our location to other species of unknown motivations and capacities.

Unfortunately, we can't count on any help from above and beyond. We shouldn't wait around for a teacher from the stars. For all we know they may be waiting to see if we can do a bit better on our own. We have no choice but to do our best to become the wise ones ourselves. It's either that or become another cosmic statistic. While we search for our brothers from another planet, why don't we just assume they are out there and act accordingly?

Will our core spiritual values of love and compassion be a part of alien religions? Will they even make the same distinction between religion and science or will it all be mixed together? Remember, they are a lot more advanced than we and they've had a long time to think about all this. If an integration of scientific and spiritual capacities is the key to long-term survival, then advanced ETs will long ago have accomplished this. Maybe, among humans, they'll relate best to Buddhist monks or Native American elders and regard SETI scientists as merely the switchboard operators who allowed them to make contact with the knowledgeable Earthlings.

As I've discussed, many arguments over the galactic prevalence of intelligent species, and the likely success of SETI, hinge on whether intelligence is an evolutionary development that, in most cases, actually aids or hinders survival. We don't really know whether our big heads will increase or decrease the longevity of our species, but one thing is clear: intelligence is obviously something you have to learn to use well. Our intelligence is a gift, but it's complicated and it came without an owner's manual. It would be nice if we could just call tech support and ask exactly how we are supposed to work this global civilization thing, but we don't know the number. So, if there is anybody out there with experience in how to get a planet to chill out and solve problems, without everybody wanting to kill everybody else, we could use a little help right about now. We've been trying it by trial and error without great success. How about some tips?

If I met an ET, that would be my first question (after the usual pleasantries). Not "How do you build your wonderful machines?" but "How did you learn to live with yourselves? How do you survive the transition to being a global, technical species? Do you have a spare manual?" Thinking about aliens in this way can help us to confront ourselves. How would we measure up? Do we have what it takes?

My belief in aliens is inseparable from a certain unavoidable, foolish, naturalistic optimism about our own ultimate prospects. Everything that I've learned about the nature of our universe and our biosphere tells me that life will find a way to thrive. Gaia, as Lynn Margulis has said, "is a tough bitch." If her noosphere chops off its head, she'll keep grooving along. In time, she may grow another noosphere, giving a different proto-intelligent species a chance at reaching the big time. I see our proud little spurt of technical invention as a little eddy in a whirling universe that is evolving, self-organizing, and moving inexorably toward more life and more intelligence. Our little whorl could wink out in an instant, or it could grow into a deeper, more stable mind-storm.

Is psychogenesis limited to Earth? I doubt it. Will there be a psychozoic age of the universe? Has it already begun? If we believe even in the possibility of the transformation to wisdom and immortality, then we must live in a universe increasingly permeated with intelligence, and suffused with love. I proved it mathematically in the last chapter, and equations don't lie. ☺

What do I really believe? I think our galaxy is full of species who have crawled up from the slime of their home worlds, evolved self-awareness and started to tinker, passed beyond the threat of technological self-extermination, and transcended their animal origins to move out into the cosmos. The vasty deep is thick with spirits. The wise ones are out there waiting for us to join them.

Notes on Sources and Suggestions for Further Reading*

SECTION I: HISTORY

In these pages I have only scratched the surface of the rich and varied history of ideas regarding extraterrestrial life. My historical account is highly selective, with examples chosen for the way in which they presage or illuminate modern thoughts and trends. For the reader wanting more, there are two comprehensive and complementary books: Michael J. Crowe's *The Extraterrestrial Life Debate, 1750–1900* (New York: Dover, 1999) gives a detailed account up to the dawn of the twentieth century. Steven J. Dick's *The Biological Universe* (Cambridge University Press, 1996) presents a thorough and thoughtful history of ideas about ET throughout the twentieth century.

Another good general history is *Planets and Planetarians: A History of Theories of the Origin of Planetary Systems* by Stanley L. Jaki (Edinburgh: Scottish Academic Press, 1978).

As for original sources from the seventeenth century, my favorite—as you might have guessed if you've read this far—is *Conversations on the Plurality of Worlds,* originally written in 1686 by Bernard le Bovier de Fontenelle and since reissued countless times in numerous languages. An English translation by H. A. Hargreaves was published by University of California Press in 1990.

A masterful and penetrating analysis of the lives and work of the Copernican revolutionaries can be found in Arthur Koestler's *The Sleepwalkers* (New York: Macmillan, 1959). If you don't want to slog through the whole thing, at least read the sections on Kepler and Galileo. Further insightful sources on Galileo are James Reston Jr.'s *Galileo: A Life* (New York: HarperCollins, 1994) and Dava Sobel's *Galileo's Daughter* (New York: Walker, 1999). If you want to dig more into an analysis of some of Kepler's lesser known theories and works, I recommend Bruce Stephenson's *The Music of the Heavens: Kepler's Harmonic Astronomy* (Princeton University Press, 1994).

Cosmology: Historical, Literary, Philosophical, Religious, and Scientific Perspectives, edited by Norriss S. Hetherington (New York: Garland, 1993),

---

*Much more extensive, irregularly updated notes can be found at funkyscience.net.

provides a varied and well-chosen selection of chapters covering the history of cosmology.

Norman H. Horowitz, one of the Viking biology investigators, gives a well-written and accessible inside account of exobiology in the 1970s and the Viking search for life on Mars in *To Utopia and Back, the Search for Life in the Solar System* (New York: W. H. Freeman, 1986). A more detailed source on the history of Mars exploration, with a focus on Viking and the early years of exobiology, is *On Mars, Exploration of the Red Planet, 1958–1978* by Edward C. Ezell and Linda N. Ezell. This book was published in 1984 as part of the NASA History Series (NASA Special Paper 4212). Recently, it was posted on the Web in its entirety by the NASA History Office. Their Web site at history.nasa.gov is a vast source of information on the history of space exploration.

Valuable primary sources for the early history of SETI include the proceedings from the First All Soviet Union Conference on Extraterrestrial Civilizations and Interstellar Communication in 1964, edited by G. M. Tovmasyan and published as *Extraterrestrial Civilizations* by the Israel Program for Scientific Translations (Jerusalem: 1967), and the proceedings of the First International Conference on Extraterrestrial Civilizations and Problems of Contact with Them in 1971, edited by Carl Sagan and published as *Communication with Extraterrestrial Life* (Cambridge: MIT Press, 1973).

A wonderful compendium of classic, seminal scientific papers and essays ranging from circa 70 B.C. to 1980 is *The Quest for Extraterrestrial Life: A Book of Readings,* edited by Donald Goldsmith (Mill Valley, Calif.: University Science Books, 1980). This book also contains a typically iconoclastic foreword by Fred Hoyle, who concludes, "It will be especially interesting to see whether it is astronomy that absorbs biology, or the other way around."*

SECTION II: SCIENCE

In my humble opinion, the best book about planetary science is my own *Venus Revealed: A New Look Below the Clouds of Our Mysterious Twin Planet* (Reading: Addison Wesley, 1997). If you liked *Lonely Planets,* you'll also enjoy *Venus Revealed.* If you hated this one, then don't bother. Actually, one book might be better: *Worlds Without End: The Exploration of Planets Known and Unknown* (Reading: Perseus, 1998) by John Lewis, my doctoral thesis adviser and scientific mentor.

An excellent collection of chapters about planetary science written by professionals in the field is *The New Solar System,* edited by J. Kelly Beatty and Andrew Chaikin (Cambridge University Press). New editions are issued every few years, so make sure you pick up the most recent one.

---

*Many of these books are out of print, but most can be purchased on the Web. If the book is still in print, please consider buying it new so that the author gets her dime.

An underappreciated gem of a book about Cosmic Evolution is *Atoms of Silence: An Exploration of Cosmic Evolution* by Hubert Reeves (translated from the original French by my old boss John Lewis and his wife, Ruth Lewis; Cambridge: MIT Press, 1984).

A masterful summary of modern cosmology can be found in Timothy Ferris's *The Whole Shebang: A State-of-the-Universe(s) Report* (New York: Simon & Schuster, 1997).

Readers wanting to learn more about the history of ideas about, and the recent discovery of, extrasolar planets should consult Ken Crosswell's *Planet Quest: The Epic Discovery of Alien Solar Systems* (New York: Simon & Schuster, 1997).

For further reading about the Gaia hypothesis, I recommend the proceedings of the first "serious" scientific conference devoted to the subject, the American Geophysical Union's Chapman Conference on the Gaia Hypothesis, held in San Diego in March 1988, published as *Scientists on Gaia,* edited by Stephen Schneider and Penelope Boston (Cambridge: MIT Press, 1991). Another excellent book presenting a "Gaian" picture of evolution is *Microcosmos: Four Billion Years of Microbial Evolution* by Lynn Margulis and Dorion Sagan (New York: Summit, 1986). Margulis and Sagan have written several other books, all recommended for their infuriatingly provocative and insightful views of evolution. Another valuable source about the important steps in biological evolution, written from a long-term global perspective, is *The Major Transitions in Evolution* by John Maynard Smith and Eors Szathmary (Oxford: W. H. Freeman, 1995).

Early roots of the Gaia concept can be found in Vladimir I. Vernadsky's fascinating and prescient *The Biosphere,* written in 1926 and finally translated in its entirety into English in 1998 (New York: Copernicus, 1998). The concept of the noosphere is elaborated in Pierre Teilhard de Chardin's brilliant and inspiring *The Phenomenon of Man* (1955; English translation, New York: Harper & Row, 1965).

Two good treatments of complexity theory are James Gleick's classic *Chaos: Making a New Science* (New York: Viking, 1987) and Stuart Kaufman's *At Home in the Universe: The Search for Laws of Self-Organization and Complexity* (Oxford University Press, 1995).

Some good recent books about exobiology and astrobiology are Amir Aczel's *Probability 1: Why There Must Be Intelligent Life in the Universe* (New York: Harcourt Brace, 1998), David Darling's *Life Everywhere: The Maverick Science of Astrobiology* (Reading: Perseus, 2001), Peter Ward and Donald Brownlee's *Rare Earth: Why Complex Life Is Uncommon in the Universe* (New York: Springer-Verlag, 2000), Paul Davies's *The Fifth Miracle: The Search for the Origin and Meaning of Life* (New York: Simon & Schuster, 1999), Robert Shapiro's *Planetary Dreams: The Quest to Discover Life Beyond Earth* (New York: Wiley, 1999), and Christian De Duve's *Vital Dust: Life as a Cosmic Imperative* (New York: HarperCollins, 1995).

Less current, but still well worth reading, are George Gamow's *Biography of the Earth: Its Past, Present and Future* (New York: Pelican, 1948), and *Life Beyond Earth: The Intelligent Earthling's Guide to Life in the Universe* by Gerald Feinberg and Robert Shapiro (New York: William Morrow, 1980).

A fascinating essay that discusses the concept of a galactic habitable zone, written by Polish science fiction writer and polymath Stanislaw Lem, is "The World as Cataclysm" in the book *One Human Minute*. An English translation was published in 1986 by Harcourt Brace Jovanovich.

SECTION III: BELIEF

The bible of books about SETI is still *Intelligent Life in the Universe* by Iosif Shklovskii and Carl Sagan, published originally in Russian as Shklovskii's *Universe, Life, Mind* (1962), and first published (to Shklovskii's surprise, as described in the text) as a dual-author, English-language book in 1966. This book has been reissued numerous times and is currently in print (Boca Raton: Emerson-Adams, 1998). The quintessential Carl Sagan book about extraterrestrial life is *The Cosmic Connection: An Extraterrestrial Perspective* (New York: Dell, 1975). Another worthy book from this era, often overlooked, is *The Galactic Club: Intelligent Life in Outer Space* by Ronald N. Bracewell (San Francisco: W. H. Freeman, 1974).

More recent good books about SETI include *Is Anyone Out There? The Scientific Search for Extraterrestrial Intelligence* by Frank Drake and Dava Sobel (New York: Dell, 1994), and Seth Shostak's *Sharing the Universe: Perspectives on Extraterrestrial Life* (Berkeley: Berkeley Hills, 1998).

Iosif Shklovskii's recollections on his involvement in SETI, and his rationale for his increasingly pessimistic beliefs about alien contact, are described in his autobiography, *Five Billion Vodka Bottles to the Moon: Tales of a Soviet Scientist* (New York: W. W. Norton, 1991).

A valuable collection of essays representing the "contact pessimist" school, fueled by the Fermi-Hart paradox, is *Extraterrestrials: Where Are They?* edited by Ben Zuckerman and Michael Hart (Pergammon, 1982).

The literature on UFOs and alien encounters is vast, and a complete bibliographic essay would alone take up a large volume. An excellent recent book that treats the subject from a religious studies perspective and contains extensive references is Brenda Denzler's *The Lure of the Edge: Scientific Passions, Religious Beliefs, and the Pursuit of UFO's* (Berkeley: University of California Press, 2001). A lighthearted and delightfully illustrated account of American UFO culture is presented in Douglas Curran's *In Advance of the Landing: Folk Concepts of Outer Space* (New York: Abbeville, 2001).

Numerous stories of alien encounters and other strange phenomena in Colorado's San Luis Valley can be read in Christopher O'Brien's *The Mysterious Valley* (New York: St. Martin's Press, 1996).

The scientific perspective on UFOs is well represented in *UFO's: A Scientific Debate,* edited by Carl Sagan and Thornton Page (W. W. Norton, 1972), in Carl Sagan's *The Demon Haunted Word* (Random House, 1996), and in the magazine *Skeptical Enquirer.*

Two good books on the "Roswell incident" are *The Roswell Report: Case Closed* by Captain James McAndrew of the United States Air Force (Washington, D.C.: Headquarters USAF, 1997) and *Roswell: Inconvenient Facts and the Will to Believe* by Karl T. Pflock, a veteran ufologist who describes his journey from Roswell believer to skeptic (New York: Prometheus, 2001).

For a seemingly endless stream of superficially convincing reports of crashed saucers, hidden alien bodies, and "black ops" government complicity with alien civilizations, read Dr. Steven M. Greer's *Disclosure: Military and Government Witnesses Reveal the Greatest Secrets in Modern History* (Charlottesville: Carden Jennings, 2001).

John Mack's work with experiencers and his interpretation of the abduction phenomenon are described in *Passport to the Cosmos: Human Transformation and Alien Encounters* (New York: Random House, 1999).

Frank Drake explains his strong belief that the aliens whom we will eventually contact through SETI are likely to come from immortal civilizations in "On Hands and Knees in Search of Elysium," *Technology Review* 78 (June 1976).

Credits and Permissions

NASA Planetary Protection bumper sticker. NASA.

Fern Fractal. Fractal generated by Harold Cooper.

Mandelbrot Haeckel similarity. Mandelbrot Set generated by Harold Cooper.

Drawing by Ernst Haeckel. *Kunstfromen der Natur*. (Leipzig & Vienna: Verlag des Bibliographischen Instituts. 1904. Courtesy Dover Pictorial Archive.)

Message to extraterrestrials. H. W. & C. W Neiman. 1920.

Drake Message to globular cluster M13. 1974. Frank Drake.

Alien virus message. Courtesy of the author.

Author and Alien. Photograph by Antony Cooper.

South Park. Cartman abducted by Aliens. By Trey Parker and Matt Stone. (Comedy Central/Viacom/Paramount.)

ET's Landing Diner. Photograph by the author. Courtesy of Tim Edwards.

Frame from Tim Edwards's video. Courtesy of Tim Edwards.

UFO Watchtower. Photograph by the author. Courtesy of Judy Messoline.

Viking Cydonia. NASA.

MGS Face. NASA/JPL/MSSS.

Triple Julia Set. Avebury Trusloe. Wheat. 29th July 1996. Photograph by Lucy Pringle.

TEXT

H. L. Mencken. Speech before the National Press Club. (New York: Random House, Inc.)

Bernard le Bovier de Fontenelle. *Entretiens sur la Pluralité des Mondes*. 1686. Translation by H. A. Hargreaves. Permission granted by the Regents of the University of California. (Berkeley: University of California Press. 1990.)

Salman Rushdie. *The Ground Beneath Her Feet*. © Salman Rushdie 1999. (Reprinted by permission of Henry Holt and Company, LLC. New York. 1999.)

Words and Music by Bart Howard. TRO—© 1954 (Renewed) Hampshire House Publishing Corp. New York, NY. Used by permission.

Lyrics and Music of Bob Dylan. © 1973,1974 by Ram's Horn Music. All rights reserved. International copyright secured. Reprinted by permission.

Index

Earth *(continued)*
  born in steam, 90–92, 167
  a giant heat-engine, 157
  holding water, 92–93, 168
  inside out, 88–89
  setting the scene, 94–96
  size of, 275, 277
Earth-centered worldview, 9, 14
Eberhart, Jonathan, 189
Edgett, Ken, 187
Edwards, Tim, 340
Edwards Air Force Base, 365
EG&G, 368
ego death, 385
Ehman, Jerry, 307
Einstein, Albert, 69, 238, 305, 312
Encyclopedia Galactica, xv
energy
  of accretion, 157
  geothermal, 123
Enlightenment era, 30, 33, 65, 256, 381
entropy, 269–270
Epicureans, beliefs about other worlds, 29
Epicurus, 7
Epsilon Eridani, 293
equilibrium, 277
Erhard, Werner, 377
ESP, 332
EST, 377
ET, *see* extraterrestrial life
eukaryotes, 119
Europa, 192–193
  biosphere of, 200–201
  first close encounters with, 240
  ocean beneath the surface of, xiii, 59, 111, 187–204
  possibility of life on, 283
  radiation on, 201
European explorers, 21
European Space Agency, xxi
evidence, philosophizing in the absence of, 8
evolution
  arguments for, 36
  chemical, 99, 165
  climate, 174
  convergent, 44–45

Darwinian, 107, 136
  thermal, 156
"evolutionary divergence," 131
exobiology, xxxi, 112, 147, 221–236, 233, 236, 419
  Cosmism, 223–227, 301
  detecting and protecting life, 221–223
  Sagan and the little green men, 233–236
  Vikings on Mars, 229–233
  Wolf Traps and setbacks, 227–229
"Exobiology: Approaches to Life beyond the Earth," 222
Exobiology Division (NASA), 223
exoplanets, 205–220
  discovery of more, 207–213
  on the edge of knowing, 216–219
  the first hundred, 213–216
  good news for modern man, 219–220
  mediocrity of Earth, 205–207
"Explanation for the Absence of Extraterrestrials on Earth, An," 313
exploration
  interstellar, 219
  planetary, biocentric, 248
extraterrestrial (ET) life
  debate over, xxvii–xxviii, 43
  resurgence of scientific interest in, xiii
"Extraterrestrial Highway," 306
"extraterrestrial hypothesis" for UFOs, 355
Extraterrestrial Retrieval Team, 367
extremophiles, 130, 139–140

"face," on Mars, 323, 358–362
Fermi, Enrico, 311
Fermi-Hart paradox, 313
Fermi's paradox, 310–333
  absence of evidence, 310–311
  evidence of absence, 311–313
  the great silence, 321–323
  Hart's answer, 313–317
  magical solution, 331–333
  message in a virus, 324–331
  where the aliens are, 317–321
51 Pegasi, 208–209

TORY READ

DAVID GRINSPOON is principal scientist in the Department of Space Studies at the Southwest Research Institute, and adjunct professor of Astrophysical and Planetary Sciences at the University of Colorado. His previous book, *Venus Revealed*, was a *Los Angeles Times* Book Prize Finalist. An adviser for NASA on space exploration strategy, he lectures widely and has appeared on numerous television and radio programs. His writing has appeared in the *New York Times*, *Slate*, *Astronomy*, *Nature*, *Science*, *Scientific American*, *Natural History*, and *The Sciences*. He maintains the Funky Science Web site at www.funkyscience.net.